SCIENCE
IN
DISPUTE

SCIENCE IN DISPUTE

Volume 1

NEIL SCHLAGER, EDITOR

Produced by
Schlager Information Group

Detroit • New York • San Diego • San Francisco • Cleveland • New Haven, Conn. • Waterville, Maine • London • Munich

SCIENCE IN DISPUTE

 Volume 1

Neil Schlager, *Editor.* Brigham Narins, *Contributing Editor.*

Barbara J. Yarrow, *Manager, Imaging and Multimedia Content.* Robyn V. Young, *Project Manager, Imaging and Multimedia Content.* Dean Dauphinais, *Senior Editor, Imaging and Multimedia Content.* Kelly A. Quin, *Editor, Imaging and Multimedia Content.* Leitha Etheridge-Sims, Mary K. Grimes, David G. Oblender, *Image Catalogers.* Lezlie Light, *Imaging Coordinator.* Randy Bassett, *Imaging Supervisor.* Robert Duncan, *Senior Imaging Specialist.* Dan Newell, *Imaging Specialist.* Luke Rademacher, *Imaging Specialist.* Christine O'Bryan, *Graphic Specialist.*

Maria Franklin, *Permissions Manager.* Margaret A. Chamberlain, *Permissions Specialist.* Kim Davis, *Permissions Associate.* Shalice Shah-Caldwell, *Permissions Associate.*

Mary Beth Trimper, *Manager, Composition and Electronic Prepress.* Evi Seoud, *Assistant Manager, Composition and Electronic Prepress.* Dorothy Maki, *Manufacturing Manager.* Rhonda Williams, *Buyer.*

Michelle DiMercurio, *Senior Art Director.* Pamela A. E. Galbreath, *Art Director.*

While every effort has been made to ensure the reliability of the information presented in this publication, the Gale Group neither guarantees the accuracy of the data contained herein nor assumes any responsibility for errors, omissions or discrepancies. The Gale Group accepts no payment for listing, and inclusion in the publication of any organization, agency, institution, publication, service, or individual does not imply endorsement of the editors or publisher. Errors brought to the attention of the publisher and verified to the satisfaction of the publisher will be corrected in future editions.

This book is printed on recycled paper that meets Environmental Protection Agency standards.

The paper used in this publication meets the minimum requirements of American National Standard for Information Sciences-Permanence Paper for Printed Library Materials, ANSI Z39.48-1984.

This publication is a creative work fully protected by all applicable copyright laws, as well as by misappropriation, trade secret, unfair competition, and other applicable laws. The authors and editors of this work have added value to the underlying factual material herein through one or more of the following: unique and original selection, coordination, expression, arrangement, and classification of the information.

Gale Group and design is a trademark used herein under license.

All rights reserved including the right of reproduction in whole or in part in any form. All rights to this publication will be vigorously defended.

Copyright © 2002
Gale Group
27500 Drake Road
Farmington Hills, MI 48331-3535

ISSN 1538-6635
ISBN: 0-7876-5765-4

Printed in the United States of America
10 9 8 7 6 5 4 3 2 1

CONTENTS

About the Series vii

Advisory Board viii

List of Contributors ix

Astronomy and Space Exploration

Is Pluto a planet? 1
Is the International Space Station the appropriate next
step for humanity's exploration of space? 8
Does the accumulation of "space debris" in Earth's orbit
pose a significant threat to humans, in space and
on the ground? 16
Historic Dispute: Is Urbain Le Verrier the true discoverer
of Neptune? 24

Earth Science

Did life on Earth begin in the "little warm pond"? 33
Did water once flow on the surface of Mars? 41
Are hazards due to earthquakes in the New Madrid
Seismic Zone overestimated? 48
Historic Dispute: Are Earth's continents stationary
throughout geologic time? 57

Engineering

Will a viable alternative to the internal combustion engine
exist within the next decade? 67
Will wind farms ever become an efficient, large-scale
source of energy? 76
Do the potential dangers of nanotechnology to society
outweigh the potential benefits? 86
Have technological advances in sports such as tennis, golf,
and track and field supplanted the athletic achievements
of the participants in those sports? 93

Life Science

Should the threat of foot-and-mouth disease be met by
the destruction of all animals that might have been exposed
to the virus? 101
Are XYY males more prone to aggressive behavior
than XY males? 111
Were dinosaurs hot-blooded animals? 119
Has DNA testing proved that Thomas Jefferson fathered
at least one child with one of his slaves, Sally Hemings? 131

Mathematics and Computer Science

Should statistical sampling be used in the United States Census? 141

Has the calculus reform project improved students' understanding of mathematics? 149

Do humans have an innate capacity for mathematics? 157

Does private strong encryption pose a threat to society? 163

Medicine

Does the addition of fluoride to drinking water cause significant harm to humans? 173

Should xenotransplants from pigs raised at so-called organ farms be prohibited because such organs could transmit pig viruses to patients—and perhaps into the general population? 184

Should the cloning of human beings be prohibited? 194

Historic Dispute: Did syphilis originate in the New World, from which it was brought to Europe by Christopher Columbus and his crew? 208

Physical Science

Is the cost of high-energy laboratories justified? 219

Historic Dispute: Do neutrinos have mass? 227

Do hidden variables exist for quantum systems? 236

Does cold fusion exist? 244

Historic Dispute: Is Earth the center of the universe? 253

General Subject Index 263

ABOUT THE SERIES

Overview

Welcome to *Science in Dispute*. Our aim is to facilitate scientific exploration and understanding by presenting pro-con essays on major theories, ethical questions, and commercial applications in all scientific disciplines. By using this adversarial approach to present scientific controversies, we hope to provide students and researchers with a useful perspective on the nature of scientific inquiry and resolution.

The majority of entries in each volume of *Science in Dispute* cover topics that are currently being debated within the scientific community. However, each volume in the series also contains a handful of "historic" disputes. These include older disputes that remain controversial as well as disputes that have long been decided but that offer valuable case studies of how scientific controversies are resolved. Each historic debate is clearly marked in the text as well as in the Contents section at the beginning of the book.

Each volume of *Science in Dispute* includes approximately thirty entries, which are divided into seven thematic chapters:

- Astronomy and Space Exploration
- Earth Science
- Engineering
- Life Science
- Mathematics and Computer Science
- Medicine
- Physical Science

The advisory board, whose members are listed elsewhere in this volume, was responsible for defining the list of disputes covered in the volume. In addition, the board reviewed all entries for scientific accuracy.

Entry Format

Each entry is focused on a different scientific dispute, typically organized around a "Yes" or "No" question. All entries follow the same format:

- **Introduction:** Provides a neutral overview of the dispute.
- **Yes essay:** Argues for the pro side of the dispute.
- **No essay:** Argues for the con side of the dispute.
- **Further Reading:** Includes books, articles, and Internet sites that contain further information about the topic.
- **Key Terms:** Defines important concepts discussed in the text.

Throughout each volume users will find sidebars whose purpose is to feature interesting events or issues related to a particular dispute. In addition, illustrations and photographs are scattered throughout the volume. Finally, each volume includes a general subject index.

About the Editor

Neil Schlager is the president of Schlager Information Group Inc., an editorial services company. Among his publications are *When Technology Fails* (Gale, 1994); *How Products Are Made* (Gale, 1994); the *St. James Press Gay and Lesbian Almanac* (St. James Press, 1998); *Best Literature By and About Blacks* (Gale, 2000); *Contemporary Novelists*, 7th ed. (St. James Press, 2000); *Science and Its Times* (7 vols., Gale, 2000-2001); and *The Science of Everyday Things* (4 vols., Gale, 2002). His publications have won numerous awards, including three RUSA awards from the American Library Association, two Reference Books Bulletin/Booklist Editors' Choice awards, two New York Public Library Outstanding Reference awards, and a *CHOICE* award for best academic book.

Comments and Suggestions

Your comments on this series and suggestions for future volumes are welcome. Please write: The Editor, *Science in Dispute*, Gale Group, 27500 Drake Road, Farmington Hills, MI 48331.

ADVISORY BOARD

Donald Franceschetti
    Distinguished Service Professor of Physics and Chemistry, University of Memphis

William Grosky
    Chair, Computer and Information Science Department, University of Michigan–Dearborn

Jeffrey C. Hall
    Assistant Research Scientist, Associate Director, Education and Special Programs, Lowell Observatory

Stephen A. Leslie
    Assistant Professor of Earth Sciences, University of Arkansas at Little Rock

Lois N. Magner
    Professor Emerita, Purdue University

Duncan J. Melville
    Associate Professor of Mathematics, St. Lawrence University

Nicholas Warrior
    Lecturer in Mechanical Engineering, University of Nottingham (UK)

LIST OF CONTRIBUTORS

Amy Ackerberg-Hastings
    Independent Scholar

Mark H. Allenbaugh
    Lecturer in Philosophy, George Washington University

Linda Wasmer Andrews
    Freelance Writer

Peter Andrews
    Freelance Writer

John C. Armstrong
    Astronomer and Astrobiologist, University of Washington

Maury Breecher
    Independent Scholar

Sherri Chasin Calvo
    Freelance Writer

Loren Butler Feffer
    Freelance Writer

Adi R. Ferrara
    Freelance Science Writer

Randolph Fillmore
    Science Writer

Maura C. Flannery
    Professor of Biology, St. John's University

Natalie Goldstein
    Freelance Science Writer

Alexander Hellemans
    Freelance Writer

Robert Hendrick
    Professor of History, St. John's University

Eileen Imada
    Freelance Writer

Anne K. Jamieson
    Freelance Writer

Brenda Wilmoth Lerner
    Science writer

K. Lee Lerner
    Professor of Physics, Fellow, Science Research & Policy Institute

Charles R. MacKay
    National Institutes of Health

Lois N. Magner
    Professor Emerita, Purdue University

M. C. Nagel
    Freelance Science Writer

Stephen D. Norton
    Committee on the History and Philosophy of Science, University of Maryland, College Park

Lee Ann Paradise
    Science Writer

Cheryl Pellerin
    Independent Science Writer

David Petechuk
    Freelance Writer

Jed Rothwell
    Contributing Editor, *Infinite Energy Magazine*

Brian C. Shipley
    Department of History, Dalhousie University

Seth Stein
    Professor of Geological Sciences, Northwestern University

Edmund Storms
    Los Alamos National Laboratory (ret.)

Marie L. Thompson
    Freelance Writer/Copyeditor

Todd Timmons
    Mathematics Instructor, University of
    Arkansas-Fort Smith

David Tulloch
    Freelance Writer

Elaine H. Wacholtz
    Medical and Science Writer

ASTRONOMY AND SPACE EXPLORATION

Is Pluto a planet?

Viewpoint: Yes, Pluto's status as a planet has been continually defended by the International Astronomical Union. Although its size and elliptical orbit do differ from those of the solar system's other planets, these criteria are arbitrary traits that do not discount its status as a planet.

Viewpoint: No, Pluto has more in common with the comets and asteroids discovered in the Kuiper Belt during the 1990s.

The controversy surrounding the planetary designation of Pluto sounds deceptively simple. While Pluto was identified as a planet upon its discovery in 1930, recent refinements in the taxonomy of orbital bodies have raised questions about whether Pluto is "really" a planet, or one of the smaller, more numerous objects beyond Neptune that orbit our Sun. Part of the controversy is essentially semantic: there is no rigid, formal definition of "planet" that either includes or excludes Pluto. The other eight planets are a diverse group, ranging greatly in size, composition, and orbital paths. Size is the primary distinction that sets them apart from the thousands of smaller objects orbiting the Sun, such as asteroids and comets. Pluto, however, is much smaller than the other planets but much larger than the bodies found in the asteroid belts. This fact alone has prompted some scientists to "demote" Pluto as a planet.

Size is not the only issue raised by astronomers who want to reevaluate Pluto's planetary status. For example, they point out that Pluto's orbit differs significantly from that of the other planets, and that its composition is more similar to comets than to the other planets. Scientific organizations, such as the International Astronomical Union, however, maintain that Pluto is a major planet, and that such distinctions are arbitrary.

Perhaps the most compelling aspect of this controversy is what it says about our understanding of the solar system. The image of our solar system consisting of one Sun and nine planets is elegant, easy to picture, and has been a staple of astronomy textbooks for more than 70 years. But as scientists learn more about the smaller bodies that orbit our Sun, and look far beyond Pluto and see a wide population of other orbital bodies, it seems simplistic and naive to view Pluto as the outer boundary of the solar system. If Pluto is re-assigned to the broader category of "Trans-Neptunian Objects," one of the small, solar system bodies orbiting beyond Neptune, it would become a recognizable exemplar of a group of far-off objects made mysterious by their distance from us, but nevertheless a part of our solar system.
—LOREN BUTLER FEFFER

**Viewpoint:
Yes, Pluto's status as a planet has been continually defended by the International Astronomical Union. Although its size and elliptical orbit do differ from those of the solar system's other planets,**

Clyde Tombaugh searches for Pluto at the Lowell Observatory in 1930.
(AP/Wide World Photos. Reproduced by permission.)

these criteria are arbitrary traits that do not discount its status as a planet.

Pluto, the last major planet of Earth's solar system, has been considered a planet since its discovery in 1930 by American astronomer Clyde W. Tombaugh at Lowell Observatory in Flagstaff, Arizona. Tombaugh was conducting a systematic search for the trans-Neptunian planet that had been predicted by the erroneous calculations of Percival Lowell and William H. Pickering. Some scientists maintain that the only reason Pluto is considered a planet today is because of the long, ongoing and well publicized search for what was then referred to as Planet X. When Pluto was discovered, media publicity fueled by the Lowell Observatory "virtually guaranteed the classification of Pluto as a major planet," according to Michael E. Bakick in *The Cambridge Planetary Handbook*.

Society of Astronomers Declare Pluto a Planet However, it is not the public opinion that determines whether a celestial body is a planet or not. That responsibility rests with a scientific body known as the International Astronomical Union (IAU), the world's preeminent society of astronomers. In January of 1999 the IAU issued a press release entitled "The Status of Pluto: A Clarification." In that document the IAU stated, "No proposal to change the status of Pluto as the ninth planet in the solar system has been made by any Division, Commission or Working Group." The IAU stated that one of its working groups had been considering a possible numbering system for a number of smaller objects discovered in the outer solar system "with orbits and possibly other properties similar to those of Pluto." Part of the debate involved assigning Pluto an identification number as part of a "technical catalogue or list of such Trans-Neptunian Objects." However, the press release went on to say that "The Small Bodies Names Committee has, however, decided against assigning any Minor Planet number to Pluto."

Notwithstanding that decision, in year 2000 the Rose Center for Earth and Science at New York City's American Museum of Natural History put up an exhibit of the solar system leaving out Pluto. That action received press coverage and re-ignited the controversy. Alan Stern, director of the Southwest Research Institute's space studies department in Boulder, Colorado, criticized the museum's unilateral decision, stating, "They are a minority viewpoint. The astronomical community has settled this issue. There is no issue."

Still, the argument continues, occasionally appearing in journal articles and in the popular press. However, for every argument against Pluto's designation as a major planet, there seems to be rational counter arguments for retaining that designation. Scientists who argue against Pluto being a major planet stress Pluto's differences from the other eight planets—the four inner planets, Mercury, Venus, Earth, and Mars, and the four giant planets, Jupiter, Saturn, Uranus, and Neptune. Supporters of Pluto as a major planet believe such arguments are fallacious because the two groups of planets could, as the Lowell Observatory put it in 1999, "scarcely be more different themselves." For instance, Jupiter, Saturn, Uranus, and Neptune have rings. Mercury, Venus, Earth, Mars, and Pluto do not. Mercury has an axial tilt of zero degrees and has no atmosphere. Its nearest neighbor, Venus, has a carbon dioxide atmosphere and an axial tilt of 177 degrees. Pluto has an axial tilt between those two extremes—120 degrees. (Earth's tilt is 23 degrees.)

Size Doesn't Matter A main argument against Pluto being a major planet is its size. Pluto is one-half the size of Mercury, the next smallest planet in our solar system. In fact, Pluto is even smaller than the seven moons in our planetary system.

"So what?" is the response of Pluto's defenders. They point out that size is an arbitrary criterion for determining the status of orbiting bodies. Mercury, for instance, is less than one-half the size of Mars, and Mars is only about one-half the size of Earth or Venus. Earth and Venus are only about one-seventh the size of Jupiter. From the standpoint of giant Jupiter, should the midget worlds of Mercury, Venus, Mars, and Earth be considered planets?

The most commonly accepted definition of a planet, according to University of Arizona educator John A. Stansberry, is that a planet is "a spherical, natural object which orbits a star and does not generate heat by nuclear fusion." For an

KEY TERMS

ASTEROIDS: Rocky, inert, irregularly shaped, airless objects found in great abundance in Earth's solar system. Most are located in an "Asteroid Belt" located between Mars and Jupiter.

COMETS: Very small, but often spectacular, visitors that create amazing celestial shows when they enter the inner solar system and come close to the Sun. The heat from the Sun releases water from their surfaces. The water carries some of the comet substances, dust or dirt, along with it and creates huge tails and streamers in the sky.

KUIPER BELT: A disk-shaped region past the orbit of Neptune roughly 30 to 100 AU from the Sun containing many small icy bodies called Trans-Neptunian Objects (TNOs), and considered to be the source of the short-period comets. (AU—astronomical unit: 1 AU is the average distance of Earth from the Sun.)

MOONS: Satellites that orbit around planets.

OORT CLOUD: Astronomers estimate that the Oort Cloud reaches about 200,000 times Earth's distance from the Sun and may contain as many as 100 trillion comets.

TRANS-NEPTUNIAN OBJECTS (TNOs): Small solar system bodies orbiting beyond Neptune, between 30 and 50 AU from the Sun. They form the Kuiper Belt.

object in space to maintain a spherical shape it has to be large enough to be pulled into that shape by its own gravity. According to that definition, "Pluto is clearly a planet," concludes Stansberry.

Detractors have also pointed out that Pluto's highly eccentric orbit has more in common with comets that originate from the Kuiper Belt than with the other eight planets in our solar system. That's true, reply Pluto's supporters, but just because Pluto has an eccentric orbit doesn't mean that it isn't a planet. Besides, Pluto's elongated orbit is only slightly more "eccentric" than Mercury's.

Another argument against Pluto being a planet is that it has an icy composition similar to the comets and other orbital bodies in the Kuiper Belt. Supporters of Pluto's planetary status argue that planets are already categorized into two unlike groups: The inner planets—Mercury, Venus, Earth, and Mars—which are composed of metals and rock, and the outer planets—Jupiter, Saturn, Uranus, and Neptune—which are, essentially, giant gaseous planets. Why couldn't there be three kinds of planets: terrestrial, giant gas planets, and icy rocks? Pluto may simply be the first planet in a new category.

Pluto: The Dividing Line Apparently Pluto is the largest of the 1,000-plus icy objects that have been discovered in the Kuiper Belt region of space. Since Pluto is the largest, that provides further support for Pluto retaining its "major planet" designation. There has to be a dividing line between "major" and "minor" planets, or as some call them, planetesimals. As Descartes reminds us, "No one of the sciences is ever other than the outcome of human discernment." In this context the quote reminds us that the distinctions we make between objects in orbit around the Sun are determined by our human decision. Planetary scientist Larry A. Lebofsky proposes that the IAU simply define Pluto's diameter of 1,413 miles (2,273 km) (almost half that of Mercury, the next smallest planet) as the smallest acceptable size for a major planet. It's an elegant solution to an argument that is fueled by differences in definitions. Let Pluto be the dividing line. In an age when we look forward to the development of technology that will soon allow us to spot orbiting bodies around other stars, it is a solution that would enable us to adopt a planetary classification scheme that would, as Stansberry puts it, provide "consistent and rational answers whether applied to our solar system, or to any other star system." —MAURY M. BREECHER

Viewpoint:
No, Pluto has more in common with the comets and asteroids discovered in the Kuiper Belt during the 1990s.

Overview In the year 2000 when Hayden Planetarium at New York City's American Museum of Natural History opened its shiny new Rose Center for Earth and Science, the astronomy hall was not the only new phenomenon. Its planetary display depicted a new view of the solar system: Mercury, Venus, Earth, and Mars—the rocky planets—were grouped together as the "terrestrials"; Jupiter, Saturn, Uranus, and Neptune as the "gas giants"; and Pluto . . . well, where *was* Pluto? Pluto was assigned to the Kuiper Belt, or perhaps more accurately, the Trans-Neptunian Belt, which the museum simplistically described as a disk of small, icy worlds beyond Neptune.

Pluto is now generally accepted among astronomers as being one of the three-to-four hundred objects discovered in the Kuiper Belt during the final decade of the 1900s. These objects are commonly known as Trans-Neptunian Objects (TNOs). Although Pluto's composition is unknown, its density indicates it is a mixture of ice and rock, as is the case with other TNOs. All the rocky planets are closest to the Sun, and those furthest from the Sun are all gaseous—except Pluto, which lies furthest of all planets from the Sun. How can this be if it is truly a planet? In light of this, can Pluto still be called a planet?

Brian Marsden, associate director for Planetary Sciences at the Harvard-Smithsonian Center for Astrophysics and director of the International Astronomical Union (IAU), the worldwide hub for recording sightings of comets and asteroids (astronomical bodies given minor planet designation), believes Pluto is a comet. In a radio interview with Robyn Williams entitled "Comets versus Planets," Marsden describes how astronomers suspect TNOs are dragged out of the Kuiper Belt (named for Gerard Kuiper who first suggested it as the source of short-period comets) when they pass close to Neptune. Neptune, in turn, may drag them somewhere else where they become influenced by the forces of other planets. Some TNOs get thrown further toward Earth to become short-period comets, circling the Sun approximately every six years. Others get thrown far out into the solar system into the region of the Oort Cloud where passing stars throw them back at us. These are the long-period comets, which often display spectacular tails when their icy component vaporizes as they approach the Sun. Others are proto comets—TNOs that remain in orbit in the solar system but have an unstable orbit. "Pluto being the biggest comet, if you like . . . Imagine bringing Pluto in to the distance of the Earth from the Sun. That would be a comet of the millennium as all the ice vaporized and made a great long tail," explains Marsden.

The Question Arises Although its planetary status has been questioned by some since its discovery in 1930, Pluto remained relatively secure in its ninth-row seat until 1992. It was then that David Jewitt of the University of Hawaii, and Jane X. Luu of Harvard, discovered "a curious object called '1992 QB1'," writes Tony Phillips in "Much Ado about Pluto." This small, icy body, Phillips explains, is about the size of an asteroid, orbiting 1.5 times further away from the Sun than Neptune, and was the first indication there might be "more than just Pluto in the distant reaches of the solar system." Phillips notes that, apart from its comparatively large size, Pluto is almost indistinguishable from other Kuiper Belt objects (KBOs) and even the short-period comets. What does distinguish Pluto from other KBOs, however, is its luminosity—a 60% higher albedo (reflective power) than anticipated for other KBOs. This phenomenon is attributed to Pluto's mass and gravitational quality, both of

**Pluto
Hubble Space Telescope · Faint Object Camera**

A view of Pluto from the Hubble Telescope.
(© NASA/Roger Ressmeyer/Corbis. Reproduced by permission.)

which are significant enough for the planet to "retain a tenuous atmosphere, from which bright frosts may be deposited on the surface," says Jewitt.

Utilizing an 8,192 x 8,192 pixel CCD camera, and experimenting with an even larger 12,000 x 8,000 pixel camera, Jewitt and colleagues have discovered KBOs one-third the diameter of Pluto—all in just the tiny portion of the sky (50 sq. degrees) examined thus far from their vantage point at the University of Hawaii. In 1996 they discovered TO66, calculated to be 497 miles (800 km) in diameter. "It would be incredible in its own right if Pluto proved to be the only 1,250-mile (2,011-km) object. I think we'll have Pluto II, Pluto III . . . within a few years," Jewitt told Phillips. Jewitt, speculating on the possibility of discovering KBOs larger than Pluto, wonders what happens then to Pluto's planetary designation.

Defining a Planet Also complicating matters is the scientific definition of a planet: There is no formal one, and therefore the question of Pluto's planetary status, like the above definition of the Trans-Neptunian Belt, also may be simplistic. In his article entitled "Is Pluto a Giant Comet?" Daniel W.E. Green, associate director of the IAU's Central Bureau for Astronomical Telegrams, writes: "The issue was/is over 'major planet' status, not 'planet' status. Indeed, with apparently nonstellar companions being discovered at an increasing rate around other Milky Way stars, the issue about how to define the word 'planet' is becoming more complex, and it is obvious that the word planet needs, in almost all cases, to have accompanying qualifier words ("major," "minor," "principal," etc.) for usage of the word 'planet' to make much sense in any given context."

"It's very difficult to come up with a physically meaningful definition under which we'd have nine planets," said Hal Levison, astronomer at the Southwest Research Institute in Boulder, Colorado, to David Freedman for his article entitled "When is a Planet not a Planet?" Freedman writes: "One generally accepted [definition] is a 'non-moon, sun-orbiting body large enough to have gravitationally swept out almost everything else near its orbit.' Among the nine planets, Pluto alone fails this test.... The second is a 'non-moon, sun-orbiting body large enough to have gravitationally pulled itself into a roughly spherical shape.' Pluto passes this test—but so do Ceres, a half dozen or so other asteroids, and possibly some other members of the Kuiper Belt." Also, the latter definition can be applied to stars, which are self-luminous—something a planet, by definition, is not.

Ceres, the small, rocky body discovered between the orbits of Mars and Jupiter in 1801, was declared a planet shortly thereafter. Another similar body, Pallas, discovered in the same orbital zone a year later, was also designated a planet. However, when these two discoveries were followed by numerous similar findings, their planetary status was changed to asteroid, even though they were certainly the largest members of what is now known as the Asteroid Belt. Similarly, Pluto is undoubtedly the largest member discovered in the Kuiper Belt thus far. However, it is just one of 60, and possibly hundreds of thousands, of comet-like objects in the fascinating Belt that extends far beyond our planetary boundary. "If you're going to call Pluto a planet, there is no reason why you cannot call Ceres a planet," Freeman quotes Marsden as saying.

Also relating to the size issue is how small can a planet be and still be called a planet? Some astronomers say anything larger than 620 miles (998 km) in diameter. This would put Pluto "in" and Ceres (at 580 miles, or 933 km) "out." However, is this arbitrary designation because Pluto was "in" long before anyone knew its true size? When first discovered, scientists believed it was about the size of Earth. By 1960 that estimate was adjusted to about half the diameter of Earth. By 1970, when its moon Charon was discovered and scientists realized the object they were seeing was not one but two, Pluto's size diminished again to one-sixth that of Earth. By this time it had become entrenched in literature and history as planet number nine, yet had diminished in size to less than half that of Mercury and smaller even than the seven satellites in our solar system commonly called moons. These include the Moon, Io, Europa, Ganymede, Callisto, Titan, and Triton. In his article "Pluto is Falling from Status as a Distant Planet," Rick Hampson quotes Michael Shara, curator of the American Museum of Natural History, as saying: "If Pluto is a planet then so is Earth's moon and hundreds of other hunks of debris floating around the Sun." Complicating the size issue is the fact that Pluto's moon is proportionately larger to Pluto than any other moon is to its respective planet. In fact, many astronomers who still consider Pluto a planet say it and Charon are, in fact, a double planet.

And what of Pluto's eccentric orbit—highly elliptical, tilted 17 degrees from that of the eight planets, and orbiting the Sun twice for every three times Neptune does? "About a third of them [TNOs] have Pluto-like orbits, and all of them appear to be, like Pluto, amalgams of ice and rock," writes Freedman. "Also, Pluto actually crosses Neptune's orbit, swapping places with Neptune every 248 years to spend 20 years as the eighth instead of the ninth planet. But does that mean it should *not* be considered a planet?"

"If Pluto were discovered today," writes Green, "it would be handled by the IAU's Minor Planet Center and given a minor-planet designation, as has happened for the hundreds of other TNOs discovered since 1992." Echoing this belief, Levison says in Kelly Beaty's article "Pluto Reconsidered," "I firmly believe that if Pluto were discovered today, we wouldn't be calling it a planet."

Changing History So, can a planet's designation be changed? Let's return to Ceres and Pallas, whose discoveries were followed by two more "planets" in 1804 and 1807. Because they floated about between Mars and Jupiter, these four bodies were designated as the fifth, sixth, seventh, and eighth planets, with Jupiter, Saturn, and Uranus moved to ninth, tenth, and eleventh positions. When, in the late 1800s, many more such bodies were discovered occupying the same area of space, astronomers reclassified the four newcomers as minor planets, or asteroids. This included Ceres, almost twice the size of the next largest asteroid.

More than 100 years later and roughly 50 years after the designation of Pluto as a major planet, discoveries of TNOs continue. While most astronomers agree Pluto is a TNO—albeit the largest one discovered thus far—the IAU officially announced in 1999 they had no intention of changing Pluto's planetary status. A formal poll taken at the 2000 General Assembly of the IAU and published in *Northern Lights* found that only 14% of astronomers from around the world now consider Pluto a major planet, 24% consider it a TNO and not a major planet, and 63% give it "dual status" as both planet and TNO. Green points out that many years went by before astronomers finally accepted Copernicus's heliocentric concept of the universe, and suggests it may be years yet before new astronomers "fully accept the poor logic behind viewing Pluto as the ninth major planet."

While the IAU sticks with tradition, the undeniable facts surrounding the dispute are leading others in a different direction. Green's article includes a small list of the plethora of scientific publications arguing against Pluto's status as a major planet. Freedman writes in his article, "Pluto doesn't need any official ruling to move into minor planethood. It could happen on a de facto basis, and it probably will. . . . Some of the newest astronomy textbooks, in fact, are already openly questioning Pluto's status."

Marsden believes that keeping Pluto as a planet misleads the public, and particularly school children, by presenting an archaic perspective of the solar system "that neatly ends in a ninth planet, rather than trailing off beyond Neptune into a far-reaching and richly populated field of objects."

In a brief article entitled "Pluto and the Pluto-Kuiper Express," Jewitt writes: "Bluntly put, one has two choices. One can either regard Pluto as the smallest, most peculiar planet moving on the most eccentric and most inclined orbit of any of the planets, or one can accept that Pluto is the largest known, but otherwise completely typical, Kuiper Belt object. The choice you make is up to you, but from the point of view of trying to understand the origin and significance of Pluto it clearly makes sense to take the second opinion. . . . Our perception of Pluto has been transformed from a singularly freakish and unexplained anomaly of the outer solar system to the leader of a rich and interesting family of Trans-Neptunian bodies whose study will tell us a great deal about the origin of the solar system. So, we have discovered -1 planets and +1 Kuiper Belt. It seems a fair trade to me." —MARIE L. THOMPSON

Further Reading

Asimov, Isaac, and Greg Walz-Chojnacki. *A Double Planet?: Pluto & Charon*. Milwaukee: Gareth Stevens Publishing, 1996.

Bakick, Michael E. *The Cambridge Planetary Handbook*. Cambridge, UK: University of Cambridge, 2000.

Chang, Kenneth. "Planetarium Takes Pluto off Planet A-List." *New York Times* (January 22, 2001).

Freedman, David H. "When Is a Planet Not a Planet?" *The Atlantic Monthly Digital Edition*. February 1998. <http://www.theatlantic.com/1>.

Harrington, R. S., and B.J. Harrington. "Pluto: Still an Enigma After 50 Years." *Sky and Telescope* no. 59 (1980): 452.

Jewitt, D, J. Luu, and J. Chen. "The Mauna Kea-Cerro Tololo (MKCT) Kuiper Belt and Centaur Survey." *Astronomical Journal* no. 112 (1996): 1238.

Marsden, B. G. "Planets and Satellites Galore." *Icarus* no. 44 (1980): 29.

Moomaw, Bruce. "Pluto: A Planet by Default." *SpaceDaily: Your Portal to Space* (June 28, 1999). <http://www.spacedaily.com/news/pluto-99a.html>.

"The Status of Pluto: A Clarification." News release from the International Astronomical Union. <http://www.iau.org/IAU/FAQ/PlutoPR.html>.

Stern, Alan A., and David J. Tholen, eds. *Pluto and Charon*. Tucson: University of Arizona Press, 1997.

———, and Jacqueline Mitton. *Pluto and Charon: Ice Worlds on the Ragged Edge of the Solar System*. New York: John Wiley & Sons, Inc., 1998.

Webster, Cynthia. "Pluto Is a Planet." News release by the Lowell Observatory, Flagstaff, Arizona (February 3, 1999). <http://www.lowell.edu/press/pluto_is_a_planet.html>.

Wetterer, Laurie A. *Clyde Tombaugh & the Search for Planet X*. Minneapolis: Carolrhoda Books Publishing Inc., 1996.

Is the International Space Station the appropriate next step for humanity's exploration of space?

Viewpoint: Yes, the ISS provides an effective platform from which manned exploration of the solar system can begin, and it represents an important model for international cooperation.

Viewpoint: No, the ISS is a poor use of valuable space exploration funds, and its low-earth orbit can do little in the way of generating creative new strategies that will make the exploration of distant locations in the solar system more feasible.

Space exploration must take place first in our imaginations. There, dreams and plans can flourish unencumbered by the drab realities of political, financial, and technological constraints. To assess the future of man's activities in space, we must ask both if we are moving toward the ultimate satisfaction of our ideal goals for exploration, as well as whether we are best using our limited resources right here and now. The International Space Station (ISS), the focus of space exploration at the start of the twenty-first century, has been a source of controversy since it was first proposed in the 1970s. The program to build the station was begun in 1984, and it was ready to begin hosting astronauts approximately 14 years later. Transformations in science, technology, medicine, and politics that took place in the intervening years have only made the International Space Station more controversial.

The proponents of the International Space Station argue that it provides an effective platform from which manned exploration of the solar system can begin. They suggest that living on the station will introduce astronauts to some of the physiological challenges likely to be presented by interplanetary visits. Experiments to be performed on the station will guide the design and production of systems for use in space environments. Advocates also hope that technological benefits, such as improved drug production, can be achieved for earthly uses as well.

The International Space Station has also produced a number of socio-political achievements. It is a fine example of successful international collaboration, involving the participation of 16 nations. Public interest in space has been stimulated by activities at the ISS, including the unintended publicity surrounding the visit of Dennis Tito, an American millionaire who—in the company of Russian cosmonauts and very much against the wishes of his countrymen at NASA—became the first space tourist in 2001.

But many interested in the future of space exploration charge that the ISS is not merely a poor way to spend limited resources. They argue that it fails the imagination as well, and that it will impede, not advance, the ultimate goal of man's exploration of the solar system. In many ways, the ISS is too similar to previous space stations Skylab and Mir, and even to the space shuttles. For example, it provides the same sort of micro gravity environment for scientific experiments that have already been provided by other space stations, and advances in our understanding of micro gravity—made during the long construction period of ISS—have rendered many of its planned experiments and projects redundant or even obsolete. Other developments, such as the evolution of a private sector-based space exploration industry, have changed the social, political, and economic context of the ISS.

Critics of the ISS also charge that it will do little or nothing to speed our access to more remote locations in space. In order for man to visit Mars or other planets in our solar system we must not only design new technological systems, we must also find strategies that change the economic constraints of space travel by taking advantage of things that can reduce the cost—for example, finding ways to utilize minerals and energy present on the Moon or Mars—while minimizing expensive efforts such as transporting heavy cargo from Earth's surface. The low-earth orbit of the ISS can do little in the way of generating creative new strategies that will make the exploration of distant locations in the solar system less expensive.

The future of the ISS is closely connected to the goals and constraints of society. Was the ISS a mistake, a timid half-step that has only taken away resources from bolder initiatives such as the colonization of Mars? Or is it a proud international achievement that has provided both a model for future international cooperation and a platform from which space exploration can advance? The answer is still being debated. —LOREN BUTLER FEFFER

Viewpoint:
Yes, the ISS provides an effective platform from which manned exploration of the solar system can begin, and it represents an important model for international cooperation.

The International Space Station, a technological marvel, is a bold and visionary achievement that has reaffirmed our commitment to making humankind's age-old dream of space exploration a reality. Furthermore, work on the station will not only expand our knowledge of medicine, physics, chemistry, and biology, but also provide insights into technological advances that will help build stronger economies on Earth.

The very existence of the ISS is a blow to arguments, dating back to the 1970s, that were voiced by critics who claimed that a space station wasn't needed, that it was an inappropriate "next step." The space station was needed then, and it is still needed now—not just for the reasons of science and national prestige that were first argued by its advocates back in the 1970s and 1980s, but for even more important reasons. The ISS has promoted international cooperation, helped ensure peaceful co-existence and prosperity among nations, and has helped keep the dream of space exploration alive.

No one can deny that the ISS is one of the most challenging technological, managerial, and political tasks ever accomplished by humankind. It has also become a public relations bonanza for those who support the space program. According to these supporters, the very existence of the space station has generated greater public interest in further space exploration. By engaging the public, the ISS helps humanity maintain its focus outward, toward the heavens. And as we live and work on the space station—and even visit like millionaire Dennis Tito—the media will continue to spotlight developments and generate interest and support for the space program. That focus may eventually cause us to take the next, "next step," whether it be a return to the Moon or a manned mission to Mars.

Critics of the ISS have stated that too much money has been spent on the station, when it could have been better spent here on Earth. But these critics fail to realize that the $50 billion-plus cost in the construction and maintenance of the ISS has been spent right here on the planet. It has helped fuel the economies in the 16 nations that cooperated to make the space station a reality. Money for the construction of the station supported the creation of thousands of jobs for highly skilled workers, including U.S. and Russian engineers who worked side by side on the space station instead of on projects involving armaments and missiles.

The argument against the ISS has always been that it cost too much, and that it was the wrong route to take to advance space exploration. Critics charged that the building of the ISS would turn out to be an orbital White Elephant, a colossal waste of time and money. Historically, such arguments have often been short-sighted. The "White Elephant" critics will be right only if they succeed in convincing citizens around the world that they should turn their backs on space exploration.

The ISS—A Platform for Further Space Exploration and Medical Research The idea of a manned orbiting structure astonishingly predates the birth of motorized flight. In 1857, author Everett Hale's science fiction narrative, "The Brick Moon," was published in the *Atlantic Monthly*. The possibility of humanity creating a station in space as a base for further exploration was first proposed by Russian rocket

The service module of the International Space Station on display in Russia prior to its launch.
(Photograph by Yuri Kadobnov. © AFP/Corbis. Reproduced by permission.)

designer Konstantin Tsiolkovsky in 1912, only nine years after the Wright brothers flew for the first time. In the 1920s and 1930s, German rocket scientists Werner Oberth and Werner Von Braun and other early rocketship pioneers also postulated a space station as a departure point for interstellar exploration. The ISS is the realization of the dreams of these early pioneers.

The ISS also provides opportunities for learning to manage extended stays in space, experience vital for exploration of the Moon, Mars, and the solar system. For proof of the role ISS will play in this new era of exploration, look no further than the multinational effort behind it. (This effort is further examined in the following section.)

"We need longer missions in space—missions that last months or even years," testified noted heart surgeon Michael E. DeBakey in 1993. His point was that space-station research on ways to combat the effects of weightlessness on the human body could yield "countermeasures that could be applicable to the aged and feeble, as well as astronauts." Better treatments for the bone loss of osteoporosis is one advance widely expected.

Undeniably, advances in biology, medicine, and physics will occur as a result of experiments that could not be conducted on Earth. Those advances will create new opportunities for financial empires just as early space research resulted in the development of new consumer products and materials such as Teflon non-stick coatings on pots and pans, heat-and cold-resistant ceramics, and solar cells that energize sun-powered calculators and automobile batteries. Future research aboard the ISS holds promise of delivering even greater consumer benefits. For instance, goals proposed for future ISS-based research include the development of purer, more powerful, but less toxic, forms of drugs that could be used to treat diseases ranging from AIDS and cancer to emphysema and high blood pressure.

As *Aviation Week and Space Technology* (February 1999) editorialized in a column entitled "The Space Station Wins on Cost-Benefit Analysis," "The station will dramatically improve our ability to live and work in space over the long-term. In the coming years, the station will also be the genesis of numerous medial, scientific, and technological discoveries that will benefit and inspire humankind for generations to come—and its critics will be long forgotten."

Those outcomes, when they occur, will convince future historians that Donald R. Beall, chairman and CEO of Rockwell International Inc., was right when he argued in a June 9, 1993, *USA Today* newspaper column that the then U.S. Space Station should be built. At that time, he called it the "next logical step in the exploration and habitation of space." He was right then and he is right today. In the future, we may very well look at the money invested on the construction of the ISS in the same way that

KEY TERMS

EMBRYOLOGY: Study of the changes that an organism undergoes from its conception as a fertilized egg (ovum) to its emergence into the world at hatching or birth.

ESCAPE TRAJECTORY: A flight path outward from a primary body calculated so a vehicle doesn't fall back to the body or orbit it.

GEOSTATIONARY ORBIT: A circular path 22,300 mi (35,900 km) above the Earth's equator on which a satellite takes 24 hours, moving from west to east, to complete an orbit, and so seems to hang stationary over one place on the Earth's surface.

GRAVITY WELL: An analogy in which the gravitational field is likened to a deep pit from which a space vehicle has to climb to escape the gravitational pull of a planetary body.

IONIZING RADIATION: Radiation that knocks electrons from atoms during its passage, leaving positively or negatively charged ions in its path.

LOW EARTH ORBIT: An orbit 93 to 186 mi (150 to 300 km) from Earth's surface.

MICROGRAVITY: Small gravity levels or low gravity.

NANOTECHNOLOGY: Building devices on a molecular scale. The idea of manipulating material on a nanometer scale—atom by atom—was first discussed by physicist Richard Feynman in 1959.

ORBIT: The path that a comet, planet, or other satellite follows in its periodical revolution around a central body. For instance, the Earth follows an orbit around the Sun. The Moon revolves around the Earth. Now, an artificial "moon," the International Space Station, also circles the Earth.

PAYLOAD: Originally, the revenue-producing portion of an aircraft load—passengers, cargo, mail. By extension, that which a spacecraft carries over and above what is needed for its operation.

PERMAFROST: A condition in which a deep layer of soil doesn't thaw during the summer but stays below 32°F (0°C) for at least two years, even though the soil above it thaws. It gives rise to a poorly drained form of grassland called tundra.

PHASE TRANSITION: A change in a feature that characterizes a system—like changes from solid to liquid or liquid to gas. Phase transitions can occur by changing variables like temperature and pressure.

SPACE STATION: An artificially constructed outpost in outer space in which humans can live and work without space suits. Salyut 1, history's first space station, was launched into orbit by the Soviet Union in April 1971.

TURBULENCE: The flow of a fluid in which the flow changes continually and irregularly in magnitude and direction. It's the opposite of streamline.

---

we now look back at the relatively paltry sums that were spent on the acquisition of the Louisiana Purchase by Thomas Jefferson or the purchase of Alaska, which at that time was called "Seward's Folly." The perspective of time has revealed that those purchases were not only excellent investments, but actual bargains!

The ISS—A Symbol of International Cooperation and Peacemaking Never before have so many space-faring nations pooled their interests, expertise, resources, and facilities to such an extent and for such an important venture. The creation of the International Space Station, the largest peacetime engineering project in history, was a 15-year struggle accomplished against enormous engineering and political odds. As such it represents a triumph of human will and cooperation.

"The ISS is the most cooperative international peacetime venture since the formation of the United Nations," wrote Fred Abatemarco in a *Popular Science* article published more than a half-year before the actual opening of the space station. As *Popular Science* also pointed out in a later article that celebrated the first visit of American and Soviet cosmonauts to the newly completed station in December 1998, "Only in war have nations come together on a larger scale."

Thus, it can be factually argued that the ISS has contributed to and will continue to contribute to World Peace, a peace that is likely to encourage further economic productivity and progress.

As the international teams of astronauts, scientists, and engineers live and work and share trials and triumphs on the ISS, most of humanity will vicariously share their grand adventure through broadcast, the Internet, and other media coverage. These explorers are likely to encounter unexpected challenges and difficulties, but those obstacles will be seen in retrospect as opportunities that produced unanticipated discoveries and rewards.

Indeed, the successful collaboration in creating and operating the ISS may, in the future, be seen as the cornerstone on which a lasting peace between Earth's suddenly not-so-far flung peoples was built. Thus, it is appropriate that the symbol of the ISS is a montage that includes four elements: a background of stars representing the conquest of space, a chain bordering an image of Earth that symbolizes the millions of people around the world who have worked to build the station, a five-pointed star representing the five national space agencies involved, and a laurel wreath symbolizing peace. —MAURY M. BREECHER

Viewpoint:
No, the ISS is a poor use of valuable space exploration funds, and its low-earth orbit can do little in the way of generating creative new strategies that will make the exploration of distant locations in the solar system more feasible.

In April 1997, physics professor Robert Park from the University of Maryland-College Park testified before the Committee on Science subcommittee on space and aeronautics on the International Space Station. He was also director of public information for the American Physical Society (APS), a 40,000-member professional organization of physicists, many of whom worked in the space program or on technologies that made it possible.

According to Park, at a time when opportunities for scientific discovery in space had never been greater, many physicists and experts in the biological sciences felt the space program's priorities are seriously misplaced.

"The space station," he said, "stands as the greatest single obstacle to the continued exploration of space." In a period of sharp NASA budget cuts, the ISS is a fixed cost that was exempted from the cuts, and whose construction cost overruns were balanced, he added, "by postponing what little science is planned for the station. The result has been the near paralysis of planetary exploration."

And Park isn't the first to say so. The Space Studies Board was established in 1958 to give external, independent scientific and program guidance in space research to NASA and other government agencies. In March 1993, Space Studies Board Chair Louis Lanzerotti sent a letter to NASA Administrator Daniel Goldin. NASA had announced it would conduct a sweeping review of the space station program—

"The current redesign efforts should be based on a realistic assessment of the depth and pace of America's commitment to human exploration of the inner solar system [Lanzerotti wrote]...If the goal of human exploration is superseded as the premise for the nation's space station program, planning and implementation of orbital research infrastructure should be adjusted to meet the requirements of the new objectives efficiently and cost-consciously. We must recognize, however, that such decisions might significantly delay the nation's option for human expansion into the solar system."

From its beginnings in 1984, when Ronald Reagan launched the program to build a permanently occupied space station in Earth orbit, the station was seen as a testbed for exploration technologies. There, fledgling spacefarers would learn to build and operate space habitats, prevent microgravity damage by establishing evidence-based medical knowledge and understanding the ecology of closed environments in space, and work together in small isolated international crews.

Now, not quite 20 years later—after many redesigns, cost overruns, and delays—the habitat is built and an international crew is working together. But as a testbed for exploration technologies, many experts consider the space station a typical U.S. space enterprise.

During a CNN Interactive special in November 1999, former NASA historian and Duke University history department chair Alex Roland described the future of space travel in the second millennium as "more of what we've had for the last 30 to 40 years, which is, at considerable risk and great expense, sending humans up into low orbit—[who] float around and look busy and come back to Earth—and not really accomplishing very much."

Scientifically speaking, according to Robert Park, in the official view of the American Physical Society, there's no justification for a permanently manned space station in Earth orbit. The reason? The International Space Station (ISS) is "yesterday's technology and its stated scientific objectives are yesterday's science."

Park cited lessons learned from *Skylab*, the first space station, abandoned in 1974 because the scientific return didn't justify its cost; 83 space shuttle missions, starting in 1981; and the Russian space station *Mir*, launched in 1986 and continuously occupied for 10 years—

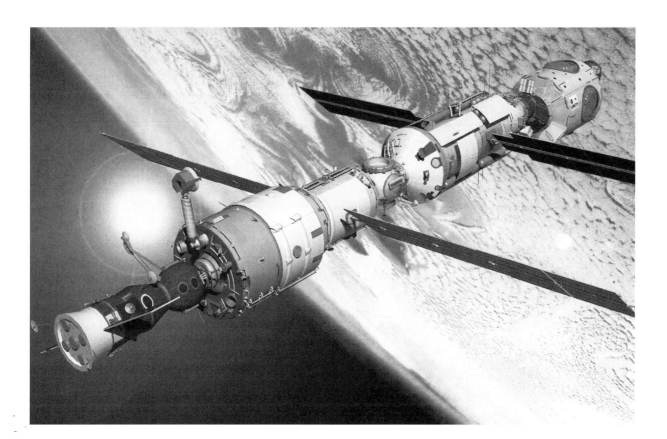

An artist's rendering of the International Space Station as it looked on November 8, 2000.
(© Reuters NewMedia Inc./Corbis. Reproduced by permission.)

- The microgravity environment is more harmful to human health than suspected—damage includes osteoporosis, muscle atrophy, diminished immune function, and chronic diarrhea.
- Astronauts in Earth orbit are exposed to cosmic and solar ionizing radiation at levels far greater than those tolerated for workers in the nuclear industry—levels high enough to raise concerns about central nervous system damage.
- The risk of a catastrophic encounter with orbital debris has grown substantially since 1984. Plans call for the station to maneuver out of the path of debris large enough to track, but smaller debris can penetrate the hull.

Microgravity is the only unique property of a space station environment, and the station was originally seen as a microgravity research lab. But, in the physical sciences, research completed on the shuttle or *Mir* produced no evidence that a microgravity environment has any advantage for processing or manufacture. A few basic experiments in areas like turbulence and fluid phase transitions might benefit from a microgravity environment, but they aren't high-priority experiments and could be conducted on unmanned platforms or on the shuttle.

And in the biological sciences—

- Experiments on the shuttle and *Mir* established that flies, plants, and small mammals can go through their life cycles in microgravity. It was valid to ask whether important biological processes like cell-cycle regulation, gene regulation, embryology, and cell differentiation could work without gravity. The answer is in: they can.
- Vestibular organs, bones, and muscles of larger mammals are affected by microgravity, but they are designed to sense or negotiate gravity so it's no surprise they're gravity-dependent. There is no evidence that studies of dysfunction in microgravity contributed to understanding how organisms function on Earth.

Where No One Has Gone Before Among the loudest protests against ISS are those who are dedicated to opening the space frontier to human settlement as quickly as possible. And except in the most indirect, long-term way, giving as many people as possible access to space isn't among the stated goals of the ISS.

For example, in a September 1999 keynote address at the *Space Frontier Conference VIII—Launching the Space Millennium*, NASA Administrator Daniel Goldin said when station is in a steady operational phase NASA hopes to turn the keys over to an entrepreneur—to make the ISS a commercial venture—and the government will be just another ISS tenant and user. If the station isn't commercialized, after 10 years of operation a national review will determine whether it still meets NASA's research needs and the needs of its international partners and the private sector. If not, it'll be shut down and de-orbited.

This decade-long window is what the late Gerard O'Neill—a Princeton professor, author of *The High Frontier: Human Colonies in Space* (1977), and founder of the Space Studies Institute—called the leisurely pace that characterized an isolated, protected program funded by a government dominant in the world. In his *Alternative Plan for US National Space Program*, O'Neill said the time scale for substantial accomplishment (measured as substantial economic or scientific return) must be shortened to outrun today's fierce global competition. That means five years maximum, rather than 20 to 40 years, for a significant return on investment.

Of course, none of the entrepreneurs who make up the nation's growing commercial space enterprise, or those who are working to open the space frontier to human settlement (the Mars Society, the Space Frontier Foundation, ProSpace, Back to the Moon, the Planetary Society), will wait 10 years—or even five—for the ISS to prove itself. And the growing number of advocates for the private use of space have their own ideas about how best to get there; most of these, in contrast to the options available with ISS, include actually *going*.

Testifying in October 1998 before the subcommittee on Space and Aviation of the House Committee on Science, former astronaut Charles "Pete" Conrad Jr. said US strategic goals over the next 40 years in space should include fostering a commercial space industry, exploring the solar system, settling the solar system, and exploring the universe.

Achieving any of these goals requires cheap access to space—transportation to low Earth orbit that's easy to operate, reliable, and inexpensive. There's no inherent technical barrier to creating such a capability. In energy terms, once you're in low Earth orbit, you're halfway to anywhere else in the solar system. Once we can put people, supplies, equipment and cargo in Earth orbit more effectively than we do now, that will open space up in a way that hasn't been possible to date.

A corollary to cheap access to space is that a lot of money can be saved and time scales for all space activities drastically shortened by making maximum use of resources that are already at the top of Earth's gravity well—like the intense solar energy available everywhere in space except in planetary shadows, and the abundant oxygen, silicon, and metals available on the Moon's surface. Up to now, all the materials used in space have been lifted from the Earth's surface—the bottom of Earth's gravity well.

According to Gerard O'Neill, the ratio of a rocket's liftoff weight (mainly propellants) to final weight (the remaining upper-stage rocket plus payload) is typically 60 to 1 for weights lifted to geostationary orbit or an escape trajectory. Deducting vehicle structure, the useful payload is typically only about 1/100th of lift-off weight. In contrast, for a rocket lifting a payload to escape from the Moon, payload plus vehicle can be as much as 60% of lift-off weight, instead of less than 2%, as in lift from the Earth. The advantage for launch from the Moon is 35-fold. And because the Moon is in vacuum, efficient launch machines there could operate on solar electric power and wouldn't need rocket fuel to launch materials into space.

The Space Studies Institute advises launching out of Earth's gravity well only items that will bring a high payback—information, intelligence (computerized or in the human brain), and sophisticated tools—not materials (like oxygen) or heavy structures. That's the way we opened the new world of the Americas, and it's the only practical way to open the new world of space.

In that sense, according to former NASA historian Alex Roland, nanotechnology represents the future of space technology. With nanotechnology, computers and machines can be built virtually at the atomic level. That means a very small package could be launched from Earth and become a very powerful instrument in space—not just a passive scientific instrument receiving information, but a machine that can do work in space. If nanotechnology proceeds at its current rate of progress, we will be able to send powerful instruments long distances at high rates of speed, and put very powerful instruments into low Earth orbit at a very slight cost. That has enormous commercial implications.

A Civilization of Spacefarers Robert Zubrin is founder and president of the Mars Society and author of several books about human space exploration, including *Mars Direct*, *The Case for Mars*, and *Entering Space*. He believes the space enterprise is a vehicle for transforming humanity into a spacefaring civilization—one that will have a virtually infinite number of new worlds to explore and settle, starting with Mars.

Like any other off-world destination, Mars travel involves developing launch systems that reduce the cost of space access. This hasn't been easy and it won't happen overnight. But government agencies like NASA can help by aiding new entrepreneurial launch companies in several ways. One is by directly contracting to develop small reusable launch systems. That's the most direct way, and it's how existing launch systems were developed.

By directing such procurement to small and medium-sized companies that are not now fielding launch systems, NASA and other agencies can create more competitors and different kinds of systems. Other subsidies include credits and tax breaks on investments. The government also could open the payload manifest for major space projects like the space station, including cargo delivery, on a competitive basis and give contingent contracts to such companies for some payloads.

To create private-sector companies to supply the medium- and heavy-lift launchers needed to build and sustain human space settlements would require a large manifest open to commercial efforts. The government could best help by establishing an initiative to permanently establish a manned base on the Moon or Mars. Once a base was established on Mars, and a large cargo requirement opened to competition, we would see the development of large and interplanetary launch systems. —CHERYL PELLERIN

Further Reading

Beall, Donald R., and Dick Zimmer. "At Issue: Should the Space Station Be Built? *USA Today* (June 9, 1993). Reprinted in *CQ Researcher* (Dec. 24, 1993): 1145.

Bergrun, Norman. "The International Space Station: A Momentous Cultural, Scientific, and Societal Undertaking." *AIAA Bulletin/The Insider News* (Sept. 1999): B18-22.

Burrows, William E. "Why Build a Space Station? *Popular Science* 252, no. 5 (May 1998 Special Issue): 65-69.

Caprara, Giovanni. *Living in Space: From Science Fiction to the International Space Station.* Buffalo, NY: Firefly Books Ltd., 2000.

Carroll, Michael, and Andrew Chaikin. "Our Man-Made Moons." *Popular Science* 252, no. 5 (May 1998 Special Issue): 101.

Ferris, Timothy. *Life Beyond Earth.* New York: Simon & Schuster, 2001.

Goldin, Daniel. "The International Space Station: Striding Toward the Future." *Popular Science* 252, no. 5 (May 1998 Special Issue): 11.

Heath, Erin. "The Greatest Show Off Earth." *National Journal* 33, no. 18 (May 5, 2001): 1336-37.

Heinlein, Robert. *The Moon Is a Harsh Mistress.* St. Martin's Press, 1997 reprint.

Kluger, Jeffrey. "Who Needs This?" *Time* 152, no. 21 (Nov. 23, 1998): 88-91.

Kraft, Christopher. *Flight: My Life in Mission Control.* New York: Dutton, 2001.

Proceedings of The Lunar Development Conference, Return To The Moon II, July 20-21, 2000, Caesar's Palace-Las Vegas, sponsored by the Space Frontier Foundation, Space Studies Institute, and The Moon Society. Available from The Space Frontier Foundation, $45. Orders, 1-800-78-SPACE, or info@space-frontier.org.

Sagan, Carl. *Pale Blue Dot: A Vision of the Human Future in Space.* New York: Random House, 1994.

Stover, Dawn. "The International Space Station." *Popular Science* 252, no. 5 (May 1998 Special Issue): 48-54.

United States Advisory Committee on the Future of the U.S. Space Program. *Report of the Advisory Committee on the Future of the U.S. Space Program.* Washington, D.C., National Aeronautics and Space Administration (distributed by the National Technical Information Service), December 1990.

Vizard, Frank. "ISS Isn't About Science." *Popular Science* 254, no. 1 (Feb. 1999): 73.

Zubrin, Robert. *Entering Space: Creating a Spacefaring Civilization.* J.P. Tarcher: 2000.

Does the accumulation of "space debris" in Earth's orbit pose a significant threat to humans, in space and on the ground?

Viewpoint: Yes, the probability of collisions between operational spacecrafts (including satellites) or between spacecraft and existing debris is increasing, and the presence of nuclear-powered spacecraft makes any collision potentially disastrous.

Viewpoint: No, evidence shows that no one has ever been killed by space debris, and satellites and space vessels have very rarely sustained serious damage from impacts in orbit.

Just as an individual ship seems tiny compared to the immensity of the sea, or an airplane relative to the entirety of the sky, man-made satellites seem very insignificant in size compared to the vastness of space. The collection of material referred to as "space debris" ranges in size from tiny paint chips to huge space stations. But relative size alone does not necessarily determine the extent of the threat to humans posed by orbital debris. After decades of space exploration and in the face of an increasing presence of space-based technology, debate rages on about how to assess and manage the risks associated with space debris.

Those who argue that space debris does not pose a significant threat to humans draw their confidence from the vastness of space itself, the ability of the atmosphere to burn up objects before they reach Earth's surface, and from ongoing efforts by engineers and scientists to minimize the proliferation of space debris and to protect against it. No one has been killed by space debris, and satellites and space vessels have very rarely sustained serious damage from impacts in orbit. Confidence that the good fortune of the past is likely to continue into the future, despite increased activity in space, is based in part on measures being taken by space agencies to improve their understanding and control of space debris.

Those who fear that space debris poses an imminent threat argue that the uncertainty surrounding the dangers of space debris is reason enough to proceed with the greatest of caution. Rather than draw comfort from the absence of any catastrophic collisions during the first four decades of space exploration, they point to a collection of observations of minor to moderate impacts as a warning sign of potential disaster. Although serious damage has been avoided, space debris has fallen in and near populated areas, injured people, killed livestock, damaged terrestrial structures, and caused observable damage to satellites and space vehicles. Any increase in space-based technology could easily lead to an increase in such events, with an accompanying threat of more serious consequences. Perhaps the greatest worry associated with space debris comes from the possibility that highly toxic nuclear and chemical fuels used in space vehicles could re-enter Earth's atmosphere and contaminate a populated area. —LOREN BUTLER FEFFER

Viewpoint:

Yes, the probability of collisions between operational spacecrafts (including satellites) or between spacecraft and existing debris is increasing, and the presence of nuclear-powered spacecraft makes any collision potentially disastrous.

The biggest problem with "space debris," also referred to as orbital debris, is that no one knows for sure how to predict the risk they present. The official word is that the chances of human casualties as a result of orbiting or reentering debris are (currently) small. These statistics are arrived at using modeling techniques based on current trends and conditions, both on Earth and in space. The models are subject to change as trends change and knowledge expands. According to NASA itself, the growth in amount of debris poses a "rapidly increasing hazard." A close inspection of our space age to date reveals many close calls and potential catastrophes. Most of them were averted as a result of plain luck, rather than engineering virtues.

Dangers of Space Debris The U.S. Space Command tracks more than 8,000 orbiting objects larger than 4 in (10 cm), of which only 7% are operational. The rest are debris—dead satellites, parts of exploded rockets, nuts, bolts, other lost hardware, etc. Most orbital debris is found in Low Earth Orbit (LEO), in altitudes below 1,243 mi (2,000 km). At these altitudes, objects collide at a speed of 6.2 mi per second (10 km/sec). In such velocities, even very small objects can inflict serious damage. It is estimated that smaller, non-tracked debris number in the millions. How much damage can an object smaller than 4 inches inflict in space? NASA documented a cracked space shuttle window from an impact with a paint chip estimated to be 0.008 in (0.2 mm) in diameter. Larger debris could cause damage up to and including a complete breakup of the spacecraft. And if a paint chip, which can't be tracked from the ground, cracked a space shuttle window, imagine what it would do to an astronaut out on a space walk. Space shuttle windows are replaced after many missions—the result of impact damage.

The size of the debris is not the only factor in the risk to space missions and the personnel aboard. In a 1997 report, NASA scientists pointed out that a single component failure due to a collision with debris could have repercussions ranging from simply functional—limited to that component without affecting the mission or craft—to catastrophic. The hypothetical example cited in the report is a perforation of an air-tank stored outside the manned module of a space station. In a worst-case scenario, the perforation could cause a pressure change in the tank that is strong enough to thrust the space station to an altitude from which orbit can't be recovered, resulting in a catastrophic failure of the space station. It will be dragged into Earth's atmosphere, where it will break up. No doubt human lives will then be at significant risk, if not lost. With the amount and sizes of debris near the space station, this scenario is not far fetched. According to the NASA report, the chances of collision increase with the size of the object and the longer it stays in orbit. The International Space Station (ISS) is very large, and is meant to stay in space for at least 10 years.

Arguably many improvements and better detection systems have been implemented on the shuttle and the ISS. But in a 1999 congressional testimony, Professor Stephen A. Book, member of the National Research Council's Committee on Space Shuttle Upgrades, had this to say about proposed upgrades aimed at protecting the shuttle from orbital debris: "Considering the predicted high level of risk from this hazard even after these modifications are made, the space shuttle upgrades program should solicit additional upgrade proposals for protecting the shuttle from meteoroids and orbital debris." And we must keep in mind that when people and machinery interact, even if protective measures are in place, the unexpected sometimes happens. In June 1999, the Air Force notified NASA that an old Russian rocket was going to pass too close to the then-empty ISS. A course alteration command that was sent to the station's computers was faulty, and the onboard computers shut ground controllers out of the steering system for 90 minutes, by which time it was too late to change course. As luck would have it, the rocket

KEY TERMS

DEBRIS SHIELDING: A lightweight protection that prevents damage to space crafts in orbit from hypervelocity impact.

LOW EARTH ORBIT: A circular orbit from about 90 to 600 mi (144 to 960 km) above Earth.

NEAR EARTH OBJECTS: Asteroids that come close to Earth's orbit.

SPACE DEBRIS (ORBITAL DEBRIS): NASA defines it as any man-made object in orbit around Earth that no longer serves a useful purpose.

TRAJECTORY: The curve that an object describes in space.

VELOCITY: The rate of motion of an object, distance per unit time, in a given direction.

Orbital debris in Saudi Arabia.
(NASA, Lyndon B. Johnson Space Center. Reproduced by permission.)

passed much further from the station than initially anticipated. A favorable miscalculation, but nevertheless it illuminates the unpredictability of space debris and our lack of reliable knowledge about them. Speaking to ABC News about the steering problem, James Van Laak, then deputy manager of space station operations, said, "This is just the first of many opportunities to be humble, and I hope that everyone will keep in mind that we're learning as fast as anybody." Can we afford this learning curve when astronauts are in the station, or in the shuttle?

When people rely on machinery to protect them in space, or trust not to fall on them from the sky, we must also ask ourselves how reliable the machinery really is. Shuttles, space stations, rockets, and satellites are only as good as their human designers and engineers, who may underestimate existing dangers during design and review stages. When Thiokol engineers sought to redesign the space shuttle O-ring, for safety reasons, NASA cut off funding, saying that the current design was good enough. Then came the *Challenger* disaster, and seven lives were lost. The official inquiry report cited poor O-ring design as a contributing factor in that tragedy.

Mother nature itself often foils even our best plans and protection measures. NASA's Skylab is a prime example. The hefty, expensive research facility was supposed to be refurbished by NASA after being boosted to a higher orbit. What no one at the time thought to consider was the increased activity of the Sun. The result of this increase was an expansion of our atmospheric region, which served to change Skylab's orbit. The orbit decayed rapidly, catching NASA unprepared. In the end, all that was left to try was to change the re-entry orbit so that it would hopefully pass over non-populated areas. These attempts failed. While many pieces of Skylab splashed into the ocean, plenty of pieces ended up on the ground, in Australia. In his book, *Collision Earth!* British astronomer Peter Grego recalls, "Ranchers in sparsely populated Western Australia found their estates strewn with hundreds of melted, twisted metallic fragments." The largest piece recovered was a 2.3-ton "fragment." One cow was reportedly killed by falling debris. Robert A. Frosch, who was NASA's administrator at the time Skylab crashed, was quoted as saying that he would rather look at chicken entrails than try to predict solar activity. Apparently he was right. On February 7, 1991, the Soviet space station Salyut 7 fell back to Earth as a result of increased solar activity. This time the people of Capitan Bermudez, Argentina, were the lucky ones, as once again debris littered a populated area. Attempts to change the reentry path were, once again, unsuccessful.

Falling Debris and Exposure to Radioactivity

Risks posed by orbital debris to people on the ground aren't limited to the heavy, twisted chunks of metal that fail to burn on reentry and fall from the sky. What very few people consider are the substances carried inside the various spacecrafts and rockets. Many times, these substances fall back to Earth. On January 24, 1978, after suffering a technical malfunction onboard, the Soviet satellite *Cosmos 954* disintegrated over the Northern Territories in Canada. Scattered across a vast area were thousands of radioactive particles, pieces of the satellite's nuclear power core that survived reentry. The Soviets were unable to predict where *Cosmos 954* might fall—they estimated somewhere between Hawaii and Africa; nor were they able to alter the satellite's flight path on reentry.

Over 60 nuclear devices were launched into orbit so far. NASA will launch three more missions involving nuclear-powered crafts in the coming years. The army isn't saying much, but nuclear militarization of space is a known trend and an ongoing debate. What happens if a nuclear space rocket is hit by space debris? Nine nuclear spacecrafts have fallen back to Earth, so far. The *Cosmos 954* incident was the worst one, and in fairness, some cores were never breached. But of the nuclear-powered spacecraft that fell to Earth, some released measurable radioactivity into the atmosphere. How do we know what effect the release of radioactivity into our atmosphere had on our health? Can we be sure that the next nuclear-powered craft that falls to Earth will once again miss a populated area, and that its core won't be breached?

Nuclear material falling from orbit is not the only hazardous substance humans are exposed to when space debris falls to Earth. The village of Ploskoye, in Siberia, is directly under the flight path of Russian launch vehicles, and has been so for 40 years. When the first stage of the rockets separates, a large amount of unused rocket fuel explodes and rains down on the village. The fuel used in some of these rockets contains a substance known to cause liver and blood problems. The fuel coats crops and contaminates the water supply. Ploskoye and its neighbors report cancer rates 15 times higher than the national average. There is also an extremely high rate of birth defects in the area. The village doctor reports a spate of new patients after each launch, and schoolteachers report that children complain of various ailments in the days following launches. In the past five years, there hasn't been a healthy newborn in the village. The traditional space debris—fragments from the rockets—has killed cattle in the area.

More often than not, accurate predictions cannot be made about landing sites of falling debris, unless the reentry is controlled from the ground. When the Russian Mars probe (another nuclear-power craft) fell to Earth in 1996, the U.S. Space Command targeted Australia in its predictions of the probe's landing site. The probe disintegrated over South America. Nor can we currently predict how much of a spacecraft will survive reentry. Intentional de-orbiting of some dead satellites, with the expectation that they will burn up on reentry, showed otherwise. According to the Center for Orbital and Reentry Debris Studies, "recent evidence shows that some portions...sometimes significant pieces, may survive reentry and pose a hazard to people and property on the ground." Some significant reentry events include a second stage of a Delta rocket that rained debris on Oklahoma and Texas on January 22, 1997. One woman was struck by a small piece of debris. Another Delta second stage reentered in early spring, 2000. Debris was found in a farm in South Africa.

Increasing Risks and Dangers When considering the risk posed by orbital debris, one must look not only at the current state of debris environment in space, but also into future conditions in the same environment. In a report issued in 1999, the UN Committee on the Peaceful Uses of Outer Space found that the probability of collisions between operational spacecrafts (including satellites) or between spacecraft and existing debris is increasing. To date, there is only one known case of a collision between cataloged man-made objects. In 1996, a fragment of an exploded Ariane upper stage rocket damaged the French satellite CERISE. As the report points out, the cause of many spacecraft break-ups in orbit is unknown, and might be the result of collision with orbiting debris. It is impossible to tell if the CERISE collision is truly a unique unfortunate event, or only the tip of the iceberg.

The continuing launches of spacecraft compounds the existing problem. More objects in orbit mean a greater chance of collision. These collisions, in turn, will generate more fragments. The end result, if significant remedial measures are not implemented, will be an exponential increase in both the number of orbital debris and the number of resulting collisions. These resulting collisions, according to NASA, will be more and more likely to happen between larger objects, compounding the problem.

Small satellites are important communication and research tools. Current planning will increase their number in LEO significantly. These satellites sometimes weigh only ounces and measure less than 1 in (2 cm). According to a recent analysis, small satellites will create constellations in orbit that are equivalent to debris clouds. Debris clouds are swarms of debris, clustered together, that are left in orbit after a break-up or explosion of space junk. They pose increased risk to operational spacecraft because they occupy a larger area than single pieces of "space junk." Owing to their design, small satellites have no collision avoidance mechanisms, they are difficult to track, and they have no means of what NASA terms "self disposal"—the ability to boost the satellite to a higher orbit or dispose of it through controlled reentry into Earth's atmosphere. In addition, these satellites are much denser than today's conventional satellite. Their higher density ensures they will stay in LEO for a very long time, but their operational lifes-pan is actually short. Thus they will be replenished at regular intervals, unlike true debris clouds that eventually disperse. What we are faced with, then, is a possible trend of increasing, hard-to-track "debris clouds" that will be replenished at regular intervals and fly blind. This situation will increase the risk to safe space flight operations. Many of these satellites are privately owned and operated, which makes them harder to regulate.

In less than half a century of space operations, we've left quite a legacy: dead cattle, contaminated environment on Earth and in space, birth defects. As more attention is paid to the issue of orbital debris, more steps are being taken to mitigate the problem. This trend should be encouraged and enforced. Currently, the hazards of orbital debris are very real. To deny that a danger exists simply because we've been lucky so far is foolhardy. —ADI R. FERRARA

Viewpoint:
No, evidence shows that no one has ever been killed by space debris, and satellites and space vessels have very rarely sustained serious damage from impacts in orbit.

The Story of Chicken Little Most of us remember the children's story about Chicken Little who ran around shouting, "The sky is falling! The sky is falling." In truth, Chicken Little had mistaken a simple acorn to be a sign of impending catastrophe. Much like this fictional character, doomsayers would argue that the sky is actually falling and that space debris threatens to destroy life as we know it both on Earth and in space. However, experts disagree and evidence indicates that the accumulation of space debris is not as significant a hazard as some people would have us believe.

At first, the very concept of space debris appears to be a tremendous risk, especially for those traveling in space. The speed of orbital debris, the term sometimes used by NASA for space debris, can be approximately 6.2 mi/second (10 km/second). At that velocity, you could drive across the United States, coast-to-coast, in about seven and a half minutes. Even something as small as a fleck of paint moving at that rate of speed could cause damage to something in its path. Then couple that knowledge with photographs that show clusters of space debris floating around Earth and it isn't hard to understand why some people might believe that a significant threat exists.

However, this risk has been overstated and sometimes even exaggerated. History has shown that even with the copious amount of space debris circling Earth, it has had very little effect on space exploration, and even less on the planet below. The reasons behind this are many, but include the vastness of Earth and space, protective measures, and early detection systems. Together these factors have reduced the possible risk dramatically.

Most space debris that falls to Earth comes from Low Earth Orbit, which is generally considered from 90 to 600 mi (144 to 960 km) from Earth's surface. The team at NASA has, however, expanded that range to include an area approximately 1,250 mi (2,000 km) from Earth's surface. How long it takes to fall can range from a few years to over a century depending on its height. Upon reentry into Earth's atmosphere, the majority of this debris is incinerated. Anything that does survive the trip down typically lands in an unpopulated stretch of the planet such as a desert or ocean. NASA's Space Science Branch at the Johnson Space Center believes that approximately one cataloged piece of debris has fallen to Earth every day for the last 40 years. Thus far, no serious injuries or property damage have occurred as a result of this falling debris. Thanks to the atmosphere and the sheer size of Earth's land mass, the risk that falling space debris poses to anyone is extremely small.

While space does not have an atmosphere to burn up space debris, Dr. Susan Holtz, a physicist and university professor, points out that "as the solar system persists, it gets cleaner." In other words, when considering the problem of space debris, it is important to look at the big picture. Often we tend to think of space as much smaller than it is in reality. One way Dr. Holtz explains the size of space to her students is to tell them the following story:

"Space is big. To give you an idea of how big it is, let's go on a space trip from the Earth and travel toward the Sun. Let's drive day and night at 100 miles an hour and not take a pit stop. It'll take us 100 years driving day and night at 100 miles an hour to get to the Sun. After 29 years we would cross Venus' orbit, and after 65 years we would cross Mercury's orbit."

Considering the small size of objects like satellites or the shuttle placed against an envi-

Delta rocket debris in Texas in 1997.

(NASA, Lyndon B. Johnson Space Center. Reproduced by permission.)

ronment as vast as space, the risk of severe collisions is minimal. Even when an object in space is hit by space debris, the damage is typically negligible even considering the high rate of speed at which the debris travels. Thanks to precautions such as debris shielding, the damage caused by space debris has been kept to a minimum. Before it was brought back to Earth via remote control, the MIR space station received numerous impacts from space debris. None of this minor damage presented any significant problems to the operation of the station or its various missions. The International Space Station (ISS) is designed to withstand direct hits from space debris as large as 0.4 in (1 cm) in size.

Most scientists believe that the number of satellites actually destroyed or severely damaged by space debris is extremely low. The Russian *Kosmos 1275* is possibly one of these rare instances. The chance of the Hubble Space Telescope suffering the same fate as the Russian satellite is approximately 1% according to Phillis Engelbert and Diane L. Dupuis, authors of *The Handy Space Answer Book*. Considering the number of satellites and other man-made objects launched into space in the last 40 years, the serious risk posed to satellites is astronomically low.

In fact, monitoring systems such as the Space Surveillance Network (SSN) maintain constant track of space debris and Near Earth Orbits. Thanks to ground-based radar and computer extrapolation, this provides an early warning system to determine if even the possibility of a collision with space debris is imminent. With this information, the Space Shuttle can easily maneuver out of the way. The Space Science Branch at the Johnson Space Center predicts the chance of such a collision occurring to be about 1 in 100,000, which is certainly not a significant enough risk to cause panic. Soon the ISS will also have the capability to maneuver in this way as well.

Understanding How a Space Debris Field Works Understanding how a space debris field works also helps to minimize the risk of harm to objects in space as well as people on Earth. Like any other physical object, the more we understand what elements affect it, the more we can anticipate change. For example, we know that over time water flowing over rock will eventually erode it. It may take a long time, but the erosion will occur. In many ways, time is on our side with regard to space debris. By understanding it better and making sure we don't add to the problem by creating more debris, we can develop simulations on how the objects in space are likely to react. Trying to describe and predict the current and future status of a space debris field is, without question, a monumental task, but scientists are working on it. By using computer programs to calculate debris characteristics such as mass, trajectory, and velocity, NASA researchers can create mod-

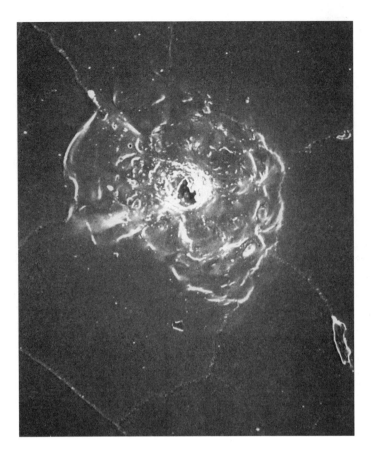

Damage to the space shuttle *Endeavor* from orbital debris during its December 2000 mission.
(NASA, Lyndon B. Johnson Space Center. Reproduced by permission.)

els that reflect the way a debris field behaves. One of the most intricate models is called EVOLVE. It tracks the progression of a space debris field, indicating how certain factors can influence it. In this manner, future complications caused by launches or accidental explosions can be predicted and/or avoided. After all, the ability to anticipate a problem and prepare for it is a large part of the equation when we evaluate risk significance.

In addition, space agencies around the world have taken steps to reduce space clutter. The United States, for example, has taken an official stand that is outlined in the 1996 National Space Policy that clearly states: "The United States will seek to minimize the creation of new orbital debris." For example, space mechanics are far more careful with regard to their tools. In the past, space mechanics sometimes let go of their tools and were unable to recover them. Strident efforts are now made to retain all objects used to repair satellites and conduct other missions. The Russians have also agreed to do their part. They used to purposely destroy their equipment in space to prevent it from falling into the wrong hands, but now refrain from that practice. Newly designed crafts and operating procedures also play a part in helping to keep space clean, while researchers continue to investigate safe ways to clean up the debris that currently exists. Everything from forcing the debris to reenter the atmosphere in a controlled manner to nudging it away from the Earth's orbit has been discussed. An activity such as collecting garbage from inside the space station and sending it back to Earth to burn up at reentry is one tangible way space explorers are helping to ensure the reduction of space clutter.

At this time there is no international treaty on how to deal with space debris; however, several nations have joined together to form the Inter-Agency Space Debris Coordination Committee (IADC). The IADC assesses the subject of space debris and how it should be handled in the future. Japan, like the United States, has developed a list of safety policies regarding space debris. Because this is ultimately a global issue, other countries such as France, The Netherlands, and Germany have jumped on the bandwagon with regard to addressing this issue.

Despite all the public discussion regarding space debris, some people fear that significant danger is being hidden from them so that space exploration can continue without interruption. However, there is no evidence that supports this theory. In fact, this theory doesn't make much sense in principle. Nonetheless, rumors concerning space debris as well as earth-shattering asteroids continue to circulate and are sometimes fueled by the media. However, perhaps Stephen P. Maran, recipient of the NASA Medal of Exceptional Achievement, puts it best. In his book entitled *Astronomy for Dummies* he says, "Conspiracy theorists think that if astronomers did know about a doomsday asteroid, we wouldn't tell. But let's face it, if I knew the world was in danger, I would be settling my affairs and heading for the South Sea, not writing this book."

Looking at the Big Picture After all, in a world inundated with health warnings and cautions, it is no wonder that people are becoming overly cautious about everything. It's getting more difficult to determine what is media hype and what things should really concern us. Many people have become either desensitized to the barrage of cautions or they have become needlessly worried. In truth, science should embrace logic and balance. Naturally, some things really do pose a significant and imminent threat to our health and lives, but we need to keep it all in perspective. Most of us stand a better chance of getting hit by lightning than we do of getting hit by space debris. So, what about the risk the astronauts face in space? While it is true that they face a greater risk of getting hit by space debris than we do on Earth, Dr. Susan Holtz puts it best when she says, "They're sitting on liquid fuel and then torching it so that the rocket will shoot into space. It is my opinion that this poses a greater risk to their well-being than space debris." In short, in the grand scheme of things, there's a lot more to worry about. Therefore, when analyzing the significance of the risk posed by the accumulation of space debris, it is important to focus on the big picture. Consider all that is being done to understand and correct the problem and keep in mind that the accumulation of space debris has not posed a significant threat thus far. Of course, to dismiss the problem of space debris altogether is certainly irresponsible, but there's no need to be a modern-day Chicken Little either. We are not in immediate danger and to be fearful that something is going to fall from the sky and kill us is overreacting to the problem. —LEE A. PARADISE

Further Reading

Center for Orbital and Reentry Debris Studies. <http://www.aero.org/cords/index.html>.

Chiles, James R. "Casting a High-tech Net for Space Trash." [Abstract]. *Smithsonian Magazine*. 1999. <http://www.smithsonianmag.si.edu/smithsonian/issues99/trash.html>.

Dunn, Marcia. "Astronauts Boost Space Station, Take Out Trash." *Nandotimes*. 2001. <http://archive.nandotims.com>.

Engelbert, Phillis, and Diane L. Dupuis. *The Handy Space Answer Book*. Detroit, MI: Visible Ink Press, 1998.

Maran, Stephen P. *Astronomy for Dummies*. Foster City, CA: IDG Books Worldwide, Inc., 1999.

NASA's Orbital Debris Web site. <http://www.orbitaldebris.jsc.nasa.gov/orbital_debris.html>.

The Orbital Debris Research Project. <http://sn-callisto.jsc.nasa.gov>.

The Star Child Team. "What Is Space Trash?" 2001. <http://starchild.gsfc.nasa.gov/StarChild/shadow/questions/question22.html>.

Wilson, Jim. "Killer Garbage in Space." *Popular Mechanics*. 2001. <http://popularmechanics.com/popmech/sci/9608STSPM.html>.

<http://www.space.com>.

Historic Dispute: Is Urbain Le Verrier the true discoverer of Neptune?

Viewpoint: Yes, Le Verrier's calculations and his tenacity in compelling others to search for Neptune make him the true discover of the planet.

Viewpoint: No, John Couch Adams is the true discoverer of Neptune.

More than subtle differences lie at the center of the debate over who is the true discoverer of Neptune. At question here is what constitutes a true discoverer. Thus, each person's definition of justice and fair play are significant factors in the way this issue is viewed. Differences of opinion abound regarding what is more important. Some people say that the title of "discoverer" should go to the first person who theorized about the existence of Neptune. Others say something more is required and that it isn't enough to have a good hunch; verification is essential. The theory has to be placed within the context of what we know to be true. At the same time, progress is often made when we are willing to color outside the lines. It is, after all, the first person to propose a new theory who gets the ball rolling. But what if that person doesn't possess all the necessary tools or skills to verify the idea? Some people say that it is the one who actually proves the theory who deserves the credit. Still others would argue that the glory should be shared.

And, of course, one must look at the reality of how things work. Regardless of how unfair it might seem, a good idea is rarely acknowledged, except in some circles. Most of the time it isn't until a scientist sinks his or her teeth into proving a theory and actually sees the process through to a tangible result that a discovery is officially recognized. For better or ill, part of this recognition involves self-promotion.

One problem that disrupts the proper attribution of scientific discoveries in all fields is timing. There are many cases where a discovery has been made, but not recognized, even by the discoverers themselves. Many times, these discoveries are disputed or ignored, sometimes for decades, only to be reaffirmed later on. In such cases, who receives the accolades, the initial founder or the person who proved the discovery? Many times, scientists must give credit for findings in a retrospective capacity, awarding acknowledgement to former researchers who could not validate their claims and/or were ignored by their peers, because their findings were too controversial at the time of the discovery.

This becomes even more convoluted when one deals with theoretical discussion—a key part of scientific fields such as astronomy. Can it be said that a theory is the guiding force behind a discovery? After all, were it not for the original source of conjecture, why would the scientists begin looking for those facts? Either because of lack of skill, technology, or resources, these theorists cannot prove their own hypotheses. However, does this negate their contribution in a particular discovery?

For these reasons, there is debate whether or not Urbain Jean-Joseph Le Verrier, a French mathematician, truly discovered the planet Neptune. During the initial study of Uranus, scientists began to question the unusual qualities of the planet's orbital behavior. As these studies progressed, a theory developed

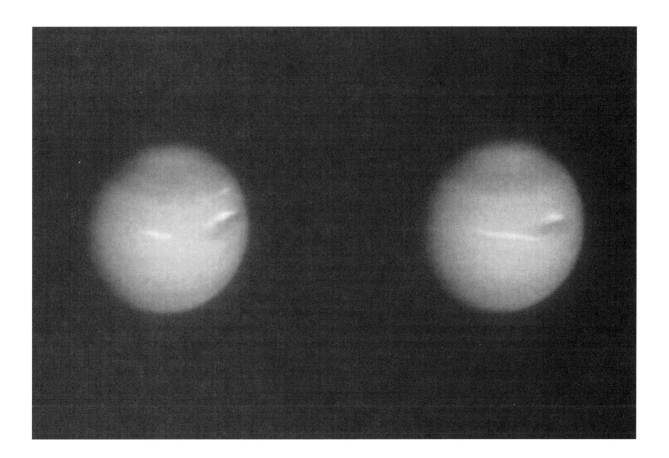

Neptune
(Neptune, photograph by U.S. National Aeronautics and Space Administration (NASA). JLM Visuals. Reproduced by permission.)

concerning the gravitational influence of an as-yet-undiscovered planet on Uranus's orbit. Two such theorists were Thomas Hussey and John Couch Adams. Both were, for the most part, ignored by their peers. Hussey was ignored because of his lack of skill in mathematics, and Adams, although skilled, was ignored because he was not well known.

Le Verrier would later independently develop his own hypotheses concerning this unseen planet. However, unlike those of his predecessors, Le Verrier's theories would be published and accepted by his fellow peers. One of these peers was George Airy, director of the Royal Observatory at Greenwich. Airy would play a key role in inspiring James Challis at the Cambridge Observatory to search for the predicted planet. Airy was also aware of, and had previously refuted, Adams's research, which was surprisingly similar to Le Verrier's calculations.

Due to poor communication within the scientific community, several independent searches began for what would later be named Neptune. Many of these searches would result in failure. However, inspired by Le Verrier's prompting, a pair of German astronomers, Johann Galle and Heinrich d'Arrest, found the elusive planet in September of 1846. What was not known at the time was that Challis had already observed the planet the month before, but had not properly recorded it due to time constraints.

So, this leaves us with the question of who should be accredited for Neptune's initial discovery. Many would say that Le Verrier's calculations and his diligence to have astronomers search for the planet should provide him with the accreditation. However, his calculations are very similar to the less experienced and tenacious Adams. And almost a decade before Adams, Hussey developed the theory but could not provide the calculations to back up his claim. We must also ask if providing the theory of an item's existence is more important than the actual proof of its existence. In Neptune's case, even this is debatable, as Challis observed Neptune before Galle and d'Arrest but didn't realize what he'd uncovered.

It cannot be disputed that Le Verrier's skill and tenacity helped guide the discovery of Neptune. However, one must also weigh the numerous factors attributable to that discovery. This includes deciding whether or not it is the development of theory or the verification of fact that determines where the final accreditation should be placed. Many times a discovery results from a combination of factors and events. Giving credit for a discovery is not always a clean-cut decision. Often it is the person demonstrating the greatest leadership qualities or the one who is most persistent who is provided the ultimate recognition. However, it can be debated whether this is the fairest assessment of accomplishment in every situation, especially with regard to the final discovery of Neptune. —LEE A. PARADISE

KEY TERMS

MATHEMATICAL ASTRONOMY: The study of astronomical bodies from a purely mathematical basis, as opposed to observational astronomy (which can still be heavily mathematical). Mathematical astronomy is often applied to difficult or impossible-to-see objects and situations, such as the location of an unseen planet or the future positions of the known planets.

PLANETARY ORBITS: Ancient astronomers assumed the planets moved in perfect circles, but in the seventeenth century Johannes Kepler showed that they moved in elliptical orbits. Issac Newton's laws of motion seemed to accurately predict the motion of the known planets. Indeed, the small deviations from the predicted paths of some of the outer planets enabled unseen planets to be discovered, further strengthening Netwon's laws. However, the orbit of the innermost planet, Mercury, could not be explained until Albert Einstein's Relativity Theory. For most practical purposes Newton's laws are still used, and elliptical orbits hold in all but the most extreme circumstances.

INVERSE PERTURBATION THEORY: The method of calculating the mass and orbital information of an unseen astronomical body by studying the deviations (perturbations) from other bodies' calculated orbits. It was successfully used to explain the perturbations in the orbit of Uranus, and thereby discover the planet of Neptune. However, attempts to calculate the position of a further planet were not successful, although eventually the ninth planet of Pluto was discovered using photographic comparisons.

TRANSURANIC PLANET: Any planet that lies beyond the orbit of Uranus. Similarly a transneptunian planet is one that lies beyond the orbit of Neptune. In the early nineteenth century the existence of any transuranic planets were generally discounted, until the problems with the observed positions of Uranus inspired some to search for a such an entity.

Viewpoint:
Yes, Le Verrier's calculations and his tenacity in compelling others to search for Neptune make him the true discover of the planet.

Urbain Le Verrier was not the first person to observe the planet Neptune. He was not even the first person to calculate the position of the planet with any accuracy. The calculations he performed, predicting the location of Neptune, were accurate only through good fortune. However, without his efforts Neptune would not have been discovered. His personal energy and mathematical skill, combined with a good deal of persistence and some luck, led to the first confirmed sightings of Neptune. For this, he and no other should be considered the true discoverer of Neptune.

Since antiquity, viewers of the night sky noticed that only five of the stars in the sky moved in relation to all the other "fixed stars." These were called the wanderers, or planets, and came to be known in the Western world as Mercury, Venus, Mars, Jupiter, and Saturn. Due to their use in navigation they became the most studied objects in science. Then in 1781 the world was shocked by William Herschel's discovery of a new planet. At first not even Herschel dared to believe the object he had found was a planet, but the evidence was overwhelming. Eventually it came to be called Uranus.

Herschel's new planet immediately began to cause problems for scientists, as its movements seemed to defy mathematical prediction. Going over old records, it turned out that other astronomers had sighted Uranus earlier than Herschel, but had failed to recognize its wandering nature. However, these older observations refused to fit into any standard planetary orbit. Many reputations were damaged, and many inaccurate tables charting Uranus's movement were made. Something appeared to be wrong with planetary physics. Some even suggested that Newton's universal laws of gravity were not that universal after all.

Theory of a Mystery Planet A number of suggestions were made to explain the strange behavior of Uranus, which lagged behind the expected positions. Comet impacts, a strange substance that existed only far out in space and slowed the planet, or mathematical and observational errors were all considered. Then in the mid-1820s Uranus began to speed up, further confusing the issue. A new theory began to gain some popularity. Perhaps Uranus was being affected by another undiscovered planet. One of the first to seriously consider finding this imaginary planet was an English clergyman, Thomas Hussey, who had a keen interest in astronomy. In 1834 he suggested that someone could calculate the position of this unseen planet, but since such mathematical endeavor was beyond his knowledge, he approached England's foremost astronomer, George Airy, but Airy dismissed the idea and Hussey made no further attempt to promote it.

In 1843 another Englishman, John Couch Adams, also theorized that another planet was responsible for the strange motions of Uranus. However, unlike Hussey, Adams was a brilliant mathematician. He began a series of complex calculations, and using some inspired guesswork, came up with an estimated location for this mystery planet. Adams's next step was to get an astronomer to search the area near his estimated location. He tried George Airy, but once again Airy was skeptical, and with good reason. Adams was a young man with no track record in astronomy, and yet he was claiming to have done something no one else had even attempted. Airy had other things on his mind, and his own calculations suggested that a new planet would not solve all of Uranus's movement problems. There was also a comedy of errors between the two men, who never managed to meet at this time, despite several attempts. If Adams could have explained his work in person, and in full, it may have convinced Airy more than did the short written notes they exchanged.

Le Verrier Estimates the Location of the Mystery Planet There the matter lay until the French mathematician Urbain Jean-Joseph Le Verrier (often written Leverrier) entered the ring. Le Verrier also independently speculated that an undiscovered planet might be the reason for Uranus's strange behavior. But whereas Adams had just graduated, Le Verrier had a distinguished academic history, held a prestigious teaching post at the Ecole Polytechnique in Paris, and had written a number of papers on mathematical astronomy. In 1845 he began to consider the motion of Uranus seriously. His first paper on the subject showed that the influence of Jupiter and Saturn were not enough to account for the inconsistencies in Uranus's motion. In England, Airy read Le Verrier's paper, and while impressed with it, still did not take Adams's calculations seriously. Then, in June 1846, Le Verrier published a second paper that showed conclusively that there was no standard elliptical orbit for Uranus that fitted the observations. Something else must be influencing the strange path of Uranus. The paper concluded that this new element must be an undiscovered planet. Le Verrier then performed an exhaustive series of calculations to determine the position of this mystery planet. Beginning with simple assumptions he slowly narrowed down the position of the new planet by more complex calculations. The paper ended with an estimated location of the unseen planet, and a promise that Le Verrier's next paper would contain more detail on the planet's orbit.

Le Verrier's estimated position was very close to that of Adams's, despite the difference in their methods. However, they were unaware of each other's work. The only person who had access to both calculations was Airy, who received Le Verrier's paper in late June. Airy began to give much more credence to Adams's work, but still had his doubts, which he communicated to Le Verrier in a short letter (although he did not mention Adams's calculations at all). Le Verrier's answer dispelled any doubts Airy had, and he finally began to press for a search for the new planet.

Although Airy was the director of the Royal Observatory at Greenwich, he did not search for the new planet himself (he had poor eyesight), or get others there to do so. Instead, he urged James Challis at the Cambridge Observatory to perform the search. Cambridge had a larger telescope than any at Greenwich, and Airy did not want to disrupt the day-to-day work of the Royal Observatory. Furthermore, Airy, like Adams, was a Cambridge graduate, and may have wanted to keep everything "in-house."

Search for Neptune Begins Challis was on holiday, but began the search when he returned on the 18th of July 1846. The only person who was told of the search was Adams, who was delighted that his work was finally being pursued. Challis chose a painstaking method to find his quarry, comparing star positions from several nights' observations to see if any had moved. However, there were too many objects in the target area to compare all of them, and so Challis only compared the brightest 39. Unfortunately for him he observed the planet in August, but it was the 49th brightest object he had recorded, and so it remained unchecked. He continued his search into September, but did not make the crucial comparison.

Meanwhile others were also looking. In France, Le Verrier's paper prompted a brief search by astronomers in the Paris Observatory, but they stopped in August. John Russell Hind, an ex-employee of Greenwich, was also inspired by Le Verrier's paper to mount his own brief search, but did not immediately find anything. At the end of August, Le Verrier presented a third paper in which he gave a predicted mass and brightness for the unseen planet, as well as a more accurate estimated position. The paper urged a full-scale search using the method of searching for the planet's disc. Stars are so far away that they do not increase in size when they are viewed at increasing magnification, whereas planets, being much closer to the observer, do. By flipping between magnifications on a telescope an observer should be able to tell a planet from the stars by the change in size of the planet's disc. However, there was a distinct lack of interest in performing such a search. Le Verrier's difficult personality may have had something to do with this. He was considered arrogant and abrasive by many of his peers, and had more enemies than friends in astronomical circles.

Le Verrier became frustrated that no one was looking for his calculated planet, unaware that in England Challis was doing just that. In September he remembered a copy of a dissertation he had been sent by a young German student at the Berlin Observatory, Johann Gottfried Galle. He quickly sent a letter to Galle asking him to search for the planet, hoping to overawe the young student enough to get him to go behind the back of his supervisor. Galle wisely discussed the matter with his superior, and was given some time on the largest telescope in the Berlin Observatory. Galle, and another student, Heinrich d'Arrest, began to look immediately. They were greatly aided by their access to a new, as yet unpublished, star map of the search area. On their first night they found a star that was not on their map. The next night they checked the position of their "star," and were delighted to find that it had moved. They had found a new planet.

Le Verrier wrote to a number of astronomers informing them of the discovery, including Airy, and suggesting that the planet be called Neptune. However, he later changed his mind, and decided that "Le Verrier's planet" would be a better name. A naming battle ensued, similar to an earlier one over Uranus. Airy and Challis finally decided to go public with the details of their search, and on the prior work of Adams. They suggested the name Oceanus for the new planet, thinking their work gave them some rights in the naming of the planet.

In France the news of Adams's work was met with suspicion. After all, if Adams had indeed made such calculations why had they not been published, or at least communicated informally to other astronomers? A war of words flared up, and an impassioned controversy based strongly on nationality began. Rumors and conspiracy theories ran wild, and many in England believed Adams to be the real discoverer of the new planet.

Airy and Adams Credit Le Verrier for Discovering Neptune However, the arguments against Le Verrier being the true discoverer of Neptune were discounted by the two Englishmen at the center of the controversy, Airy and Adams. Airy noted in a letter to Le Verrier shortly after the discovery: "You are to be recognized, beyond doubt, as the real predictor of the planet's place. I may add that the English investigations, as I believe, were not quite as extensive as yours." Adams also acknowledged Le Verrier's claim at a meeting of the Royal Astronomical Society, on November 13, 1846. He stated that Le Verrier had "just claims to the honours of the discovery, for there is no doubt that his researches were first published to the world, and led to the actual discovery of the planet by Dr. Galle."

Over the years a number of other arguments have been made against Le Verrier taking full credit for the discovery. Le Verrier's (and Adams's) calculations were based on much guesswork, particularly in regards to the orbital distance of Neptune from the Sun. They were also both extremely fortunate that Uranus and Neptune were close together in the period from 1800–1850, giving rise to very obvious changes in the speed of Uranus. While it was true there was a great deal of luck and guesswork involved, this should not diminish Le Verrier's work. Many other discoverers in science have been as lucky.

While Adams deserves some credit for his work, it was not at his urging that a search for the new planet was begun or maintained. Adams's work was less rigorous, less accurate, and above all less persuasive than that of Le Verrier. Adams himself did not think he deserved the credit for the discovery, and while he may have been a more modest man than the pompous Le Verrier, he was not merely being polite. Le Verrier had done more to find the new planet, both mathematically and in urging the search. While Galle and d'Arrest also deserve credit for sighting the planet, they would not have been looking without Le Verrier's information. And while Challis may have been the first to see the planet while searching for it, he did not realize its nature. Although Adams's calculations could have led to the discovery, the secrecy kept by Airy and Challis, and the lack of analysis of star sighting by Challis, "robbed" Adams of the claim of co-discoverer. For even if Challis had first sighted the new planet, Adams's work was taken seriously only because of Le Verrier's work.

Le Verrier deserves to be recognized as the true discoverer of Neptune, as without his efforts the search would not have been made at that time. He calculated the correct place in the sky, and no matter how much luck was involved, the planet was where he said it would be. Had Le Verrier not published his calculations, no astronomer would have looked for Neptune in the proper area, and the discovery of the planet would have been delayed by many years. —DAVID TULLOCH

Viewpoint:
No, John Couch Adams is the true discoverer of Neptune.

John Couch Adams (1819–1892) cannot be credited with the "optical" discovery of the transuranic planet Neptune nor can Urbain Jean Joseph Le Verrier (1811–1877). A cursory acquaintance with the facts suffices to show that priority in this matter unquestionably belongs to the German astronomer Gottfried Galle (1812–1910) and his assistant Heinrich Louis d'Arrest (1822–1875). Nevertheless, ever since Galle and d'Arrest identified Neptune on September 23, 1846, the priority dispute over the planet's "discovery" has invariably, and quite rightly, focused on Le Verrier and Adams. A close look at the facts, though, indicates that priority belongs to Adams alone.

Before proceeding, a few remarks regarding the nature of scientific discovery are in order. Traditionally, discovery has been viewed as a unitary event, occurring at a specific time and place and attributable to an individual or small group of individuals. Thomas Kuhn rejected this monolithic account. He argued that many scientific discoveries—such as oxygen, electric charge, etc.—cannot be isolated in this way because the discovery processes itself is instrumental in creating the identity of that which is discovered. Accordingly, assigning priority in such cases is never manifest and often results in disputes even when all relevant facts are known. Kuhn claims this is not typically so for traditional or unitary discoveries. However, priority disputes can arise in traditional cases when information is incomplete. The discovery of Neptune is a paradigmatic example of this latter sort.

Manifold reasons exist for the controversy surrounding Neptune's discovery. The most important revolves around two main issues. These are (1) the initial obscurity—some would say secrecy—surrounding Adams's researches as compared with the very public pronouncements of Le Verrier and (2) the revolutionary nature of inverse perturbation theory and the difficulties involved in solving such problems. Since knowledge of the latter is useful for understanding the former, it will be advantageous to begin with a discussion of the inverse perturbation problem associated with the anomalous motions of Uranus.

Inverse Perturbation Theory In 1832, then Plumian Professor of Astronomy at Cambridge, George Biddell Airy (1801–1892) addressed the British Association for the Advancement of Science (BAAS) on the progress of nineteenth-century astronomy. In his report, Airy documented the latest research on the anomalous motion of Uranus and the many attempted explanations of this behavior in terms of either (a) a modification of the law of gravity or (b) by appeal to a purturbing planet beyond the orbit of Uranus. Increasing confidence in the validity of Newton's gravitational theory gradually focused research on the hypothesis of a transuranic planet. However, predicting the location of such a planet required accomplishing what most scientists at the time deemed impossible—solving an inverse perturbation problem.

Until the mid-nineteenth century, theoretical work in astronomy exclusively employed forward modeling. This involves deducing consequences from known conditions. For example, traditional perturbation theory seeks to determine the observable effects of certain known bodies on other known bodies. In contrast, inverse perturbation theory attempts to infer the mass and orbital elements of an unknown body by reference to the observed behavior of a known body. (In their simplest versions, inverse problems amount to reconstructing the integrands used to produce an integral from knowledge of just the resulting integral or reconstructing individual summed items from their resulting sum.) Successful application of inverse perturbation theory to the problem of Uranus held forth the promise of not only discovering a new planet but also opening up a whole new branch of theoretical astronomy dealing with unobservable phenomena.

Unfortunately, progress in developing inverse perturbation methods was forestalled for two very important reasons. First, it was correctly perceived that the inverse perturbation problem pertaining to the anomalous motions of Uranus was underconstrained by the data—meaning the available data were insufficient for determining a unique set of orbital elements for the reputed transuranic planet. The general consensus, to which Airy subscribed, was that many more years of Uranian observations would have

John Couch Adams
(Adams, John Couch, photograph. © Hulton-Deutsch/Corbis. Reproduced by permission.)

to be taken before any successful attempt could be made to resolve the problem. Second, and just as important, few individuals possessed the requisite mathematical skills to carry-through the calculations, and fewer still the patience and resolve necessary to actually see the work through to completion. Consequently, any proposed inverse perturbation solution would be highly suspect, especially if proffered by a young, unproven researcher.

Adams Begins His Study John Couch Adams was just such an individual. He stumbled upon the problem quite by chance. During a visit to a Cambridge book shop on June 26, 1841, Adams came across a copy of Airy's 1832 report on the progress of astronomy. On July 3, 1841, Adams confided to his notebook that he had "[f]ormed a design, at the beginning of this week, of investigating as soon as possible after taking my degree, the irregularities in the motion of Uranus, which are as yet unaccounted for; in order to find whether they may be attributed to the action of an undiscovered planet beyond it; and if possible thense to determine the elements of its orbits &c approximately, which wd. probably lead to its discovery."

After graduating in 1843 from St. John's College, Cambridge, Adams devoted all his spare time to computing the elements of the transuranic planet. By October he had obtained an approximate solution assuming a circular orbit. In February 1844, James Challis (1803–1882), Professor of Astronomy at Cambridge, wrote on Adams's behalf to Airy, then Astronomer Royal, requesting a full-set of Greenwich Uranian observations. Armed with this data, Adams refined his solution; and by September had deduced the orbital elements for an elliptical path of the perturbing planet. He also made the first reasonably accurate prediction of its location.

Though Adams made no secret of his researches, his shy and retiring nature together with the general skepticism toward inverse perturbation methods conspired against him. When he presented his results to Challis in September 1845, he clearly hoped a search would be made with Cambridge Observatory's 11.75-inch (30-cm) Northumberland Refractor. This instrument would easily have resolved the planet's disk. Challis was dubious, though, of undertaking a search for a planet based solely on theoretical calculations, and Adams did not press the issue. Nevertheless, Challis wrote a letter of introduction to Airy for the young Cambridge Fellow. Adams attempted to see Airy twice at Greenwich but failed. On his second visit he left a summary of his results. Perhaps feeling slighted, Adams returned to Cambridge feeling he had advanced his case as far as he could.

Airy did give Adams's results due consideration. Nevertheless, he was skeptical for a number of reasons. First and foremost, he was still convinced inverse perturbation problems were impossibly complex. Second, with only a summary of his results, Airy was unaware of the prodigious computations underlying Adams's conclusions. Third, he believed Saturnian perturbations were responsible for the anomalous behavior of Uranus. Though dubious of Adams's results, Airy was sufficiently interested to write a reply. Specifically, he asked whether Adams's theory could account for errors in the radius vector (distance from the Sun) of Uranus. As Adams recounted in later years, he did not respond to Airy because he considered this a trivial question.

The standard astronomical tables for Uranus had been calculated by Alexis Bouvard (1767–1843). They predicted both the longitude and radius vector for Uranus. Over time, Bouvard's predictions increasingly failed to agree with observations. While most astronomers had focused on the errors apparent in the longitude vector, Airy had gone to great lengths calculating the errors in the radius vector. Though Adams's results clearly explained the former, Airy wanted to know if he could account for the latter. Given that Airy doubted the existence of a transuranic planet, this seemed a reasonable question. However, according to Adams, since Bouvard's orbits were wrong in the first place, both the longitude and radius vectors would necessarily be in error. Adams considered the question trivial and did not respond. This failure to reply did nothing to alleviate Airy's concerns. Consequently, Airy was not motivated to act.

Le Verrier Presents His Findings Le Verrier had begun his own investigations during the summer of 1845 at the instigation of Dominique François Jean Arago (1786–1853). Within a year he completed three Memoirs. The first, presented before the Académie des Sciences in Paris on November 10, 1845, established the exact perturbing influences of Jupiter and Saturn on Uranus. The second, presented on June 1, 1846, demonstrated that the anomalous behavior of Uranus could not be the result of "the actions of the sun and of other planets acting in accordance with the principle of universal gravitation." He concluded that a transuranic planet must be responsible and by means of inverse perturbation theory provided an approximate position for this planet. In the third Memoir, presented August 31, 1846, he provided the mass and orbital elements of the transuranic planet as well as a more accurate prediction of its position.

After reading the second Memoir, Airy realized Le Verrier's results were in close agreement with those of Adams from the year before. He promptly drafted a letter on June 26 in which he queried Le Verrier about the errors in the radius vector—the same question Adams failed to answer. Le Verrier's prompt reply of June 28 satisfied Airy. This, together with the concordance of results and Le Verrier's considerable reputation, removed Airy's last remaining doubts about the existence of a transuranic planet.

In his reply, Le Verrier further suggested that if Airy were confident enough in his results to undertake a search, then he would send a more accurate position as soon as he finished his calculations. Airy clearly thought there was a very good chance of discovering a new planet and that every effort should be made to do so. Nevertheless, he did not respond to Le Verrier. However, he did write Challis at the Cambridge Observatory on July 9 urging him to initiate a search immediately and stating that "the importance of this inquiry exceeds that of any current work, [and] is of such a nature [as] not to be totally lost by delay." Challis somewhat reluctantly began his search on July 29, 1846.

Much of the groundwork for the later controversy was now in place. Of particular note was Airy's failure to mention, in his June 26, 1846, letter to Le Verrier, Adams's earlier researches into this matter. In fact, Adams's work at this time was known only to a small circle of scientists, primarily from Cambridge. Furthermore, though the Northumberland refractor at Cambridge was the largest in England, there were any number of smaller instruments just as capable of resolving the disk of the planet. Significantly, Airy's efforts were focused solely on Cambridge. In fact, Robert W. Smith convincingly argues that, among those few in England aware of both Adams's and Le Verrier's work, there was a feeling that Cambridge was the appropriate place for the optical discovery since it was there that the mathematical discovery was made.

Galle and d'Arrest Observe Neptune Impatient for observational confirmation, Le Verrier wrote to Johann Gottfried Galle at the Berlin Observatory, on September 23, 1846, entreating him to search for the transuranic planet. Galle, together with his assistant d'Arrest, prepared to search the skies that very night. d'Arrest suggested comparing their observations with the positions recorded in the newly drawn Hora XXI star chart. As Galle observed through Berlin's 9-inch (23-cm) Fraunhofer refractor, he called out coordinates to d'Arrest, who checked them against the chart. Within minutes Galle noted the position of an 8th-magnitude star. When he called out its position, d'Arrest exclaimed, "That star is not on the map!" As news of the discovery spread, Le Verrier was lavished with praise and congratulated for establishing once and for all the validity of universal gravitation.

In the ensuing priority dispute, three factors were emphasized by those supporting Le Verrier: First, Le Verrier published first. Second, the whole course of his inquiry was available to the entire scientific community from the beginning whereas Adams's research only became widely known after the planet was observed. Third, the optical discovery by Galle and d'Arrest proved the planet existed.

Publication is certainly important in resolving priority disputes. In lieu of hard evidence to the contrary, publication date weighs heavily. However, publication itself does not establish priority. As in this situation, well-established facts can and should trump publication date. In this case, as the narrative makes clear, it is beyond dispute that Adams produced a fairly accurate prediction of Neptune's position a full-year before Le Verrier. With respect to the obscurity of Adams's work, the charge has never been made that his results were completely secret. Though it has been shown that there was no serious attempt to promulgate the results, there is no indication Adams was intentionally secretive. While the former behavior was criticized by the scientific community, the latter fact should militate against any unqualified condemnation. More telling than either of these is the perceived relationship between the optical discovery and theoretical prediction of Neptune.

John Herschel's (1792–1871) view is fairly representative. He believed the discovery was the property of Le Verrier for the simple reason that he had proved the planet existed. The proof Herschel and others accepted was not his mathematical calculations, but rather the optical discovery made by Galle and d'Arrest. Herschel wrote, "Until the planet was actually seen and shown to be a planet—there was no discovery." Yet, if the optical discovery proved the planet existed by confirming Le Verrier's prediction, it did so no less for Adams's prediction. Given these circumstances, the only way of assigning Le Verrier priority was by conjoining his mathematical prediction with the optical discovery. However, doing so would diminish the work of Galle and d'Arrest.

If one gives Galle and d'Arrest their due credit, then the optical discovery of Neptune can only be seen as confirmation of both Le Verrier and Adams's work. Consequently, Adams and Le Verrier would reign as co-discoverers. If, in addition, one strictly adheres to the methodological imperatives of publication, full-disclosure, etc., then the laurels go to Le Verrier. The problematic nature of this move, though, dictates due emphasis be given to Adams's earlier work. Once this is done, Adams alone retains priority for the discovery of Neptune. —STEPHEN D. NORTON

Further Reading

Fernie, J. Donald. "The Neptune Affair." *Scientific American* 83, 2 (Mar.–Apr. 1995): 116–19.

Grosser, Morton. *The Discovery of Neptune.* New York: Dover Publications, Inc., 1962.

Harrison, H. M. *Voyager in Time and Space: The Life of John Couch Adams.* Sussex, England: The Book Guild Ltd., 1994.

Hoyt, William Graves. *Planets X and Pluto.* Tucson: University of Arizona Press, 1980.

Kuhn, Thomas. "The Historical Structure of Scientific Discovery," in *The Essential Tension.* Reprinted from *Science* 136 (1962): 760-764. Chicago: University of Chicago Press, 1977: 165-177.

Moore, Patrick. *The Planet Neptune: An Historical Survey Before Voyager.* 2nd ed. Chichester, England: Praxis Publishing, 1996.

Sheehan, William. *Worlds in the Sky: Planetary Discovery from the Earliest Times through Voyager and Magellan.* Tucson and London: University of Arizona Press, 1992.

———, and Richard Baum. "Neptune's Discovery 150 Years Later." *Astronomy* 24 (Sept. 1996): 42–49.

Smith, Robert W. "The Cambridge Network in Action: The Discovery of Neptune." *Isis* 80 (Sept. 1989): 395–422.

Standage, Tom. *The Neptune File: Planet Detectives and the Discovery of Worlds Unseen.* London: Penguin Press, 2000.

Tombaugh, Clyde W., and Patrick Moore. *Out of the Darkness: The Planet Pluto.* Guildford and London: Lutterworth Press, 1980.

Whyte, A. J. *The Planet Pluto.* Toronto: Pergamon Press, 1980.

EARTH SCIENCE

Did life on Earth begin in the "little warm pond"?

Viewpoint: Yes, the theory that life began in the "little warm pond" has supporting evidence from a number of experiments, and competing theories are more problematic.

Viewpoint: No, either life began on the surface during the period known as the late heavy bombardment, or it began in a protected environment away from the surface and the devastating impacts.

The origin of life on Earth is one of the most important, and elusive, problems in science. Efforts to understand the origin of life have been frustrated by lack of evidence. In the face of this fundamental difficulty, the search for a scientific explanation for the origin of life has relied upon speculative hypotheses, observations from present conditions on Earth and elsewhere in our solar system, and laboratory experiments that seek to simulate the conditions of the earliest period of Earth's history.

While earlier natural philosophers considered the problem of the origin of life, it was Charles Darwin who first posed an explanation for life's origin that is consonant with the larger picture of the evolution of life on Earth. Darwin suggested that simple chemicals in small or shallow bodies of water might spontaneously form organic compounds in the presence of energy from heat, light, or electricity from lightning strikes. These organic compounds could then have replicated and evolved to create more complex forms.

Darwin's "little warm pond" remains one of the most suggestive explanations for the origin of life. A classic experiment performed by Harold Urey and Stanley Miller in the 1950s brought the problem of the origin of life, and the "little warm pond," into the laboratory. Urey and Miller filled a flask with the gases they believed were present in the atmosphere of the ancient Earth, and suspended it over a small pool of water. They applied electrical sparks to the system, and observed that complex organic compounds, including amino acids, formed abundantly in the water. Amino acids are the most basic components of life on Earth. Their production in this experiment suggested that the beginnings of life could indeed have formed in appropriate settings on ancient Earth.

The Urey-Miller experiments have remained the paradigmatic explanation for the origin of life in school textbooks, but they leave many issues unresolved, and scientists have raised serious questions about nearly every aspect of the "little warm pond" model. For example, the atmosphere of ancient Earth may well have been made primarily of carbon dioxide and nitrogen—which do not produce amino acids—rather than the more hospitable mixture of methane, hydrogen, and ammonia used by Urey and Miller. And even if amino acids were abundant in the ancient seas, it is not understood how they could have evolved into more complex forms, including the proteins that make up the genetic code of DNA that is found in all life forms today.

Scientists who believe that ancient Earth did not provide conditions similar to those of the Urey-Miller experiments have looked elsewhere to try to

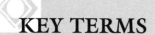

KEY TERMS

AMINO ACIDS: Amino acids are complex carbon compounds that make up proteins and other molecules important to life.

CARBONACEOUS CHONDRITES: The oldest meteorites in the solar system, believed to be remnants of the pristine building blocks of planets. They also can be rich in amino acids.

HYDROTHERMAL VENTS: Locations on the sea floor near spreading centers where magma beneath the surface heats water to high temperature and jets to the surface. They support thriving ecosystems on the otherwise barren deep sea floor.

MARIA: The dark areas on the Moon that are large crater basins filled with cooled lava. Comes from the Latin for "Sea."

TREE OF LIFE: Tool used by microbiologists to study the genetic relationships between organisms. It is composed of three branches—the eukarya, bacteria, and archaea.

find alternative scenarios to explain the origin of life. Observations of the Moon and other planets in the solar system have suggested that during the first billion years or so of Earth's history, the surface of our planet was extremely volatile and even hostile, bombarded by meteorites and intense solar radiation. These conditions make the formation of stable, warm, shallow ponds unlikely, but they suggest other interesting possibilities to explain the origin of life on Earth.

For example, the constant bombardment of ancient Earth by meteorites and comets may itself explain the source of amino acids—perhaps they formed elsewhere in the solar system and were carried here by these interplanetary travelers. Hypothetically, amino acids and other organic compounds could survive a trip from Mars or another planet to Earth, and although no undisputed evidence of such a transfer has been identified, scientists continue to study meteorites and comets to better understand the effect of these impacts on the formation of life.

Another promising focus of study is on those areas of Earth that come closest to the harsh conditions likely to have been present on early Earth. By investigating regions near volcanic rifts in the ocean floor and in deep underground wells, scientists have found microorganisms capable of surviving at the very high temperatures, which probably characterized the surface of ancient Earth.

Scientists have also tried to understand what the earliest life on Earth might have been like by studying genetic differences between organisms. These differences can be used to construct a genetic "tree of life," showing how groups of organisms are related and how they might have evolved from common, simpler ancestors. The oldest branch of life appears to include the thermophilic organisms that thrive in the hot environments on today's Earth.

Although evidence about the conditions of ancient Earth remains elusive, by bringing together observations from elsewhere in the solar system, genetic studies, and contemporary life in the strange micro-environments close to the heat of Earth's crust, scientists have begun to put together a picture of what the first billion tumultuous years of Earth's history might have been like. While this evidence challenges the simple idea of life emerging from a "little warm pond," primarily by suggesting that ancient Earth's surface lacked the stability to support such ponds, the fundamental suggestion that some combination of water, energy, and atmospheric chemicals produced simple organic compounds that ultimately gave rise to complex life remains a central idea in the efforts to explain the origin of life. —LOREN BUTLER FEFFER

Viewpoint:
Yes, the theory that life began in the "little warm pond" has supporting evidence from a number of experiments, and competing theories are more problematic.

How Did Life Begin? The best theory we currently have regarding the origin of life on Earth is that it first originated as the accumulation of organic compounds in a warm body of water. This hypothetical "warm little pond" has sup-

porting evidence from a number of experiments. However, the origin of life is clouded in uncertainty, and the precise mechanisms by which basic chemicals came together to form complex organisms is not known. The lack of evidence from ancient Earth means we may never know precisely how life began. Nevertheless, of all the speculative theories, the warm little pond remains the most promising.

In some ways the nature of this question means that a simple "yes and no" debate is of little value. To begin with, arriving at a satisfactory definition of "life" has proved difficult. While a number of attempts have been made, some definitions are so broad as to include fire and minerals, while others are so narrow they exclude mules (which are sterile).

Another major problem with determining how life began on Earth is the lack of evidence. The fossil record is limited by the fact that almost all rocks over three billion years old have been deformed or destroyed by geological processes. In addition to debating the issue of when life emerged, scientists also debate the conditions of ancient Earth. Some theories posit that early conditions on our planet was extremely cold, while other theories suggest that it was warm and temperate, and even boiling hot. Computer models have suggested that a variety of temperature ranges are possible, but without further evidence there is little consensus.

Life on Earth may have had a number of false starts. Early Earth was subjected to massive geological upheavals, as well as numerous impacts from space. Some impacts could have boiled the ancient oceans, or vaporized them completely, and huge dust clouds could have blocked out sunlight. Life may have begun several times, only to be wiped out by terrestrial or extra-terrestrial catastrophes.

Any theory on the origin of life must contain a great deal of speculation. What scientists can agree upon are the general characteristics that define life from non-life. Early life must have had the ability to self-replicate, in order to propagate itself and survive. Self-replication is a tricky process, implying a genetic memory, energy management and internal stability within the organism, and molecular cooperation. Just how the ingredients of the "primordial ooze" managed to go from simple chemical process to complex self-replication is not understood. Moreover, the process of replication could not have been exact, in order for natural selection to occur. Occasional "mistakes" in the replication process must have given rise to organisms with new characteristics.

Life from a Chemical Soup The modern debate on the origin of life was inaugurated by Charles Darwin. In a letter to a fellow scientist he conjectured that life originated when chemicals, stimulated by heat, light, or electricity, began to react with each other, thereby generating organic compounds. Over time these compounds became more complex, eventually becoming life. Darwin imagined that this process might occur in shallow seas, tidal pools, or even a "little warm pond." Later theorists have suggested variations on this theme, such as a primordial ocean of soup-like consistency, teeming with the basic chemical ingredients needed for life. While Darwin and his contemporaries saw life as a sudden spontaneous creation from a chemical soup, modern theories

tend to regard the process as occurring in a series of small steps.

In the early 1950s the "little warm pond" theory of life was given strong experimental support by the work of Harold Urey and Stanley Miller. Miller, a student of Urey, filled a glass flask with methane (natural gas), hydrogen, and ammonia. In a lower flask he placed a small pool of water. He then applied electric shocks to mimic lightning. The results were more than either scientist had hoped for—within a week Miller had a rich reddish broth of amino acids. Amino acids are used by all life on Earth as the building blocks for protein, so Miller's experiment suggested that the building blocks of life were easy to make, and would have been abundant on early Earth.

Further experiments by Sidney W. Fox showed that amino acids could coagulate into short protein strands (which Fox called proteinoids). It seemed that scientists were on the verge of creating life from scratch in a test tube. However, Fox's work now appears to be something of a dead end, as there is no further step to take after proteinoids. Proteins and proteinoids are not self-replicating, and so either there are missing steps in the process, or something altogether different occurred. Miller's work, too, has lost some of its shine, as there are now strong doubts that the atmosphere of ancient Earth contained the gases he used in his experiment. It is possible that rather than methane, hydrogen,

Charles Darwin
(Darwin, Charles, photograph. The Library of Congress.)

The phylogenic tree of life.
(Reproduced by permission of Norman Pace.)

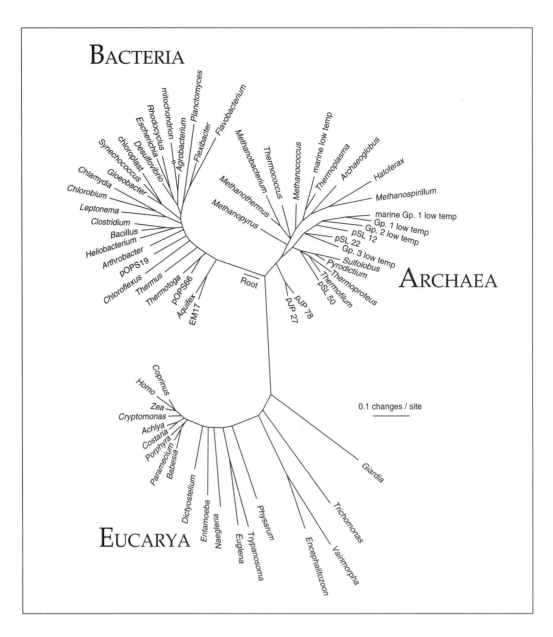

and ammonia the early atmosphere was rich in carbon dioxide and nitrogen.

Even if amino acids were common on early Earth there is still the question of how these simple compounds gave rise to the complexity of life, and to DNA, the double helix that contains the genetic code. DNA cannot replicate without catalytic proteins, or enzymes, but the problem is that the DNA forms those proteins. This creates something of a chicken-and-egg paradox. One possible explanation is that before the world of DNA there was an intermediate stage, and some scientists have suggested that RNA is the missing gap. RNA is similar to DNA, but is made of a different sugar (ribose), and is single-stranded. RNA copies instruction from DNA and ferries them to the chemical factories that produce proteins in the cell. RNA may have come first, before DNA. The RNA world may have provided a bridge to the complexity of DNA. However, RNA is very difficult to make in the probable conditions of early Earth, and RNA only replicates with a great deal of help from scientists. Some theorists think there was another, more simple, stage before RNA, but again, no evidence has been found.

Other Theories on the Origin of Life Because of the difficulties with the warm little pond theory and its variants a number of new theories have recently emerged to challenge it. Many of these theories are interesting, intriguing, and even possible. However, they all have unanswered questions, making them even more problematic than the idea of the "little warm pond."

Several decades ago scientists were amazed to discover organisms that live in very hot conditions. Dubbed thermophiles, these hot-living bacteria have been found in spring waters with temperatures of 144°F (80°C), and some species near undersea volcanic vents at the boiling point of water. There is even some evidence of underground microbes at even higher temperatures (336°F [169°C]). The discovery of such hardy organisms has led some to speculate that life originated not in a warm pond, but in a very hot one. Perhaps ancient Earth was peppered with meteor and comet impacts, raising temperatures and boiling oceans, but also providing the necessary chemical compounds to create life. Or possibly hot magma from volcanic sources provided the vital gases and compounds, and the energy, to assemble the first living organisms.

A variant of this theory considers the undersea volcanic vents as the birthplace of life, with the chemical ingredients literally cooked into life. There are even those who champion a deeper, hotter, underground origin for life. Underwater and underground origins have some advantages over other theories. Such depth might make early life safe from the heavy bombardment of material from space the planet received, depending on the size of the object striking Earth and the depth of the water. They would also be safe from other surface dangers, such as intense ultraviolet radiation. There is even some genetic evidence to support these hot theories, as thermophiles do seem to date back to near the beginnings of the tree of life. However, whether they were the trunk of the tree, or merely an early branch, is not known. There is also the question of how these hot organisms could have moved into cooler areas. Some theorists argue that it is easier to go from cool to hot, not the other way around. Also, environments such as undersea volcanic vents are notoriously unstable, and have fluctuations that can cause local temperature variation that would destroy rather than create complex organic compounds.

Some theorists have gone to the other extreme of the temperature scale, and envision life beginning on a cold, freezing ancient Earth. Just as hot microbes have been discovered, so have organisms capable of surviving the Antarctic cold. Some suggest these as our common ancestors. Again, there are some advantages to such a theory. Compounds are more stable at colder temperatures, and so would survive longer once formed. However, the cold would inhibit the synthesis of compounds, and the mobility of any early life. Also, the premise that ancient Earth was a cold place is not widely accepted.

Others have looked to the heavens for the origins of life. The early solar system was swarming with meteors and comets, many of which plummeted to Earth. Surprisingly there are many organic compounds in space. One theory suggests that the compounds needed to form the primordial soup may have arrived from space, either from collisions, or just from near misses from comet clouds. Even today a constant rain of microscopic dust containing organic compounds still falls from the heavens. Could the contents of the little warm pond have come from space?

There are also suggestions that life may have arrived from space already formed. Living cells could possibly make the journey from other worlds, perhaps covered by a thin layer of protective ice. The recent uncovering of a meteorite that originated on Mars has leant support to this theory. There is some suggestion that the meteorite contains fossilized microorganisms, but most scientists doubt this claim. However, the collision of comets and meteors is far more likely to have hindered the development of life than help create it. Objects that would have been large enough to supply a good amount of organic material would have been very destructive when they hit. It seems probable that life began on Earth, rather than in space somewhere. Also, the idea that life may have traveled to Earth does not help explain its origin; it merely transposes the problem to some distant "little warm pond" on another world.

There are a number of other theories proposing various origins of life that have appeared in recent years. Gunter Wachtershauser, a German patent lawyer with a doctorate in organic chemistry, has suggested that life began as a film on the surface of fools gold (pyrite). Some small experiments have given it some credence, but the idea is still at the extreme speculative stage. Sulfur is the key ingredient in some other theories, such as the Thioester theory of Christian R. de Duve. Thioesters are sulfur-based compounds that Duve speculates may have been a source of energy in primitive cells. In the primal ooze thioesters could have triggered chemical reactions resembling those in modern cellular metabolism, eventually giving rise to RNA. However, again there is a lack of supporting experimental evidence.

All of these new theories suffer from the same problems that beset the standard interpretation. That is, the difficulty of going from simple chemical process to self-replicating organisms. Many of these new theories are merely new twists on the original warm little pond concept. Some are boiling ponds, others are cold, but only a few offer completely different ways of viewing the origin of life. While some of these theories have some strong points, they have yet to provide the hard evidence to support the speculation. None of them has gained enough support to topple the "little warm pond" from its place as the most likely theory we have. There is much supporting evidence for the standard theory, in the form of Miller's experiments and the work on RNA. Darwin's throw-away comment in a letter may have led to more than he bargained for, but his theory on the origin of life still remains the best and most useful theory we currently have. —DAVID TULLOCH

Viewpoint:

No, either life began on the surface during the period known as the late heavy bombardment, or it began in a protected environment away from the surface and the devastating impacts.

Scientists know little about the origin of life—except that it happened. Earth formed 4.5 billion years ago from innumerable collisions of smaller rocks. The leftover debris from planet formation careening through the solar system bombarded Earth even after it achieved its full size. During this period, known as the late heavy bombardment, the surface was routinely devastated by large impacts. This period ended about 3.8 billion years ago. Oddly enough, the oldest known microfossils date back 3.5 billion years ago, and there is tantalizing evidence of life as early as 3.8 billion years ago. It seems that, as soon as life could survive on the surface of Earth, it began—in a very short amount of time. But many scientists believe the origin of the first life forms from non–life forms must have taken more than the few million years between the end of the late heavy bombardment and the first evidence of life. Therefore, either life began on the surface during the late heavy bombardment, or it began in a protected environment away from the surface and the devastating impacts.

The Miller-Urey Experiment In 1953, Stanley Miller and Harold Urey performed a radical new experiment. They generated a spark in a mixture of water, methane, ammonia, and hydrogen, a composition believed to contain the major components of Earth's early atmosphere. After only one week, Miller and Urey succeeded in forming complex organic carbon compounds, including over 20 different types of amino acids, the building blocks of life. The Miller-Urey experiment, which has been replicated thousands of times in high schools and colleges around the world, stands as a proof of the concept that, in the "warm little ponds" believed to pepper the surface of early Earth, the beginnings of life could form.

This perhaps occurred in a "warm little pond" on Earth 4 billion years ago. But long before the amino acids collected themselves into the organized structures known as life, one of the frequent large impacts of leftover solar system debris smacked into Earth, devastating the surface—boiling seas, heating the atmosphere enough to melt silica—and effectively sterilizing the "warm little pond," along with the remainder of Earth's exposed surface. The Miller-Urey experiment proved that the beginnings of life could assemble in a simple environment comprised of basic molecules. But it failed to insure that such an idyllic environment existed on early Earth. Amino acids are but one prerequisite to life. The others are a source of food or energy and—more importantly—a location stable enough to allow the organization of molecules into life.

Much of what scientists know about the conditions on early Earth comes from the Moon. Lacking any atmosphere or oceans, events recorded on the surface do not erode away over time. The Moon, in essence, acts as a geological recording device, writing its history on its surface for all to see. The lunar record can be read on any clear night on the face of the full Moon: the dark areas, or maria, are large basins excavated by massive impacts and later filled with lava. A simple pair of binoculars reveals even more craters, created by smaller impactors. Overall, the oldest features on the Moon, dating back 3.9 to 4.0 billion years ago, tell a story of frequent, massive impacts in the Earth-Moon system. Four billion years ago, the surface of the Moon—and, by extension, Earth—was a hostile environment.

Another factor rendered the surface of Earth inhospitable to life. Four billion years ago, Earth's atmosphere was largely made of nitrogen and carbon dioxide (oxygen, a product of photosynthesis, didn't arise in appreciable quantities until after microbial life flourished on Earth for over 2 billion years). An oxygen atmosphere, illuminated by radiation from the Sun, produces O_3, or ozone. Therefore, without oxygen, Earth lacked an ozone layer capable of protecting life from harmful and even fatal ultraviolet radiation. To make matters worse, the Sun was far more active in the past, producing tremendous solar flares and incredible bursts of ultraviolet radiation. Any organism on the surface of the Earth would be damaged by the intense radiation—a fate ill-suited to further reproduction.

Lastly, the surface of Earth was much different four billion years ago. The total amount of exposed land area, or continents not covered by water, was much less than today. Therefore, any potential microbial environments were most likely covered in a layer of water, perhaps 3,300 ft (1 km) thick or more. This reduces the chance of the proper compounds being confined to a "warm little pond," as in the Miller-Urey experiment.

But scientists know one thing for certain— life did begin. Somehow, the proper building blocks such as amino acids were formed on, or delivered to, Earth. Other than the "warm little pond," there are several other sources of amino acids in the solar system. The oldest meteorites, known as carbonaceous chondrites, represent the pristine material from which the planets were formed. These objects routinely rain down on Earth even today, and represent a potential source of amino acids for life. In addition to meteorites, comets deliver a large amount of material to

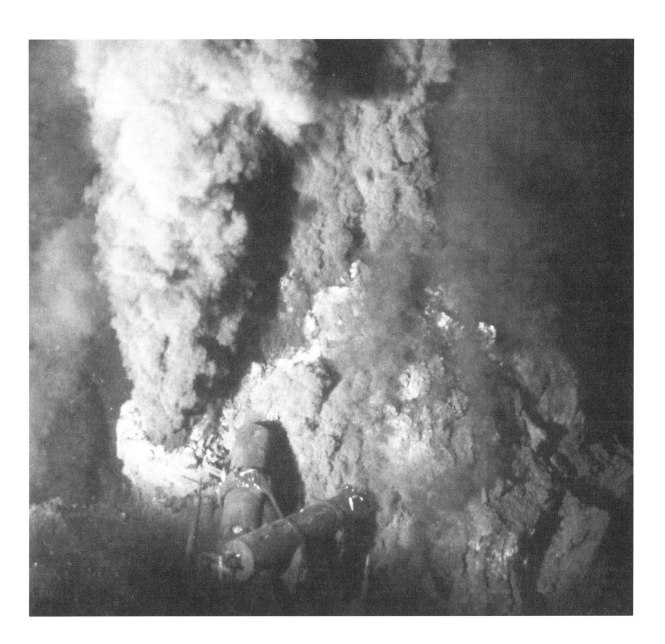

A hydrothermal vent.
(© Ralph White/Corbis. Reproduced by permission.)

Earth's surface. While no samples of comets have been examined to date, NASA's *Stardust* mission plans to return a sample of comet Wild-2 in 2006, and scientists expect to find complex carbon compounds, including amino acids. Even if the surface of Earth did not allow for the formation of amino acids, and later life, ample sources of raw materials exist.

Life Found in Extreme Heat The surface of Earth is not the only place where life could have started. Today, two unique environments support extensive microbial ecosystems, in what many would consider inhospitable locations. Near volcanically active rifts on the sea floor, known as hydrothermal vent systems, microorganisms not only exist, but also thrive and support a dynamic ecosystem of macroscopic life forms as well. Volcanically heated water leaves the hottest vents, known as black smokers, with a temperature in excess of 750°F (400°C). This water interacts with the seawater at 28°F (-2°C). Some microorganisms found in this environment thrive at temperatures in excess of 212°F (100°C), or the boiling point of water. The organisms get the energy by chemical reactions with the materials released by the black smokers. On early Earth, where submarine volcanism was widespread, these environments were abundant. They have the advantage of being located thousands of feet underwater, and are well protected from both smaller impacts and ultraviolet radiation.

Still, some of the largest impactors to hit Earth during the heavy bombardment had enough energy to boil the oceans completely, destroying even the submarine environments. Therefore, scientists have looked deep underground for habitats capable of supporting life. Recently, sediments retrieved from deep wells near the Hanford Nuclear Reservation in Washington State show evidence of microbial activity. Some of these organisms are from samples originally at depths of almost 10,000 ft (about 3 km).

The organisms thrive on water from deep aquifers and chemical reactions in the subsurface. Due to their depth, these environments would also be protected from impacts and UV radiation.

Both of these high-temperature environments exist—over 212°F (100°C) in the hydrothermal vents, and as high as 167°F (75°C) in the subterranean environments. Organisms whose optimum growth temperature is above 113°F (45°C) are known as thermophiles, and organisms with an optimum growth temperature above 176°F (80°C) are known as hyperthermophiles. Clearly, the first organisms to live in such environments would have to be thermophiles or hyperthermophiles. Therefore, if the organisms living on early Earth were tolerant to heat, chances are they originated in one of these environments.

Scientists have no samples of life from early Earth, but the genetic code of all organisms contains remnants from the first organisms. By studying the differences in the genes between organisms, scientists can determine how one species relates to one another—which organism evolved from which, and how closely they are related. By charting the genetic differences between organisms, scientists construct a genetic "tree of life" relating all the species to one another. Study of all the life on Earth reveals several major classes of organisms. First of all, there are the eukarya, containing all the microorganisms whose cells have a nucleus, as well as the familiar macroscopic life like fungi, oak trees, and humans. Next are the bacteria (monara), single-cell organisms without cell nuclei. Lastly, the oldest organisms are the archaea, which are closest to the "root" of the tree, and presumably most closely related to the common ancestor that first evolved on early Earth. Study of the archaea reveal a large number of thermophilic and hyperthermophilic organisms, indicating the common ancestor was also likely to be thermophilic, making the subterranean and hydrothermal vents appealing locations for the evolution of life.

Did Life Begin in Space? Another intriguing idea is that life did not begin on Earth at all, but was later brought to Earth during the heavy bombardment by interplanetary meteorites. To date, there are at least a dozen known Martian meteorites, launched from the surface of Mars after a large impact and later retrieved on Earth's surface. In the summer of 1996, David McKay and other researchers at NASA released preliminary evidence of microbial activity in Martian meteorite ALH84001. Since that time, much of their evidence has been discounted as a misinterpretation or contamination by terrestrial sources. However, the discovery opened up the question of interplanetary "seeding" of biological material. Amino acids and other compounds can survive the trip from Earth to Mars, and it may be possible that microorganisms can as well. It is possible that life swapped worlds several times, as sterilizing impacts rendered one planet inhospitable for a time, and the other planet acting as a temporary safe haven until the next large impact. If this is the case, the question "where did life begin?" becomes even more complex.

Scientists have learned, from studying the Moon, that Earth's surface was far too hostile to support "warm little ponds" stable enough for life to form unless life began frequently—and very rapidly—between impacts, or was delivered to another world for safe keeping. It appears as if sterilizing impacts forced life's origin to sites deep underground or underwater. In the near future, the new science of astrobiology may be able to answer the question "where did life begin?" Two locations in the solar system, other than Earth, are likely places to look for life today. Our neighboring planet Mars may have subterranean environments similar to those found on Earth. Jupiter's moon Europa is composed of a rocky core surrounded by a shell of liquid water and ice. Hydrothermal vents may exist below the surface of the kilometer-thick ice sheet on the ocean floor. If life is found in either of these locations, it is likely that life began in hydrothermal or subterranean environments on Earth as well.
—JOHN ARMSTRONG

Further Reading

Davies, Paul. *The Fifth Miracle: The Search for the Origin of Life*. Penguin Press, 1998.

Horgan, John. "In the Beginning..." *Scientific American* (February 1991):100-109.

Kump, Lee R., and James Kasting. *The Earth System*. New Jersey: Prentice Hall, 1999.

Maddox, John. *What Remains to Be Discovered*. New York: Simon & Shuster, 1998.

Madigan, Michael T., John M. Martinko, and Jack Parker. *Brock Biology of Microorganisms*. 9th ed. New Jersey: Prentice Hall, 2000.

"Microbes Deep Inside the Earth." <www.sciam.com/1096issue/1096onstott.html>.

Mojzsis, S. J., G. Arrhenius, K. D. McKeegan, T. M. Harrison, A. P. Nutman, and C. R. L. Friend. "Evidence for life on Earth Before 3.8 Billion Years Ago." *Nature* no. 384 (Nov. 1996): 55-59.

Monastersky, Richard. "The Rise of Life on Earth." *National Geographic* (March 1998): 54-81.

Ward, Peter D., and Donald Brownlee. *Rare Earth: Why Complex Life is Uncommon in the Universe*. Copernicus Books, 2000.

Did water once flow on the surface of Mars?

Viewpoint: Yes, evidence from various Mars missions indicates that water once flowed on Mars and helped shape the current landscape.

Viewpoint: No, the topographical features suggesting that water once flowed on Mars lack crucial elements necessary to support the theory that water flowed on Mars.

People have wondered about the planet Mars for hundreds of years. Mars is Earth's closest neighbor in the solar system, and it is close to Earth in size and appearance as well. While Mars presents no obvious bodies of water, polar ice caps on the planet grow and shrink with the seasons. Yet more intriguing to observers are the mysterious channels visible on the Martian surface. The appearance of these channels have led to speculations about the presence of water on Mars. While modern telescopes and exploratory missions have shown conclusively that there is no water flowing on the Martian surface today, speculation about a watery past on Mars has continued.

Mars presents a range of geological features and formations that could have been formed by water and accompanying erosion. Valleys, gullies, riverbeds, and flood plains appear in images of the Martian landscape. But is a Martian "riverbed" really the same geological feature recognized on Earth, or is it just a superficial resemblance? Critics of the claim that Mars once contained large quantities of water argue that it is much more likely that forces of wind and volcanism, rather than water, shaped the surface of Mars.

Why does it matter whether water ever flowed on the surface of Mars? Mars is Earth's closest neighbor in the solar system, and has long seemed like the best planet to support some form of life. From the wildest fictional speculation about "Martians" to earnest research efforts at NASA, interest in the possibility of life on Mars has always been high. If it can be proved that Mars once had seas, rivers, or lakes of water on its surface, then it would seem much more likely that it also supported some form of life as well. —LOREN BUTLER FEFFER

Viewpoint:

Yes, evidence from various Mars missions indicates that water once flowed on Mars and helped shape the current landscape.

Sometimes referred to as the Red Planet and named after the God of War, Mars has captured the attention of scientists and storytellers alike for centuries. Many have believed, and hoped, that it could be Earth's twin—a world capable of sustaining life like our own planet. However, recent observation has proven that Mars is quite different from Earth—the key difference being that it seems to lack a significant amount of surface water.

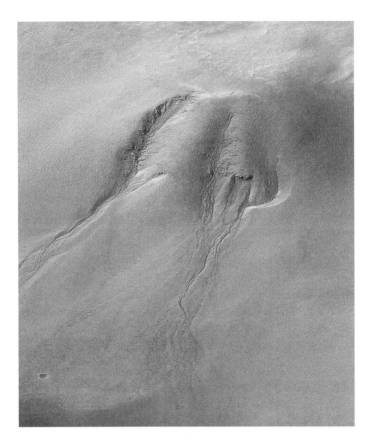

This image from the *Mars Global Surveyor* shows gullies eroded into the wall of a meteor impact crater in Noachis Terra.
(© AFP/Corbis. Reproduced by permission.)

But was Mars always so water-poor in nature? Could this hostile world have once possessed flowing rivers and vast lakes? After much debate, only one conclusion can be reached: Yes, water once flowed freely on Mars's rugged surface.

Water Features on the Martian Landscape

Astronomers have studied the existence of water on Mars for 300 years, watching the seasonal changes in the planet's polar icecaps. The first evidence of flowing water came almost a century before the first person even entered space. Giovanni Schiaparelli, an Italian astronomer, discovered the presence of canali (channels) on Mars's surface. Because some people misunderstood the word "canali" to mean "canals," the belief that the channels were built by intelligent beings grew. While this theory made for wonderful fiction, it did little to further the true understanding of Mars. Later *Viking* missions would dispel this theory as pure imagination (JPL Mars Program, 2001). However, it did leave a major question: Just where did these channels come from?

As the geography of Mars was documented through tens of thousands of photos during numerous missions, one thing became increasingly clear. Water, and copious amounts of it, had helped shape the Martian landscape. During an earlier period, perhaps several billion years ago, Mars was probably far warmer and wetter than it is today. Lakes, rivers, and maybe even oceans covered the surface and slowly left their mark over the millennia. Schiaparelli's channels were not the result of an alien hand, but erosion. The data obtained from photo mosaics of the Martian surface helped support this hypothesis.

Outflow Channels and Valley Networks

There are two specific types of water-related geological features that appear in several areas on Mars's surface—outflow channels and valley networks. Outflow channels come from chaotic terrain and can be well over 62 miles (100 km) wide and 1,240 miles (1,995 km) in length. Their characteristic flow features, like islands and craters, are typically formed from catastrophic flooding. Valley networks, on the other hand, can be broken up into two categories—long, winding valleys with few tributaries, and smaller valley networks, which can have numerous and complex tributaries. These are shaped through a more peaceful process than outflow channels, and are similar to river valleys found on Earth.

Ken Edgett, a staff scientist at Malin Space Science Systems (MSSS), stated, "... the *Mariner 9* spacecraft found evidence—in the form of channels and valleys—that billions of years ago the planet had water flowing across the surface." This hypothesis was also supported by Gerald A. Soffen, Chief Scientist for the *Viking* missions, who stated that "... the presence of braided channels suggests to many geologists that they are the result of previous periods of flowing water, suggesting a very dynamic planet."

Photos of different sections of Mars during the *Viking* missions also suggest the earlier existence of flowing water. In the Ravi Vallis, catastrophic outflow channels were revealed. Likewise, evidence was found through NASA's *Mariner 9* that rivers and/or flooding most likely formed a complex system of channels in the Chryse Palanitia. And, as discussed by Paul Raeburn in his book entitled *Uncovering the Secrets of the Red Planet*, cratered terrain with fine channels indicate the former presence of a lake in the Margaritifer Sinus region of Mars.

Data obtained later from the study of sedimentary rock suggest that lakes once existed on the planet surface. One such example exists in the Valles Marineris, a 3,700-mile (5,953-km) long canyon. Images indicate layered terrain, much like that created by shallow seas. Similarly, Jim Head and his colleagues at Brown University found evidence of a former ocean while reviewing elevation data collected by the *Mars Global Surveyor* (*MGS*).

Carl Allen, a planetary scientist at the Johnson Space Center, discussed one of *MGS*'s key findings concerning a possible ancient ocean in the Martian northern plains. He indicated that "this part of the planet is incredibly flat and smooth, and it is a set of concentric layers, all exactly the same altitude." An ocean evaporating and/or receding beneath the planet's surface could have left these rings, which are similar to wave-cut shorelines.

The *MGS*'s instruments also revealed other evidence of the existence of ground water. Allen said, "The exciting result from the spectrometer is several large deposits, some hundreds of miles across, of the iron oxide mineral hematite. Deposits of this material probably require large amounts of water from lakes, oceans, or hot springs." And naturally, liquid water is fluid by nature, which implies movement or flow.

Data also revealed that a great deal of activity occurred in the northern lowlands, suggesting that titanic floods of an unimaginable scale carved outflow channels. To put the power of these outflows into perspective, it would take approximately 10,000 Mississippi rivers to carve out some of these channels. Valley networks discovered in the southern hemisphere resemble those formed by rain, sapping (collapses caused by soil softened by groundwater), or glacial run-off.

On July 4, 1997, the Mars *Pathfinder* landed on the Red Planet to begin its exploration of the surface. Its landing-zone was in the middle of an ancient flood plain. Photos from the Mars *Rover* revealed that a Martian rock (named Prince Charming by scientists) may be a conglomerate, created by flowing water that has acted to "round" pebbles, which later are compacted and cemented to form rock.

Mounting New Evidence Today's technology has allowed us to get a far better look at the surface of Mars than ever before. Because of this technological advance, scientists have been able to discover further evidence of water on Mars, not just from the ancient past, but also in recent times. Andrew Chaikin, the Executive Editor of *Space & Science*, explained, "The camera that spotted evidence of water gullies on Mars has a set of eyes that boggles the imagination. Even from an orbital height of 235 miles (378 km), the camera can spot objects as small as 3 yards (2.7 m) across."

The camera, built into the *MGS* spacecraft, captured the images of what appeared to be gullies or washes on the Martian surface. "We see features that look like gullies formed by flowing water and the deposits of soil and rocks transported by these flows," said Michael Malin, Chief Investigator for the *Mars Orbital Camera* at MSSS. "The features appear to be so young that they might be forming today. We think we are seeing evidence of a ground water supply, similar to an aquifer. These are new landforms that have never been seen before on Mars." Of course, time in space is a relative thing. However, many scientists agree that although it is possible that the gul-

KEY TERMS

CANALS: An artificial (man-made) waterway, which is usually tubular in shape that allows the passage of water from one place to another. Usually used for irrigation purposes.

CARBONATE ROCKS: A type of rock formed when water, slightly carbonated by CO_2 gas, flows over surface rock. This material is formed on Earth when rainwater runs over the surface, and the resulting carbonate rocks are deposited as runoff in the oceans.

CHANNELS: The bed where a natural stream of water runs or the deeper part of a river, or harbor. The term can also be used when referring to a narrow sea between two close land masses.

CONGLOMERATE ROCK: A rock that contains large amounts of gravel that is rounded. The rounding of the gravel usually takes place in a stream or river.

DEBRIS FAN: Material at the end of a ravine or gully deposited by the movement of material from the upper slope.

GULLIES: Trenches in the earth that are formed by running water.

SUBLIMATE: To cause a solid to change into a gas (or vice versa) without becoming a liquid. CO_2 does this under normal Earth conditions.

TRIBUTARIES: Streams that feed a larger stream or lake.

lies were formed long ago, it is equally possible that they were formed quite recently.

These newly discovered gullies may indeed be the evidence required to answer the age-old question scientists have been asking about Mars. Edgett explains, "Ever since [*Mariner 9*], Mars science has focused on the question, 'Where did the water go?' The new pictures from the *Global Surveyor* tell us part of the answer—some of that water went underground, and quite possibly is still there."

Edgett also explains why it is difficult for water to currently exist on the surface of Mars. Because of the planet's low atmospheric pressure, almost 100 times less than Earth, liquid water would instantly boil. But Edgett continues, "We've come up with a model to explain [the gullies] and why the water would flow down the gullies instead of boiling off the surface. When water evaporates it cools the ground—that would cause the water behind the initial seepage site to freeze. This would result in pressure building up behind an 'ice dam.' Ultimately, the dam would break and send a flood down the gully."

In a recent study, Laurie Leshin, a geochemist at Arizona State University, claims that

Mars could be more water-rich than previously suspected. After examining the water-bearing crystals in the 3 million-year-old Martian meteoroid QUE94201 (discovered in Antarctica in 1994), she discovered that the crystals were rich in deuterium, a heavy form of hydrogen. Furthermore, the ratio of the deuterium in the crystals is surprisingly similar to the ratio of the deuterium in the planet's current atmosphere. This suggests that the planet has lost two to three times *less* water than originally believed, indicating there may be more water beneath the Martian soil than suspected.

In conclusion, it seems evident that water flowed in some quantity on the surface of Mars. How much, if any, still flows there today is up for debate, but there is certainly compelling new evidence to inspire further inquiry. It is clear that water has helped shape the Martian landscape. For proof, one only needs to look at the planet's scarred surface. —LEE A. PARADISE

Viewpoint:

No, the topographical features suggesting that water once flowed on Mars lack crucial elements necessary to support the theory that water flowed on Mars.

Of all the planets, Mars bears the most striking resemblance to Earth. Like Earth, it has a hard rocky surface and experiences changes in seasons due to the tilt of its axis. It has an atmosphere—albeit 1,000 times less dense than Earth's—that further modifies the climate. The temperature, while cold enough for CO_2 to freeze on the surface, is not outside the realm of human experience, with the daily equatorial temperature variation on Mars similar to the seasonal variation in Vostok, Antarctica. Mars even displays seasonal ice and solid CO_2 caps on the poles. But unlike Earth, Mars today lacks large amounts of liquid water on the surface.

The lack of water, coupled with Mars's tenuous CO_2 atmosphere, means that erosion of surface features is much slower than on Earth. Looking at the Martian surface, geologists can see features that date back 3.9 to 4.0 billion years ago, only 500 million years after the formation of the planet. Compared to Earth, where landforms are quite young because of the erosion at Earth's surface, nearly the entire geological history of Mars is available for examination. Interpretation of imagery from remote spacecraft yields several observations supporting the hypothesis that water flowed on the surface in the distant past. Craters in the southern highlands appear to require flowing water to explain the high erosion rates. Valley networks in the older terrain (2.0 to 3.8 billion years old) look like Earth valleys carved by water. Outflow channels, similar to the catastrophic flood features created on Earth during the last ice age, exist on some of the younger terrain. Finally, gully formations with debris fans covering fresh dunes indicate some kind of fluid flow in the very recent past (100 to 10,000 years ago). All of this evidence suggests a rich water history for Mars.

Little Evidence of Water However, as one looks deeper for supporting observations, very little evidence for water can be found. High erosion rates like those found in the southern highlands would leave behind huge expanses of carbonate rocks, the weathering product between water and the CO_2–rich atmosphere. To date, detailed spectroscopic analysis of the surface reveals no such deposits. In fact, deposits of the mineral olivine, indicating dry conditions, are widespread. The valley networks lack the fine tributary structure seen in similar Earth formations, indicating the source of the flowing liquid came from a few discrete sources rather than flowing over the surface. The outflow channels are of such huge scale that accounting for the shear volume of water released by the floods—enough water to cover the surface of Mars to a depth of 3,300 feet (1,006 m)—becomes difficult. And the gullies, the most compelling geological evidence of recent water modification, are located primarily on the central peaks of craters or on the pole facing slopes of cliffs, making it even more difficult to explain the localized melting of water.

Most researchers in the scientific community agree that the current Mars climate cannot support large amounts of liquid water on the surface. Today, Mars is drier than any desert on Earth. If all the water in the atmosphere condensed on the surface, its depth would be less than the width of a human hair. There is enough water in the polar caps to cover the planet to a depth of 33 feet (10 m), but it is locked in a frozen state. And thanks to the thin CO_2 atmosphere, water that leaves the poles does so through sublimation, turning from a solid directly to gas. Also, due to Mars's distance from the Sun—1.5 times as far away as Earth—the surface temperatures, overall, are far below freezing. Since most of the purported water features occur on older terrain, however, scientists believe that the early climate of Mars was much warmer than it is today.

The average surface temperature of a planet is determined primarily by two factors: its distance from the Sun, and the constituents of the atmosphere. For example, if Earth had no atmosphere, the surface temperature would average -40°F (-18°C), far colder than the freezing point of water. With the addition of

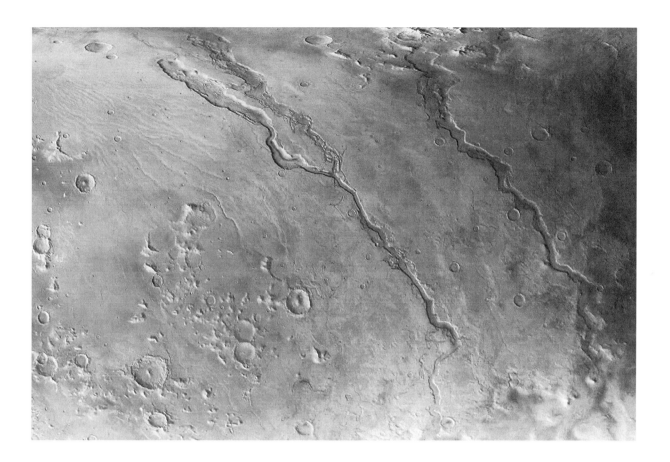

An image taken by the Mars Global Surveyor showing valley systems east of the Hellas plains.
(© AFP/Corbis. Reproduced by permission.)

the atmosphere, the global average temperature rises to 59°F (15°C). This additional warming by Earth's atmospheric "blanket" is known as the greenhouse effect. Trace gases in Earth's atmosphere, such as CO_2 and water vapor, cause this warming of the planet by allowing visible radiation from the Sun to pass through the atmosphere, and absorbing infrared radiation emitted from the planet's surface. Currently, the thin CO_2 atmosphere on Mars creates a negligible greenhouse effect, but if the Martian atmosphere were as dense as 10 Earth atmospheres (or 10 bars), the amount of greenhouse warming afforded by the increased CO_2 would be enough to warm the planet sufficiently to allow stable liquid water on the surface. This assumes, of course, that the amount of sunlight reaching Mars was the same in the past as it is today.

Since its formation, however, the Sun has been getting steadily brighter as it fuses hydrogen to helium in its core. Detailed astrophysical models of the Sun's interior indicate that the Sun is some 30% brighter today than it was 3.8 billion years ago—when water was supposedly flowing on Mars. In this case, even a 10 bar atmosphere is insufficient to warm the planet under the cooler Sun. The Sun was so faint, in fact, that it was even difficult to support liquid water on Earth at this time. But there is strong evidence for liquid water on Earth 3.8 billion years ago—problem known as the Faint Young Sun paradox.

Clearly, through some method or another, Earth managed to get enough early greenhouse gases in the form of CO_2 methane, or water vapor to support liquid water. The geological evidence on Mars, if interpreted as being formed by liquid water, means Mars must have found a way as well. In order to do this, Mars would have required a much more substantial atmosphere than it has today.

Loss of Atmosphere on Mars Over geological time, Mars has lost much of its atmosphere to space. The losses are due to its small size, and hence its lack of gravity to retain its atmosphere. Noble gases, such as argon and xenon, act as tracers of the history of the atmosphere. Since they are chemically inert, the original inventory stays in the atmosphere rather than reacting with the surface. Any loss of atmosphere results in a loss of noble gases. Since many of these gases have stable isotopes of different masses, measuring the ratio of the lighter isotope to the heavier isotope reveals something about the loss of the atmosphere. If the ratio is enriched in heavy isotopes compared to the original inventory, then a significant part of the atmosphere has been lost. Scientists measure the original inventory from meteorite fragments indicating the conditions in the solar system when Mars was formed, and they measure the current inventory with spacecrafts like *Viking* and *Pathfinder* on the surface of the planet. The result: Mars has lost a significant amount of atmosphere in the last 3.8 billion years.

Using the current atmospheric and polar inventories of CO_2 and water on Mars, it is possible to estimate the initial inventory of water and CO_2 on the planet. Unfortunately, even though 95–99% of Mars's atmosphere has been lost to space, less than 1 bar of CO_2 and less than 328 feet (100 m) of water covering the entire planet can be accounted for with atmospheric loss. While this is a great deal of water, it is insufficient to explain the geological features, and it is also an insufficient amount of CO_2 to account for the greenhouse effect needed to warm the planet to the melting point of water under the faint young Sun. Other effects are currently being explored, such as the warming and cooling properties of CO_2 and water clouds. At the present time, the addition of these complex processes make modeling the early climate even more difficult.

Recounting the evidence for an early warm and wet Mars, it seems much of the arguments for flowing water on the surface rest on geological observations from orbit. This being the case, there is nothing to be done but compare them to the only known analogs—the geological features of Earth. Numerous satellite images of Earth indicate that the dominant erosion mechanism in 99% of the images is from water. However, out of 70,000 high-resolution images taken of the surface of Mars, less than 10% show any evidence of water modification. Almost every other image, on the other hand, shows evidence of wind erosion, CO_2 frost, and additional processes never seen on Earth at the same scale. Therefore, a growing number of researchers are attempting to find alternative explanations for the "water" features that are more in sync with the Martian climate.

Alternative Explanations for Mars's "Water" Features Ample evidence of wind erosion exists on the surface of Mars. The erosion rates in the southern highlands can't be explained by wind erosion in the current climate regime, but just a slight increase in pressure creates a large increase in wind's ability to scour the terrain. Three-and-a-half billion years ago, the atmospheric pressure was perhaps high enough to support the erosion rates seen in the southern highlands. And, since wind erosion doesn't require water and removes rather than deposits material, no special deposits should be found in the area. Wind not only erodes the surface, but transports dust on a global scale. The atmosphere, laden with dust, acts more like a huge ocean carrying sediment over the landscape. In terrestrial oceans, such flows carve channel-like features, and it may be possible that some similar process is acting on Mars. Unlike the ocean, the atmosphere is very thin, but the winds on Mars also move at high speed, which may make up for the lack of density in forming these features.

Since CO_2 is the dominant constituent of the atmosphere, it may play a large role in the formation of the flow features. In particular, the outflow channels may result from a catastrophic release of CO_2 gas from layered deposits beneath the surface. The cold CO_2 gas scours out the terrain much like dense gas flows from volcanoes on Earth remove the sides of mountains. While these hypotheses are still being explored and tested, they represent some new ideas in understanding Mars.

It is clear that Mars, while similar to Earth in general, is radically different in detail. Features formed on the surface are subject to a range of conditions never experienced by similar landscapes on Earth. Liquid water at the surface exists only briefly while wind erosion on a planet-wide scale occurs even today. Therefore, researchers must be careful to remove their "water glasses" and explore Mars in the context of its own environment. Each of the above interpretations—including that of flowing water—requires more information to resolve the debate. To date, researchers have access only to images from the surface. Earth geologists are suspicious of arguments based solely on how something "looks." They prefer to gather more field observations and samples before drawing a conclusion. Mars geologists lack that ability to gather samples from the planet, but until they do so, the true nature of the "water" features will be debated. —JOHN ARMSTRONG

Further Reading

Boyle, A. "Mars Gullies Hint at Liquid Water." *MSNBC* June 22, 2001. <http://www.msnbc.com/news/423452.asp>.

Bridges, A. "Mars Hides Much More Water, Study Suggests." 2000. <http://www.space.com/scienceastronomy/solarsystem/mars hiding 000628.html>.

Caplinger, M. "Channels and Valleys." *Malin Space Science Systems.* 1995. <http://www.msss.com/http/ps/channels/channels.html>.

Carr, Michael H. *Water on Mars.* Oxford: Oxford University Press, 1995.

Chaikin, A. "An Eye for Mars: The Camera that Found the Watery Evidence." 2000. <http://www.space.com/scienceastronomy/camera_technology_000623.html>.

Davis, K. E. *Don't Know Much About the Universe: Everything You Need to Know About the Cosmos but Never Learned.* New York: Harper Collins Publishers, Inc., 2001.

Hoffman, Nick. "White Mars: A New Model for Mars' Surface and Atmosphere Based on CO_2." *Icarus* no. 146 (2000): 326–42.

JPL Mars Program. 2001. "Mars Polar Lander: Searching for Water on Mars." <http://mpfwww.jpl.nasa.gov/msp98/why.html>.

Keiffer, Hugh H., Mildred S. Mathews, and Bruce M. Jakosky. *Mars*. Tucson: University of Arizona Press, 1997.

Maran, S. P. *Astronomy for Dummies*. Foster City, CA: IDG Books Worldwide, Inc., 1999.

Mars Global Surveyor. <mars.jpl.nasa.gov/mgs>.

Milan, W. "Water on Mars: Back to the Future?" 2000. <www.space.com/news/spacehistory/water_mars_history_000626.html>.

NASA. "The Case of the Missing Mars Water." 2001. <http://science.nasa.gov/headlines/y2001/ast05jan_1.htm>.

———. "Liquid Water on Mars." 2001. <http:// liftoff.msfc.nasa.gov/news/2000/news-MarsWater.asp>.

———. "Mars Surprise." 2000. <http://www.spacescience.com/headlines/y2000/ast22jun_2.htm>.

"New Evidence Suggests Mars Has Been Cold and Dry." <www.spacedaily.com/news/mars-water-science-00l.html>.

Raeburn, P. *Uncovering the Secrets of the Red Planet*. Washington, DC: National Geographic Society, 1998.

Sagan, C. *Cosmos*. New York: Random House, 1980.

Taylor, M. R. "Water on Mars: Is the Proof Out There?" 2000. <http://www.discovery.com/news/features/marswater/marswater.html>.

Are hazards due to earthquakes in the New Madrid Seismic Zone overestimated?

Viewpoint: Yes, major earthquakes in the Midwest are rare, and the geological features of the NMSZ are quite different from the earthquake-prone regions of the West Coast.

Viewpoint: No, scientific understanding of earthquake phenomena is still limited, and the large earthquakes of 1811 and 1812 suggest that the threat is genuine.

People have a special dread of earthquakes. They strike without warning, and can in the worst circumstances in a few moments kill thousands of people and leave thousands of others injured and homeless. Hurricanes can be equally devastating, but their arrival is usually predicted hours or even days in advance, giving people the chance to flee or prepare themselves. Tornadoes give less warning than other storms, but their damage is typically very localized. Floods, avalanches, forest fires, and other natural catastrophes wreak havoc with lives and property, but do not present the special risks—or generate the fear—associated with earthquakes. In the developed world, earthquakes are treated as a unique risk to life and property, and in the regions most vulnerable to damaging seismic events, construction of buildings, roads, and other structures are strictly regulated to minimize damage and loss of life in the event of a severe earthquake.

Quakeproof construction comes at significant cost. Cities where the earthquake risk is highest—Tokyo, San Francisco, Los Angeles—have had the greatest success in regulating new construction to high safety standards and in retrofitting older structures to improve safety. In the immediate aftermath of an event such as the earthquake that struck near San Francisco in 1989, resistance to the costs of improving safety in the effected area is naturally diminished. But even in the places where earthquakes are a constant, present danger, funds for quakeproof construction are far from unlimited. Engineering decisions are always made in the context of balancing costs and benefits, and protection from earthquakes is no exception.

At the opposite extreme from the quake-sensitive cities of California are large urban centers such as New York and Chicago. A significant earthquake near either of these cities would create a catastrophe far greater than any in recorded history. Both cities house millions of people in unreinforced masonry buildings that are very likely to collapse during an earthquake. Bridges, tunnels, and mass transportation systems are also not constructed to meet earthquake standards. Large residential and commercial neighborhoods rest upon reclaimed land, or "landfill," likely to liquefy during a quake. But while the risk presented by an earthquake in these cities is enormous, the chance of an earthquake striking either place is, while not zero, very low, and the governments and citizens of New York and Chicago have chosen to treat the threat as negligible.

The region in the Mississippi River Valley known as the New Madrid Seismic Zone presents a more complicated equation of risk and benefit. In the early nineteenth century, New Madrid, Missouri, was the epicenter of some of the largest earthquakes ever recorded in the continental United States. The region was sparsely populated at that time, but historical accounts describe houses falling down in New Madrid, chimneys toppling near St. Louis, and—perhaps apocryphally—church bells ringing as far away as Boston. Because the region is much more heavily populated today, a large earthquake there could cause great death and destruction.

But engineers, scientists, and civic planners have not fully accepted the conclusion that the New Madrid region faces the same earthquake threat as the more seismically active cities on the West Coast. There were no major seismic events in the region during the twentieth century, and those lesser events that occurred (mostly earthquakes of magnitude 5 or less) did virtually no damage. While the quake-prone regions of California rest on or near active fault lines, the geology of the New Madrid region is more complicated, and scientists disagree about its risk profile for the future. More scientific study is underway, and will no doubt continue into the future. In the meantime, decisions about how to incorporate seismic risks into decisions about construction and civic planning will have to be made with incomplete scientific evidence, and the worrisome historical record of a massive quake long ago will have to be measured against the complacency inspired by nearly two hundred quake-free years. —LOREN BUTLER FEFFER

Viewpoint:
Yes, major earthquakes in the Midwest are rare, and the geological features of the NMSZ are quite different from the earthquake-prone regions of the West Coast.

Scientists face a challenge when discussing natural hazards with the public. On the one hand, experts want to warn to public about the hazard and convince citizens to take appropriate steps to mitigate it. On the other hand, overestimating the hazard can have either of two undesirable effects. If the general public does not find the warning credible, the warning will be ignored. Alternatively, accepting an overestimate of the hazard may divert time and resources from other goals that could result in more societal good.

The Cost of Stringent Building Codes An example of this challenge is illustrated by attempts to estimate the seismic hazard for parts of the central United States due to earthquakes in the New Madrid Seismic Zone (NMSZ). Recent United States Geological Survey maps predict that the seismic hazard in the area is surprisingly high, in some ways exceeding that found in California. Because most earthquake-related deaths result from the collapse of buildings—a principle often stated as "earthquakes don't kill people; buildings kill people"—the primary defense against earthquakes is designing buildings to ensure the safety of the inhabitants. As a result, authorities in Memphis, Tennessee, and other portions of the area are currently considering adopting new building codes that require that buildings be built to the stringent standards of earthquake resistance used in California. Given that these requirements are expensive and might raise building costs by 10 to 35%, it is natural to ask whether the hazard estimate is realistic. Similarly, it is important to consider whether these costs are justified by the resulting increases in public safety or whether alternative, less expensive strategies might better balance costs and benefits.

To consider these questions, it is important to recognize that a community's choice of building codes in earthquake-prone areas reflects a complicated interplay between seismology, earthquake engineering, economics, and public policy. The issue is to assess the seismic hazard and choose a level of earthquake-resistant construction that makes economic sense given that such a design raises construction costs and diverts resources from other purposes. For example, money spent making schools earthquake-resistant is therefore not available for hiring more teachers, improving their pay, or providing better class materials. Similarly, funds spent making highway bridges earthquake-resistant cannot be used for other safety improvements addressing more common problems. Thus, ideally building codes should not be too weak, permitting unsafe construction and undue risks, or too strong, imposing unneeded costs and encouraging their evasion. Deciding where to draw this line is a complex policy issue for which there is no definite answer.

Hazards and Risks In assessing the potential danger posed by earthquakes or other natural disasters it is useful to distinguish between hazards and risks. The hazard is the intrinsic natural occurrence of earthquakes and the resulting ground motion and other effects. The risk is the danger the hazard poses to life and property. The hazard is a geological issue, whereas the risk is affected by human actions.

> ## KEY TERMS
>
> **ALLUVIAL PLAIN:** An assemblage of sediments marking the place where a stream moves from a steep gradient to a flatter gradient and suddenly loses transporting power. Typical of arid and semi-arid climates but not confined to them.
>
> **ASTHENOSPHERE:** The weak or soft zone in the upper mantle of the earth just below the lithosphere, involved in plate movement and isostatic adjustments. It lies 70 to 100 km below the surface and may extend to a depth of 400 km.
>
> **EPICENTER:** The point on the earth's surface that is directly above the focus (center) of an earthquake.
>
> **GLOBAL POSITIONING SYSTEM (GPS):** An array of military satellites in precisely known orbits. By comparing minute differences in arrival times of a satellite's radio signal at two sites, the distance between markers tens of kilometers apart can be determined to within a few millimeters.
>
> **LITHOSPHERE:** The rigid outer shell of the earth. It includes the crust and upper mantle and is about 100 km thick.
>
> **MAGNITUDE:** A measure of the strength of an earthquake based on the amount of movement recorded by a seismograph.
>
> **PLATE TECTONICS:** A theory of global tectonics according to which the lithosphere is divided into mobile plates. The entire lithosphere is in motion, not just the segments composed of continental material.
>
> **RICHTER SCALE:** A commonly used measure of earthquake magnitude, based on a logarithmic scale. Each integral step on the scale represents a 10-fold increase in the extent of ground shaking, as recorded on a seismograph. Named after American seismologist Charles Richter.
>
> **RIFT:** A valley caused by extension of the earth's crust. Its floor forms as a portion of the crust moves downward along normal faults.
>
> **SEDIMENT:** Any solid material that has settled out of a state of suspension in liquid.

Unfortunately, earthquake risks are not well understood: earthquake risk assessment has been described by one of its founders as "a game of chance of which we still don't know all the rules." Nowhere is this more the case than in the New Madrid Seismic Zone, where the situation is quite different from the familiar case of California, as shown in Figure 1. Earthquakes in California occur as part of the plate boundary zone that spans most of the western United States and that accommodates most of the approximately 45 mm/year net motion between the Pacific and North American plates. (A plate is a layer of rock mass both within the earth and on its crust.) In contrast, the NMSZ is within the generally stable interior of the North American plate, which measurements from the Global Positioning System (a collection of military satellites in known orbits) show deforms by less than 2 mm/year. Hence, major earthquakes are far more common in California: large earthquakes (magnitude 7 or greater) taking up the interplate motion typically occur on major faults in California about every 100 to 200 years, whereas the small intraplate deformation in the NMSZ appears to give rise to earthquakes of this size about every 500 to 1,500 years.

As a result, high-magnitude earthquakes in the Midwest are relatively rare. Since 1816, there are thought to have been 16 earthquakes with magnitude greater than 5 (about one every 10 years), and two with magnitude greater than 6, or about one every 100 years. These have caused some damage but have been considered more of a nuisance than a catastrophe. For example, the largest NMSZ earthquake in the past century, the November 9, 1968 (magnitude 5.5) southern Illinois earthquake, was widely felt and resulted in some damage but caused no fatalities. However, a repetition of the large earthquakes that occurred from 1811 to 1812 in the NMSZ would be very damaging and likely cause deaths. Historical accounts show that houses fell down in the tiny Mississippi river town of New Madrid and that several chimneys toppled near St. Louis. This data implies that these earthquakes had magnitude about 7.2 and provide insight into the effects of future ones.

The risk that earthquakes pose to society depends on how often they occur, how much shaking they cause, and how long a time interval we consider. Figure 2 illustrates this idea for California and the New Madrid Seismic Zone, using circles representing approximate areas within which an earthquake might significantly damage some buildings. Earthquakes of a given magnitude occur 100 times more frequently in the NMSZ than in California. However, because seismic energy is transmitted more efficiently by the rock in the Midwest, NMSZ earthquakes shake about the same area as California earthquakes one magnitude unit smaller. Thus, in 100 years much of California will be shaken seriously, whereas only a much smaller fraction of the NMSZ would be. After 1,000 years, much of the NMSZ has been shaken once, whereas most of the California area has been shaken many times.

Guarding Against the Greatest Possible Hazard—Too Costly? Typically, buildings have useful lives of about 50 to 100 years. Thus, buildings in the Midwest are unlikely to be shaken during their lifetimes, whereas buildings in California are likely to be. Given this difference, several approaches are possible. One, to

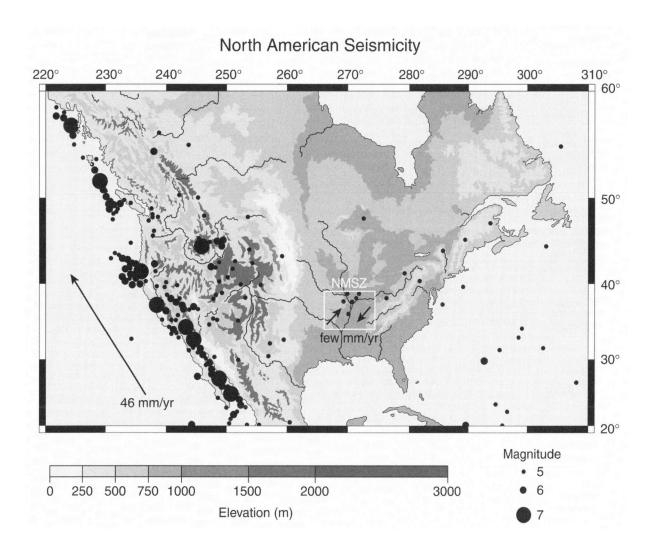

Figure 1: This map shows earthquake activity in the New Madrid Seismic Zone.

(Electronic Illustrators Group.)

assume the highest possible hazard, is taken by the United States Geological Survey seismic hazard maps, which show the New Madrid zone as the most seismically hazardous in the nation. This estimate assumes that the 1811–1812 earthquakes were, and hence the largest future earthquakes will be, magnitude 8 earthquakes, much larger than other lines of evidence suggest. The maps further assume that ground motions from these earthquakes would significantly exceed that expected from previous models based on recorded shaking. Hence, the maps predict that the ground motion expected in 50 years at 2% probability, or roughly every 2,500 (50/0.02) years, for the New Madrid zone exceeds that for San Francisco or Los Angeles. These estimates have considerable uncertainty because the underlying physical cause of the earthquakes is unclear, the magnitudes and recurrence times for the largest earthquakes are difficult to infer, and the likely ground motion from such earthquakes is essentially unconstrained. Nonetheless, these estimates have led to the proposal that buildings in the Memphis area be built to California codes. This option maximizes safety at considerable cost because buildings in Memphis are much less likely to be seriously shaken. Given that this option is based on a very high estimate of the hazard, it seems desirable to explore alternatives. For example, the code might require that buildings be built to an intermediate standard, perhaps the ground motion expected once in 500 years. The values are less uncertain than the 2,500 year ones because they are closer to what has actually been seismologically recorded. This approach seems likely to be more realistic, given how rare major earthquakes are, and gives reasonable seismic safey at significantly lower cost. As a result, it seems premature to adopt California-style codes before a careful analysis of costs and benefits.

Inform the Public with Clear and Accurate Data This discussion indicates that deciding how to address earthquake hazards in the Midwest is best done by taking a balanced approach of recognizing the hazard while not exaggerating it. Although the public and news media are drawn by images of destruction during earthquakes, it is useful to bear in mind that in the United States earthquakes have claimed an average of nine lives per year nationally, putting earthquakes at the level of risk equal with in-line skating or football, but far less than bicycles. To date, the deaths are primarily in the western United States: earthquakes in the central United States have caused no fatalities for more than 150 years. Unfortunately, official statements and publications intent upon trying to interest the public in NMSZ seismic hazards often use inflated language, for example referring to future magnitude 6 earthquakes, which are unlikely to cause fatalities or major damage, as "devastating". Most such pronouncements do not convey how rare major Midwest earthquakes are (a major earthquake anywhere in the six-state seismic zone is expected less than once every 100 years on average) and do not explain how different the situation is from California. As a result, the impressions given are misleading and may lead to poor policy choices. This is especially unfortunate because the public, media, and authorities have less sophistication in understanding earthquake issues than their counterparts in California. A more sensible way to present seismic hazards would be to be candid with the public so that evaluation of the costs and benefits of alternative policies can be done with as much information on the issues and uncertainties as possible. It would be wise to follow American physicist Richard Feynman's 1988 admonition after the loss of the space shuttle *Challenger*: "NASA owes it to the citizens from whom it asks support to be frank, honest, and informative, so these citizens can make the wisest decisions for te use of their limited resources. For a successful technology, reality must take precedence over public relations, because nature cannot be fooled." —SETH STEIN

Viewpoint:
No, scientific understanding of earthquake phenomena is still limited, and the large earthquakes of 1811 and 1812 suggest that the threat is genuine.

There is only one thing wrong with the notion that the earthquake hazard in the New Madrid Seismic Zone (NMSZ) is overestimated—geophysicists, geologists, seismologists, and other experts in this field do not know enough about how to predict earthquakes even to make this statement. This is especially true of earthquakes that happen in the Mississippi River Valley. No one knows this better than Arch Johnston, director of the Center for Earthquake Research and Information at the University of Memphis, and coordinator for the Hazards Evaluation Program at the Mid-America Earthquake Center. According to Johnston, New Madrid released more seismic energy in the nineteenth century than did the entire western United States. "In the twentieth century," however, as Johnston reported in *Seismicity and Historical Background of the New Madrid Seismic Zone,* "the fault zone has been relatively quiescent, with only a few minor-damage events exceeding magnitude 5. . . . Understanding which century is more representative of normal New Madrid behavior is perhaps our greatest current challenge."

Plate Tectonics One reason it's a challenge is that the New Madrid Seismic Zone is not like other earthquake-prone regions. Most earthquakes on the planet happen at the boundaries between tectonic plates, or land masses. This plate tectonics model of the earth's surface emerged in the 1960s and gave geoscientists a way to understand previously baffling things like the origin of mountain ranges and oceans, volcanic eruptions, and earthquakes. According to the model, the massive, rigid plates are 30 to 90 miles thick and 12 to 124 miles wide. They move at different speeds and directions on a hot, soft layer of rock called the asthenosphere, whose currents move the plates a few inches a year. Continents and oceans hide plate borders, but it is not hard to pinpoint their locations. Where plates ram together, mountains and volcanoes form; where they crack apart into yawning rifts, oceans are born; where they grind past each other, the land is racked by earthquakes. Though the plate tectonics model has been widely accepted by geophysicists, it does not explain everything about earthquake behavior and predictability.

Questions about Seismic Activity in the NMSZ In the winter of 1811–1812, three massive earthquakes struck the Mississippi River Valley, which is more than 1,000 miles from the nearest plate boundary. These intraplate (or, within plate) earthquakes were magnitude 8.2, 8.1, and 8.3, making them the most energetic earthquakes ever recorded in the contiguous 48 states. The quakes were not an isolated event. Researchers now have evidence that strong earthquakes occurred in the NMSZ repeatedly in the geologic past. And smaller earthquakes still shake the region—two to three earthquakes a week, with one or two a year large enough for residents to notice—making the central Mississippi Valley the most seismically active region east of the Rocky Mountains. "According to current scientific understanding, these were earthquakes where they shouldn't be," Johnston wrote. "They are outstanding examples of rare major-to-great earthquakes that happen remote from the usual tectonic plate boundary or active intraplate seismic zones. . . . Our understanding of the faulting process and repeat times of New

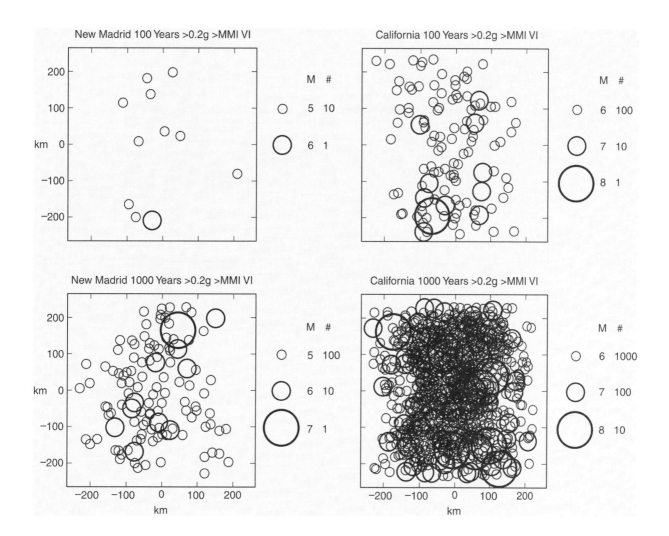

Figure 2: Charts comparing seismic activity in California and New Madrid Sesimicity Zone.
(Electronic Illustrators Group.)

Madrid characteristic earthquakes has greatly improved recently," he added, "but major questions remain."

One big—and fundamental—question is why an area in an otherwise stable continent would release such massive amounts of seismic energy? Another is, why are faults in an ancient rift system in the crust under the Mississippi River valley, inactive for millions of years, now becoming active? The proposed answers to these questions are varied and speculative.

Perhaps the biggest question is when will the next large New Madrid quake strike? A study in 1992 suggested that it might occur within the next few hundred years. By December of 1998, however, at a meeting of the American Geophysical Union, researchers who had monitored the New Madrid region for signs of strain said the next significant earthquake would not occur for another 5,000 or 10,000 years, if then.

Complicating matters is the fact that geophysicists do not understand why the New Madrid faults ruptured in the first place, leading geophysicist Paul Segall of Stanford University to tell *Science* magazine that another magnitude 7 to 8 "can't be dismissed at this point." Segall said strain may have reaccumulated rapidly in the first century after 1812 and is now slowly building toward another quake in the next century or two. Future Global Positioning System (GPS) surveys should show whether the land is motionless, or is slowly inching toward the next big earthquake. "The simplest assumption is, if [big quakes] happened in the past, they can happen in the future," Segall said.

In 1992, Segall, geophysicist Mark Zoback of Stanford, and colleagues compared a 1991 GPS survey with a nonsatellite 1950s land survey and calculated that ground near the fault was moving 5 mm to 8 mm a year. Later, Zoback reported in *Science* magazine that he and his colleagues "made a detailed study in 1991 of crustal strain with the use of a dense concentration of geodetic stations located astride a single major fault. Our repeated GPS measurements of this network in 1993 and 1997 appear to indicate lower rates of strain accumulation than we originally reported." But lower strain rates do not necessarily imply lower seismic hazard for the region, Zoback said, adding that it is possible that the strain energy released in the storm of large earthquakes in the New Madrid area over the past thousand years took hundreds of thousands, or even millions of years to accumulate. "If this is the case, a slow rate of strain accumulation over the past 6 years does not imply low seismic hazard," Zoback said. "The persistently high rate of seismic activity in the New Madrid seismic zone over the past few thousand years implies high seismic hazard in the foreseeable future. To communicate any other message to the public would seem to be a mistake."

Can Earthquakes Really Be Predicted?
Such uncertainty about earthquake behavior and prediction extends well beyond the New Madrid Seismic Zone. In February 1999, the journal *Nature* sponsored a seven-week online debate among top researchers in the field about whether the reliable prediction of individual earthquakes was a realistic scientific goal. According to moderator Ian Main—a reader in seismology and rock physics at the University of Edinburgh and associate editor for the *Journal of Geophysical Research*—all the debate contributors who expressed an opinion "agree that the deterministic prediction of an individual earthquake, within sufficiently narrow limits to allow a planned evacuation program, is an unrealistic goal." Another contributor, Max Wyss of the Geophysical Institute at the University of Alaska-Fairbanks, said, "The contributions to the debate about earthquake prediction research in *Nature* so far clearly show that we have hardly scratched the surface of the problem of how earthquake ruptures initiate and how to predict them.... If the current knowledge of the earthquake initiation process is so poorly founded that experienced researchers can maintain the profound differences of opinion present in this debate," he added, "we are in desperate need of the major research effort that is not at present being made."

Hazards Emanating from Outside the NMSZ
There is one more problem with minimizing the earthquake hazard in the New Madrid Seismic Zone—this zone may be the most recognized source of earthquakes in the central Mississippi Valley, but it is not the *only* source of seismic hazards. Some scientists estimate a 9-in-10 chance that a magnitude 6 or 7 earthquake will occur in the New Madrid Seismic Zone in the next 50 years. Others say a large earthquake will not strike the region for at least 1,000 years, or even 5,000 or 10,000 years. Even the new United States Geological Survey National Seismic Hazard Maps are based on a 1,000-year cycle for magnitude-8 earthquakes in the New Madrid Seismic Zone. But earthquakes occur in a range of magnitudes, and the New Madrid Seismic Zone is not the only source of earthquakes in the Mississippi River valley. In fact, the most recent earthquakes have come from disturbances outside that zone.

In 1968, a magnitude 5.5 earthquake struck whose epicenter was 120 miles southeast of St. Louis. In 1978 another earthquake rattled St. Louis, this one measuring 3.5 on the Richter scale. The source was the St. Louis Fault, discovered in 1948, a 45-mile-long fault running from Valmeyer, Missouri, to Alton, Illinois. In 1990 a magnitude 4.6 earthquake originated south of Cape Girardeau, Missouri, in the Benton Hills, a five-mile-wide forested incline in an otherwise flat alluvial plain. It rattled seven states, and the epicenter was near New Hamburg in Scott County, Missouri, which is outside the New Madrid Seismic Zone. After the earthquake, the Missouri Department of Natural Resources (DNR) and the USGS decided to investigate the source of seismicity in the Benton Hills.

One of the few places in the central Mississippi Valley where faults break the earth's surface is near the town of Commerce, Missouri. Here, starting in 1993, geologist David Hoffman, an earthquake specialist in the Division of Geology and Land Survey for the Missouri DNR, excavated a dozen or more 15-foot-deep trenches in the Benton Hills uplands and uncovered folds in the soil layers. The folds told Hoffman that seismic activity had occurred there within the last 10,000 years and suggested that the faults were still active. The site lies along a linear magnetic feature called the Commerce geophysical lineament. Some geophysicists believe this northeast-trending lineament may represent a fault zone more than 150 miles long. It is still poorly understood, but this geologic feature could pose another large seismic threat to the region. "There are lots of faults in the area, and it might [indicate] the potential for earthquakes in the future," Hoffman told the *St. Louis Riverfront Times* in 1999. "It takes studies of lots of these kinds of places and features and compiling all the data and analyzing it . . . to get the whole picture. We . . . have one piece of the puzzle with this spot. We're the primary ones that have even looked here. Then there's the rest of the state that hasn't been looked at."

Still Much to Learn Assessing the seismic risk in the Mississippi Valley is tricky because faults and other geologic structures were deeply buried over hundreds of millions of years by thick layers of sediment, and no clues to the area's seismicity exist at the earth's surface. Other seismic authorities agree with Hoffman that there is still a great deal to be learned about the structure of the earth's crust under the Mississippi River valley—and everywhere else. Even in northern California, where the faults are much closer to the surface, seismologists did not know about the fault that caused the magnitude 6.8, 1994 Northridge earthquake—57 dead; 9,300 injured; 13 billion dollars in damages;

50,000 homeless; 13,000 buildings destroyed—until the ground started moving.

Using geophysical techniques and a growing array of tools, researchers have gained substantial knowledge over the years about the nature of earthquakes and the processes that drive them. Beginning in 1995, the National Science Foundation's (NSF) Division of Earth Sciences announced that it would fund research opportunities in a special area of emphasis—tectonically active systems of the earth's continental crust. NSF describes the initiative, called Active Tectonics, as "a concerted attack on issues pertaining to the connectedness of individual tectonic elements and physical-chemical processes within a system, and how these relationships evolve through time." The research is supposed to cross traditional geological and geophysical disciplinary boundaries and link fields such as earthquake seismology, basin analysis, geodetics, structural geology, paleoseismology, geomorphology, reflection seismology, petrology, regional tectonics, solid-earth geophysics, geochronology, geomagnetics, rock mechanics, hydrology, tectonophysics, Quaternary studies, paleomagnetics, volcanology, and more. The result, in the future, will be better ways to define problems and approaches, better characterization of the dynamic state of lithosphere and asthenosphere, better 3-D resolution of regional tectonic architecture, integration of historical geological facts and interpretations of a region's tectonic evolution, and better ways to mitigate earthquake hazards.

In the meantime, earthquakes generated by hidden and poorly understood geologic features will keep shaking the central Mississippi Valley. Those who presume to know such things have different opinions about when they strike—in 50 or 1,000 years? Never, or tomorrow? At the Seismological Laboratory at the California Institute of Technology-Pasadena, geophysicist Hiroo Kanamori likes to say that there is only one known method to date for testing earthquake predictions—wait and see. —CHERYL PELLERIN

Further Reading

Bolt, Bruce. *Earthquakes and Geological Discovery*. Scientific American Library, 1993.

Earthquake and New Madrid Seismic Zone Information Pages. Southern Illinois University-Carbondale, Illinois <http://www.science.siu.edu/geology/quakes/>. Near-time displays of central United States seismic data at <http://www.eas.slu.edu/Earthquake_Center/NM/EW/GIFS/welcome.html>.

Hildenbrand, Thomas, Victoria Langenheim, Eugene Schweig, Peter Stauffer, and James Hendley II. *Uncovering Hidden Hazards in the Mississippi Valley*. USGS Fact Sheet 200–96, 1996.

Hough, S., J. G. Armbruster, L. Seeber, and J. F. Hough. "On the Modified Mercalli Intensities and Magnitudes of the 1811/1812 New Madrid, Central United States, Earthquakes." *Journal of Geophysical Research* 105, No. 23 (2000): 839.

Incorporated Research Institutions for Seismology (IRIS) <http://www.iris.washington.edu/HQ/iris.html>. Also the IRIS Education and Outreach Page <http://www.iris.washington.edu/EandO/>.

Inside Geology. Houghton Mifflin Co. <http://www.geologylink.com/toc/>.

Kanamori, Hiroo. "Initiation Process of Earthquakes and Its Implications for Seismic Hazard Reduction Strategy." *Proceedings of the National Academy of Sciences* 93 (April 1996): 3726–31.

Kerr, Richard. "Meeting: American Geophysical Union: From Eastern Quakes to a Warming's Icy Clues." *Science* 283, No. 5398 (January 1, 1999): 28–29.

Knopoff, L. "Earthquake Prediction: The Scientific Challenge." *Proceedings of the National Academy of Sciences* 93 (April 1996): 3719–20.

Malone, Stephen. *Surfing the Internet for Earthquake Data*. Department of Earth and Space Sciences, University of Washington-Seattle <http://www.geophys.washington.edu/seismosurfing.html>.

Nature Debates. Ian Main, moderator. February 25, 1999 <http://www.nature.com/nature/debates/earthquake/equake_contents.html.>

Newman, Andrew, Seth Stein, John Weber, Joseph Engeln, Ailin Mao, and Timothy Dixon. "Slow Deformation and Lower Seismic Hazard at the New Madrid Seismic Zone." *Science* 284, No. 5414 (April 23, 1999): 619–21.

———, J. Schneider, Seth Stein, and A. Mendez. "Uncertainties in Seismic Hazard Maps for the New Madrid Seismic Zone." *Seismological Research Letters* 72 (November 2001).

NOAA, National Geophysical Data Center, Earthquake Data <http://www.ngdc.noaa.gov/seg/hazard/earthqk.shtml>.

Nuttli, Otto. "The Mississippi Valley Earthquakes of 1811 and 1812: Intensities, Ground Motion and Magnitudes." *Bulletin of the Seismological Society of America* 63, No. 1 (February 1973): 227–48 <http://www.eas.slu.edu/Earthquake_Center/SEISMICITY/Nuttli.1973/bssa.html>.

Schweig, Eugene, Joan Gomberg, and James Hendley II. "The Mississippi Valley: 'Whole Lotta Shakin' Goin' On'." USGS Fact Sheet 168–95 (1995) <http://quake.wr.usgs.gov/prepare/factsheets/NewMadrid/>.

Stelzer, C. D. "On Shaky Ground." *St. Louis Riverfront Times* (December 16, 1999) <http://www.riverfronttimes.com/issues/1999-12-15/feature.html>.

Stover, Carl, and Jerry Coffman. *Largest Earthquakes in the United States*. Abridged from *Seismicity of the United States, 1568–1989*. USGS Professional Paper No. 1527. Washington, D. C.: United States Government Printing Office, 1993.

United States Geological Survey Center for Earthquake Research and Information. University of Memphis-Tennessee <http://www.ceri.memphis.edu>.

USGS National Earthquake Information Center. World Data Center for Seismology, Denver <http://earthquake.usgs.gov/>. New Madrid information at <http://wwwneic.cr.usgs.gov/neis/new_madrid/new_madrid.html>.

USGS National Earthquake Information Center (NEIC) <http://neic.usgs.gov>. Also United States National Seismic Net <http://neic.usgs.gov/neis/usnsn/usnsn_home.html> and Earthquake Hazards Program <http://earthquake.usgs.gov>.

Zoback, Mark. "Seismic Hazard at the New Madrid Seismic Zone." *Science* 285, No. 5428 (July 30, 1999): 663.

Historic Dispute:
Are Earth's continents stationary throughout geologic time?

Viewpoint: Yes, the best available and most widely accepted models of Earth's structure once precluded large-scale horizontal motion of continents.

Viewpoint: No, geologic evidence shows that the continents had once been in very different positions, and thus must have moved great distances across Earth's surface over time.

Scientific ideas and theories change through gradual evolution and sudden revolution. Sometimes the systematic exploration of shared principles leads to an expansion and accumulation of knowledge about a subject; other times, competing concepts about the most fundamental hypotheses of a field clash for acceptance. For several decades during the twentieth century, earth scientists wrestled with an idea at the very foundation of their field: had the continents we observe on Earth today been stationary throughout history, or had subterranean forces reshaped the pattern of land and sea throughout time? The reluctance of many scientists to accept the concept of continental drift was rooted not only in reservations they had about the nature of the evidence cited to support the novel theory, but also in concerns about the very kind of theory it was.

To understand why the theory of continental drift seemed so unacceptable to earth scientists during the twentieth century, it helps to understand the guiding methods of geology that had been developed during the previous two hundred years. One important principle that geology shared with other sciences was a commitment to simplicity—scientists sought to find the simplest theory to explain their observations. Extraneous assumptions, elaborate constructions, and obscure interpretations would all be challenged if simpler, more obvious explanations could account for the same data. This more or less philosophical preference for simplicity has been an important element of the history of science and has often played a role in scientific controversies.

A methodological or philosophical concept that is special to the earth sciences is called uniformitarianism. It is related to the general commitment to simplicity mentioned above, but arises more directly from the subject matter of Earth's history. Uniformitarianism suggests that scientists may invoke no forces to explain geologic events of the past that are not observed on Earth today. Earth has changed through history, but always and only as a result of observable transformations—not through the work of invisible forces or mysterious conditions. Uniformitarianism as a methodology itself changed somewhat during the nineteenth and twentieth centuries, but it was still a valuable concept to geologists at the time they were confronted with the unorthodox concept of continental drift.

The suggestion that Earth's continents had shifted over time, breaking off from an original supercontinent into drifting configurations that gradually came to the array of continents observed today, was extraordinarily novel. Even the theory's proponents were initially unable to explain how the continents moved, and to many scientists it seemed that for every observation that could be explained by continental drift, there were several—from geology, paleontology, and physics—that could not. Over time, new discoveries and new interpretations of controversial observations accumulated in support of the idea that the continents are not stationary. Today, we know that the brittle outer portion of Earth, called the lithosphere, is in motion. We have sophisticated ways of directly measuring this motion using lasers and satellites. The modern theory of plate tectonics accommodates the idea that the continents are in constant motion. This theory states that Earth is divided into a few large, thick slabs of rock called lithospheric plates that are slowly moving and changing in size. Continental crust comprises the upper portion of some of these plates, ocean crust comprises the upper portion of other plates, and still others have both continental and oceanic crust comprising their upper portions. Plate tectonics is a unifying theory, or paradigm, in the earth sciences. It holds such an important place in our study of Earth because it explains the formation of mountain belts, the occurrence and distribution of volcanoes and earthquakes, and the age and distribution of material in Earth's crust. The saga of the long dispute over accepting the idea that continents move is a paradigmatic story of a revolutionary transformation in scientific ideas. —LOREN BUTLER FEFFER

Viewpoint:
Yes, the best available and most widely accepted models of Earth's structure once precluded large-scale horizontal motion of continents.

Existing Theories and Their Problems One of the major geological problems in the second half of the nineteenth century was the question of how mountains originate. Although it was well-known that the rocks which formed mountains had originally been deposited in shallow coastal seas, opinions varied as to the nature of the forces and processes that caused these thick sediments to be thrust upward, tens of thousands of feet above sea level. In Europe, where the mountains best known to geologists were the inland Alps, the most influential theorist was Austria's Eduard Suess. He postulated that Earth's crust experienced irregular periods of collapse, causing a global redistribution of water and land, and changing the patterns of erosion and sediment deposition. This explained how the Alps could contain rocks that were once on the ocean floor. Suess also coined the term "Gondwanaland" to describe a now-sunken continent that he estimated had once existed in the southern hemisphere. In his theory, oceans and continents were not permanent: they could move up and down, continents becoming oceans and vice versa. However, they certainly could not drift sideways.

In the United States, geologists' experience came primarily from the Appalachians, a coastal mountain chain. American scientists, led by Yale professor James Dwight Dana, therefore, favored a different theory: continental permanence. Dana and his followers believed that the oceans and continents had been in roughly the same position throughout Earth history. They thought that mountains were raised up as a result of sedimentary accumulation along the coast in a trough called a "geosyncline." The American theory also recognized that continents and ocean floors are made of two fundamentally different materials, with distinct chemical compositions and densities. In contrast to Suess's theory, the two were not interchangeable. As the saying went, "once a continent, always a continent; once an ocean, always an ocean."

What theories like these had in common was the assumption that Earth was gradually cooling, and therefore contracting. This ongoing shrinkage was imagined to be the ultimate driving force behind subsidence, mountain building, and other geological processes. By the end of the first decade of the twentieth century, however, many realized that no theory relying on contraction could possibly work, for two reasons. First, the discovery of radioactivity meant that there was a new source of heat to account for within Earth. The Irish geologist John Joly calculated that Earth was not cooling, and thus not shrinking. However, he argued, the heat released by radioactive decay deep inside Earth could be the energy source that powered geological forces.

The second problem with the idea of a contracting Earth was equally fundamental. Surveyors in India and the United States had made very detailed measurements of the force of gravity in many different places, to correct for instrumental error. They believed that enormous mountain ranges like the Himalayas would exert a gravitational pull on the plumb lines used to level astronomical instruments. To the geophysicists' surprise, the effective mass of mountains was much less than predicted. The only possible solution was "isostasy," the idea that Earth's crust is flexible and that its pieces are in balance with each other. This made it impossible that continents (which are made of lighter material than the ocean floors) could sink to become oceans, as Suess had argued. Dana's theory that continents were permanently fixed

KEY TERMS

CONTINENTAL CRUST: The outer most layer of the lithosphere comprising the existing continents and some subsea features near the continents. Continental crust is composed of lower density rocks such as granite and andesite.

CRUST: The outermost layer of Earth, 3–43 mi (5–70 km) thick and representing less than 1% of Earth's total volume.

GONDWANALAND: An inferred supercontinent in the southern hemisphere, used to explain why the fossils and rock formations of today's southern continents are so similar. According to Suess, it existed between Africa and India before sinking to become part of the ocean floor; according to Wegener it broke apart into the continents we now see.

HOMOLOGY: A structural similarity between two objects from different places (e.g., rock formations, fossils), which suggests, but does not prove, that they have a common origin.

ISOSTASY: The "floating" behavior of components of Earth's crust. Just as icebergs float because ice is less dense than water, mountains are buoyed up by deep "roots" that are of lower density than the crustal material they displace.

LITHOSPHERE: The near-surface region of the crust and upper mantle that is fractured in plates that move across a plasticine mantle.

OCEANIC CRUST: The outer most layer of the lithosphere lying beneath the oceans. Oceanic crust is composed of high-density rocks, such as basalt.

POLAR WANDERING: Earth's magnetic field is like a dynamo driven by electrical currents in the Earth's liquid outer core. Igneous (volcanic) rocks contain small amounts of magnetic minerals that align themselves with Earth's magnetic field as they cool. These rocks provide a paleomagnetic record. Earth's magnetic poles undergo regular changes (reversals or flips) exemplified by the Jaramillo Event, a reversal of Earth's magnetic field approximately 900,000 years ago. Apparent polar wandering also is the result of restricted magnetic pole movement and continental drift.

UNIFORMITARIANISM: Traditionally the most important principle of geology, it states that "the present is the key to the past." The history of Earth must be explained entirely according to processes actually occurring and observable today; references to speculative or unknown external forces are not allowed.

---

in place seemed closer to the truth, but it would soon be challenged by Alfred Wegener.

The Initial Response to Wegener's Challenge
Although the American theory of continental permanence agreed with geophysical evidence about Earth's structure and composition, Wegener pointed out that it failed to account for a wide range of geological evidence about the actual history of Earth. In the 1920s, scientists who rejected his hypothesis had to provide other explanations for the observed correlation in fossils, stratigraphy, and glaciation, especially between South America, Africa, and India. European geologists had formerly relied on rising and sinking continents to account for these similarities, and indeed Wegener obtained his concept of Gondwanaland from Suess. But if Suess's theory was physically impossible, what response could American geologists, who insisted that these southern continents had always been widely separated, offer?

Ever since Charles Darwin showed that species are related by descent, it had been highly implausible that the same groups of plants and animals could have arisen twice in two different places. "Land bridges" were one of the most popular solutions to the problem of how species could cross oceans. Even if isostasy made it impossible for whole continents to rise and fall, it was still possible for small areas of the ocean floor to be lifted up temporarily, creating a thin pathway for migration. Charles Schuchert, the most influential American paleontologist of his time, favored this hypothesis. He believed that land bridges had connected the continents of the southern hemisphere at certain points in the past, much as the isthmus of Panama connects North and South America today. This model had the added advantage of explaining the evidence for widespread simultaneous glaciation in parts of the southern continents, because it implied that the Antarctic Ocean had been completely cut off from external circulation, leading to much colder than normal temperatures.

James Dwight Dana
(Dana, James D., photograph.
© Bettmann/Corbis.
Reproduced by permission.)

Schuchert was very aware that his theory of land bridges had to be in agreement with geophysical principles. He enlisted Bailey Willis, another eminent American geologist, to help describe the processes that could have formed these inter-continental links. By the early 1930s, they had successfully proposed an alternative to Wegener's theory that was accepted by many American paleontologists and subsequently appeared as a standard model in textbooks. Schuchert took Wegener's arguments very seriously and was even willing to accept that continents had moved horizontally over relatively short distances, providing the forces to build mountains. However, he adamantly rejected drift on a global scale because it seemed to threaten the very science of paleontology, which depended on continents remaining at the same latitude so that their past climate could be inferred. If the continents had wandered, and been at different latitudes in the past, Schuchert's interpretation of the principle of uniformitarianism would be jeopardized. Schuchert was very concerned drift denied geologists' essential ability to understand the past in terms of present conditions. Although Wegener offered a powerful solution to certain problems in paleontology and historical geology, the idea of drift also required major sacrifices. Since there was more than one way to explain the homologies between the southern continents, most American geologists followed Schuchert and stuck to the established theory of continental permanence.

If geologists were resistant to Wegener's suggestions, geophysicists had no trouble rejecting drift outright. Although Wegener had a good understanding of the problems facing existing models of Earth's structure and of the implications of isostasy for new models, his proposal that continents underwent horizontal displacement (drifting) met widespread disbelief. Willis demanded to know how the ocean floor could be soft enough to allow continents to pass through it, while simultaneously strong enough to force mountains to rise up on land. Moreover, the prevailing mathematical treatment of isostasy, the "Pratt model," implied that pieces of Earth's crust could not move horizontally at all. The leader of American geophysical surveying, William Bowie, believed that the Pratt model's practical success in dealing with field data meant that it accurately represented reality. Like other scientists, he refused to consider changing this model since it was clearly adequate and reliable.

The Pratt model required a very weak crust, one that would be incapable of lateral motion. (An equivalent version of isostasy, the "Airy model," would have let continents move horizontally, but was mathematically more complex and thus less popular.) Bowie repeatedly argued that continental drift was simply an impossibility. As a geophysicist, he was unconcerned with the masses of historical evidence that suggested the existence of Gondwanaland in the past. Another physicist, British mathematician Harold Jeffreys, argued that measurements of earthquake vibrations proved the solidity of Earth and thus provided another reason why drift was impossible. Because Wegener's theory violated the laws of physics, geologists would have to provide some other explanation for their own data. Paleontologists and historical geologists could not refute physical calculations (hence Schuchert's need to rely on Willis), while geophysicists had no interest in accounting for the details of Earth history. Because of these divisions between the various professional communities, the greatest virtue of Wegener's approach—that it drew together data from many different fields—was ineffective.

It is important to realize that geologists and geophysicists alike did not reject Wegener's idea simply because they refused to acknowledge the data. Wegener's hypothesis was one of many new theories competing in the early twentieth century to explain a wide range of confusing geological data. His theory of continental drift was disadvantaged from the start by his choice of methodology. American Earth scientists were firmly committed to the method of "multiple working hypotheses," as expounded by the senior geologist T. C. Chamberlin. According to this philosophy, a proper geological report should begin with objective data, interpretations should be left until the conclusion, and more than one hypothesis should be considered as a possible explanation. By these standards, Wegener's writings were completely unacceptable. He began with his all-embracing theory, and then presented selected facts as demonstrations of its unquestionable validity. His enthusiasm made it very difficult for many scientists to take him seriously. Because it did not meet American standards of scientific logic, Wegener's theory of drift did not receive a full hearing. At a famous symposium held by the American Association of Petroleum Geologists in 1926 (two years after Wegener's book was translated into English), nearly all the speakers rejected continental drift decisively, criticizing Wegener's dogmatic writing style as much as his ideas.

A Continuing Rejection Once they had rejected continental drift, Earth scientists considered the matter closed and were unreceptive to new evidence in its favor. The highly detailed comparison of South American and African rock formations made by South African geologist Alexander du Toit, and his strong advocacy of continental drift through the 1930s and 40s, had little impact. His writing was unfortunately just as zealous as Wegener's and his carefully gathered data in support of drift were dismissed by Willis as a "fairy tale." Geophysicists, and a younger generation of geologists who were increasingly turning to quantification and laboratory experimentation, saw no need to revise their models of Earth's crustal structure in light of old-fashioned, field-based observations.

The most striking demonstration that the permanence of continents was a scientific nonissue in the 1930s–50s comes from the work of British geologist Arthur Holmes. One often-repeated criticism of Wegener's model was that there was no adequate force to move the continents. Yet while Wegener did not live long enough to solve this problem, Holmes developed a sophisticated model of continental motion based on convection (heat-driven) currents below the crust. This model, similar in many ways to our present understanding, finally demonstrated that continental movement was physically possible. Nevertheless, Holmes's work, which was well-known from the 1930s onward in Europe and the southern hemisphere, did not change American minds. Holmes did have a persistent critic in the influential Cambridge physicist Jeffreys, who continued to argue for a solid, contracting Earth even after the general acceptance of plate tectonics. But the rapid change in most scientists' views during the 1960s indicates that the problem up until that time was not a lack of evidence or of a plausible mecha-

nism: it was a lack of interest. Even in light of Holmes's work, most Earth scientists were unwilling to reconsider the assumption of continental permanence because they saw no pressing need to do so. —BRIAN C. SHIPLEY

Charles Schuchert
(Courtesy of the Peabody Museum of Natural History, Yale University, New Haven, Connecticut.)

Viewpoint:
No, geologic evidence shows that the continents had once been in very different positions, and thus must have moved great distances across Earth's surface over time.

Continental Drift and the Theory of Plate Tectonics Continental drift, in the context of the modern theory of plate tectonics, is explained by the movement of lithospheric plates over the asthenosphere (the molten, ductile, upper portion of the Earth's mantle). Precisely used, the term "continental drift" is actually rooted in antiquated concepts regarding the structure of the Earth. Today, geophysicists and geologists explain the movement or drift of the continents within the context of plate tectonic theory. The visible continents, a part of the lithospheric plates upon which they ride, shift slowly over time as a result of the forces driving plate tectonics. Moreover, plate tectonic theory is so robust in its ability to explain and predict geological processes that it is equivalent in many regards to the fundamental and unifying principles of evolution in biology, and nucleosynthesis in physics and chemistry.

The theory of plate tectonics gained general acceptance between 1940 and 1980. Although first advanced in the early part of the twentieth century, less than one out of 10 professional geologists accepted arguments of continental drift prior to the Second World War. The resistance on the part of geologists was well founded because the original theory of continental drift asserted that the continents moved through and across an underlying oceanic crust much as ice floats and drifts through water. In contrast, according to a survey of professional geologists conducted in the 1970s, four out of five geologists accepted the arguments of plate tectonics enough to characterize the theory as well established by fact. Although the ultimate validity of any theory is not ultimately dependent upon authority (i.e., the popularity of the theory with experts), the acceptance of a hypothesis—enough to regard it as a scientific theory—within the scientific community is usually a result of the theory's ability to better explain and make predictions based upon existing data. During the twentieth century, the weight of evidence, collected from multiple lines of inquiry, made it clear that only the theory of plate tectonics could explain all the existing data related to the apparent drift of the continents.

Based upon centuries of cartographic depictions that allowed a good fit between the Western coast of Africa and the Eastern coast of South America, in 1858, French geographer Antonio Snider-Pellegrini, published a work asserting that the two continents had once been part of larger single continent ruptured by the creation and intervention of the Atlantic Ocean. In the 1920s, German geophysicist Alfred Wegener's writings advanced the hypothesis of continental drift depicting the movement of continents through an underlying oceanic crust. Wegener's hypothesis met with wide skepticism but found support and development in the work and writings of South African geologist Alexander Du Toit who discovered a similarity in the fossils found on the coasts of Africa and South Americas that derived from a common source. Other scientists also attempted to explain orogeny (mountain building) as resulting from Wegener's continental drift. The technological advances necessitated by the Second World War made possible the accumulation of significant evidence regarding Wegener's hypothesis, eventually refining and supplanting Wegener's theory of continental drift with modern plate tectonic theory.

Plate tectonic theory asserts that Earth is divided into core, mantle, and crust. The crust is subdivided into oceanic and continental crust. The oceanic crust is thin (3–4.3 mi [5–7 km]), basaltic (<50% SiO_2), dense, and young (< 250 million years old). In contrast, the continental crust is thick (18.6–40 mi [30–65 km]), granitic (>60% SiO_2), light, and old (250–3,700 million years old). The outer crust is further subdivided by the subdivision of the lithospheric plates, of which it is a part, into 13 major plates. These lithospheric plates, composed of crust and the outer layer of the mantle, contain a varying combination of oceanic and continental crust. There is a compositional change from crust material to mantle pyriditite called the Mohovisic discontinuity, and the lithospheric plates move on top of mantle's asthenosphere.

Boundaries are adjacent areas where plates meet. Divergent boundaries are areas under tension where plates are pushed apart by magma upwelling from the mantle. Collision boundaries are sites of compression either resulting in subduction (where lithospheric plates are driven down and destroyed in the molten mantle) or in crustal uplifting that results in orogeny. At transform boundaries, exemplified by the San Andreas fault, the continents create a shearing force as they move laterally past one another.

New oceanic crust is created at divergent boundaries that are sites of sea-floor spreading. Because Earth remains roughly the same size, there must be a concurrent destruction or uplifting of crust so that the net area of crust remains the same. Accordingly, as crust is created at divergent boundaries, oceanic crust must be destroyed in areas of subduction underneath the lighter continental crust. The net area is also preserved by continental crust uplift that occurs when less dense continental crust collides with continental crust. Because both continental crusts resist subduction, the momentum of collision causes an uplift of crust, forming mountain chains. A vivid example of this type of collision is found in the ongoing collision of India with Asia that has resulted in the Himalayan mountains that continue to increase in height each year. This dynamic theory of plate tectonics also explained the formation of island arcs formed by rising material at sites where oceanic crust subducts under oceanic crust, the formation of mountain chains where oceanic crust subducts under continental crust (e.g., Andes mountains), and volcanic arcs in the Pacific. The evidence for deep hot convection currents combined with plate movement (and concurrent continental drift) also explained the mid-plate "hot spot" formation of volcanic island chains (e.g., Hawaiian islands) and the formation of rift valleys (e.g., Rift Valley of Africa). Mid-plate earthquakes, such as the powerful New Madrid earthquake in the United States in 1811, are explained by interplate pressures that bend plates much like a piece of sheet metal pressed from opposite sides.

Evidence in Support of the Theory of Plate Tectonics Wegener's initial continental drift assertions were based upon the geometric fit of the displaced continents and the similarity of rock ages and Paleozoic fossils in corresponding bands or zones in adjacent or corresponding geographic areas. Wegener also argued that the evidence of Paleozoic glaciation in South Africa, South America, India and Australia—sites far removed from estimates of the geographical extent of glaciation—argued strongly for continental drift. Although Wegener's theory accounted for much of the then existing geological evidence, Wegener was unable to advance a verifiable or satisfying mechanism by which continents—with all of their bulk and drag—could move over an underlying mantle that was solid enough in composition to be able to reflect seismic S-waves.

Modern understanding of the structure of Earth is derived in large part from the interpretation of seismic studies that measure the reflection of seismic waves off features in Earth's interior. Different materials transmit and reflect seismic shock waves in different ways, and of particular importance to the scientific debate regarding continental drift is the fact that liquid does not transmit a particular form of seismic wave known as an S wave. Because the mantle transmits S-waves, it was long thought to be a cooling solid mass. Geologists later discovered that radioactive decay provided a heat source with Earth's interior that made the asthenosphere plasticine (semi-solid). Although solid-like with regard to transmission of seismic S-waves, the asthenosphere contains very low velocity (inches per year) currents of mafic (magma-like) molten materials.

Although also explained by theories of crustal upthrusting and collapse, another line of evidence in support of plate tectonics came from the long-known existence of ophiolte suites (slivers of oceanic floor with fossils) found in upper levels of mountain chains. The existence of ophiolte suites are consistent with the uplift of crust in collision zones predicted and explained by continental drift and plate tectonic movements.

As methods of dating improved, an important line of evidence in support of plate tectonics derived from the dating of rock samples. Highly supportive of the theory of sea floor spreading (the creation of oceanic crust at a divergent plate boundary such as the Mid-Atlantic Ridge) was evidence that rock ages are similar in equidistant bands symmetrically centered on the divergent boundary. More importantly, dating studies show that the age of the rocks increases as their distance from the divergent boundary increases. Accordingly, rocks of similar ages are found at similar distances from divergent boundaries, and the rocks near the divergent boundary where crust is being created are younger than the rocks more distant from the boundary. Eventually, radioisotope studies offering improved accuracy and precision in rock dating also showed that rock specimen taken from geographically corresponding areas of South America and Africa showed a very high degree of correspondence, providing strong evidence that at one time these rock formations had once coexisted in an area subsequently separated by movement of lithospheric plates.

Similar to the age of rocks, studies of fossils found in once adjacent geological formations showed a high degree of correspondence. Identical fossils are found in bands and zones equidistant from divergent boundaries. Accordingly, the fossil record provides evidence that a particular band of crust shared a similar history as its corresponding band of crust located on the other side of the divergent boundary.

The line of evidence that swayed most geologists to accept the arguments in support of plate tectonics derived from studies of the magnetic signatures or magnetic orientations of rocks found on either side of divergent boundaries. Just as similar age and fossil bands exist on either side of a divergent boundary, studies of the magnetic orientations of rocks reveal bands of similar magnetic orientation that were equidistant and on both sides of divergent boundaries (e.g., the Mid-Atlantic Ridge). Tremendously persuasive evidence of plate tectonics is also derived from correlation of studies of the magnetic orientation of the rocks to known changes in the Earth's magnetic field as predicted by electromagnetic theory. Paleomagnetic studies and discovery of polar wandering, a magnetic orientation of rocks to the historical location and polarity of the magnetic poles as opposed to the present location and polarity, provided a coherent map of continental movement that fit well with the present distribution of the continents.

Paleomagnetic studies are based upon the fact that some hot igneous rocks (formed from volcanic magma) contain varying amounts of ferromagnetic minerals (e.g., Fe_3O_4) that magnetically orient to the prevailing magnetic field of Earth at the time they cool. Geophysical and electromagnetic theory provides clear and convincing evidence of multiple polar reversals or polar flips throughout the course of Earth's history. Where rock formations are uniform or not grossly disrupted by other geological processes, the magnetic orientation of magnetite-bearing rocks can also be used to determine the approximate latitude the rocks were at when they cooled and took on their particular magnetic orientation. Rocks with a different orientation to the current orientation of the Earth's magnetic field also produce disturbances or unexpected readings (anomalies) when scientists attempt to measure the magnetic field over a particular area.

Additional Support for Plate Tectonics

Overwhelming support for plate tectonics came in the 1960s in the wake of the demonstration of the existence of symmetrical, equidistant magnetic anomalies centered on the Mid-Atlantic Ridge. Geologists were comfortable in accepting these magnetic anomalies located on the sea floor as evidence of sea floor spreading because they were able to correlate these anomalies with equidistant radially distributed magnetic anomalies associated with outflows of lava from land-based volcanoes.

Additional evidence continued to support a growing acceptance of tectonic theory. In addition to increased energy demands requiring enhanced exploration, during the 1950s there was an extensive effort, partly for military reasons related to what was to become an increasing reliance on submarines as a nuclear deterrent force, to map the ocean floor. These studies revealed the prominent undersea ridges with undersea rift valleys that ultimately were understood to be divergent plate boundaries. An ever-growing network of seismic reporting stations, also spurred by the Cold War need to monitor atomic testing, provided substantial data that these areas of divergence were tectonically active sites highly prone to earthquakes. Maps of the global distribution of earthquakes readily identified stressed plate boundaries. Improved mapping also made it possible to view the retrofit of continents in terms of the fit between the true extent of the continental crust instead of the current coastlines that are much variable to influences of weather and ocean levels.

Also damaging to older theories of an undulating crust theory were compositional chemical studies that showed that the oceanic crust was substantially younger than the continental crust. Not only were there symmetrical bands of rock on either sides of divergent boundaries with similar dating, but no rocks older than 250 million years old were ever discovered in oceanic crust. Compositional studies also allowed plate tectonic theory to explain isostacy (buoyant characteristics) and provided gravitational data showing less dense material in mountainous areas.

In his important 1960 publication, "History of Ocean Basins," geologist and U.S. Navy Admiral Harry Hess (1906–1969) provided the missing explanatory mechanism by suggesting that the thermal convection currents in the asthenosphere provided the driving force behind plate tectonics. Subsequent to Hess's book, geologists Drummond Matthews (1931–1997) and Fred Vine (1939–1988) at Cambridge University used magnetometer readings previously collected to correlate the paired bands of varying magnetism and anomalies located on either side of divergent boundaries. Vine and Matthews realized that magnetic data reveling strips of polar reversals symmetrically displaced about a divergent boundary confirmed Hess's assertions regarding seafloor spreading.

The degree with which the geological community resisted acceptance of Wegener's theory of continental drift is clearly demonstrated by the fact that Hess's assertion of thermal currents was drawn from work done by Arthor Holmes in the 1930s. The fact that there was also initial resistance to plate tectonic theory is evidenced by the fact that Vine and Matthews's publication of their findings in *Nature* came on the heels of the rejection a few months prior by the same editors of a paper by Canadian geologist L.W. Morley proposing a similar plate tectonic hypothesis. Eventually, however, the evidence for plate tectonics became overwhelming and, although murkier as to insights into earlier continental configurations, plate tectonics powerfully explains the forces that caused the continents to break apart form and drift from a common Pangaea supercontinent to their present configuration on the ever-evolving face of Earth. —K. LEE LERNER

Further Reading

Greene, Mott T. *Geology in the Nineteenth Century: Changing Views of a Changing World*. Ithaca, NY: Cornell University Press, 1982.

Hallam, A. *A Revolution in the Earth Sciences: From Continental Drift to Plate Tectonics*. New York: Oxford University Press, 1973.

Hess, H. H. "History of the Ocean Basins." *Petrological Studies: Geological Society of America*, no. 559 (1962).

Le Grand, H. E. *Drifting Continents and Shifting Theories*. New York: Cambridge University Press, 1988.

Matthews, Drummond H., and Simon L. Klemperer. "Deep Sea Seismic Reflection Profiling." *Geology* 15 (March 1987): 195–98.

McPhee, John. *Basin and Range*. New York: Farrar, Straus, & Giroux, 1980.

Oreskes, Naomi. *The Rejection of Continental Drift: Theory and Method in American Earth Science*. New York: Oxford University Press, 1999.

Sager, W., and A. Koppers. "Late Cretaceous Polar Wander of the Pacific Plate: Evidence of a Rapid True Polar Wander." *Science* 287 (January 2000): 455–59.

Stewart, John A. *Drifting Continents and Colliding Paradigms: Perspectives on the Geoscience Revolution.* Bloomington: Indiana University Press, 1990.

United States Geological Survey. "This Dynamic Earth." August 2001. <pubs.usgs.gov/publications/text/dynamic.html>.

University of California, Museum of Paleontology. "Plate Tectonics." <www.ucmp.berkeley.edu/geology/tectonics.html>.

Vine, F. J. "Spreading of the Ocean Floor: New Evidence." *Science* 154, no. 3775 (December 1966): 1405–1515.

Wegener, Alfred (trans. J. Biron). *The Origin of Continents and Oceans.* New York: Dover Publications, 1966.

ENGINEERING

Will a viable alternative to the internal combustion engine exist within the next decade?

Viewpoint: Yes, battery-powered, fuel-cell electric, and hybrid vehicles are technologically viable alternatives to the internal combustion engine now, and they are likely to be economically viable within the decade.

Viewpoint: No, current internal combustion engine technology has many advantages over its potential competitors—lower costs of production and operation, a longer driving range before refueling, and better overall performance—that will ensure its dominance for several decades to come.

The internal combustion engine is a versatile power source, used in everything from lawn mowers to rockets. However, it is most commonly associated with the car and other road transport vehicles, and it is this association that has come under increasing criticism and scrutiny in recent years. With a growing awareness of the risks of pollution to both health and the environment, the internal combustion engine has become less desirable, and practical alternatives are being sought by governments, corporations, and environmental groups worldwide.

Essentially, an internal combustion engine works like a cannon. A flammable substance, such as gasoline, is ignited in a small, enclosed space, and the resulting explosion releases energy in the form of expanding gas that can propel an object with great force. A typical car engine has hundreds of such explosions a minute, and harnesses some of the energy produced to turn the drive shaft. As the name implies, the combustion takes place inside the engine, as opposed to an external combustion engine, where the fuel burns outside of the engine, such as in a steam-powered train.

Today, the internal combustion engine is the dominant vehicle engine the world over. However, there were many other alternatives proposed and produced in the early days of automobiles, including electric, steam, and even liquid air–powered cars. The first electric motor was produced in 1833, and many models of electric cars were available from the 1880s to the turn of the century. Steam cars were also produced in the 1880s. The first gasoline-fuelled, internal combustion engine cars were built in 1891 and quickly made an impact. They had a much longer range than electric cars, and competed well against steam cars in endurance races.

Early electric and steam cars had many advantages over rival internal combustion models, including speed. In 1898 an electric car achieved the record speed of 39.25 mph (63 km), and a year later another electric model went over 65 mph (104.6 km). That record held until 1902, when a steam power vehicle reached 75.06 mph (121 km). Electric cars were also quieter—some were almost silent—and did not produce unpleasant exhaust fumes.

However, the internal combustion engine, while noisier, hotter, and dirtier than electric motors, began to dominate the car market. Other types of cars had their own problems. Early electric batteries were heavy and corroded quickly, needing to be replaced every two years, and there were many cases of battery leaks producing noxious fumes. Gasoline became cheaper; the speed, performance, range, and durability of the internal combustion engine were improved; and consumers began to prefer the noise and power of the internal combustion engine–driven car. The electric engine came to be associated with senior citizens, while the petrol engine was seen as progressive, reliable, and perhaps most importantly, cheaper to buy and run. The internal combustion engine thus prevailed.

However, while the internal combustion engine–driven car dominated the twentieth century, the twenty-first seems likely to see the reintroduction of electric cars. There are compelling reasons to find alternatives, from a greater awareness of the effects of pollution and fears over global warming, to concerns that the supply of oil is drying up, or at least becoming more expensive to find and extract.

Many different types of cars are being developed. Most of the problems that plagued the battery-powered car of the nineteenth century have been resolved. Modern car batteries are safer, last longer, and provide more power than previous-generation batteries. Another alternative electric car type uses fuel cells, which produce electricity rather than storing it as long as there is fuel in the cell. The standard fuel for fuel cells is hydrogen, which is a cheap and plentiful gas but which is also extremely flammable, and is often linked in the public mind with the *Hindenburg* airship disaster.

Hybrid engines have also been proposed. These use either a combination of two alternative technologies, such as fuel cells and batteries, or one alternative technology with a standard internal combustion engine. While many of these combinations result in a longer drive range and improved performance, the need to have two engines adds weight and size to the vehicle.

There is much debate over which alternative engine or combination is best able to reduce car emissions and still provide the consumer with the necessary power and range. Currently, there are a bewildering choice of options, each requiring its own infrastructure of refuelling stations and service industries. A number of different alternative vehicles are in use all over the world, and are particularly successful in niche areas, such as inner-city transport. However, it seems likely that only a few, possibly just one, will emerge as a serious challenger to the internal combustion engine.

The same concerns over the environment that have given rise to alternative engine research have also resulted in many recent improvements to the internal combustion engine. Modern cars have significantly lower emissions of undesirable gases than those of a few decades ago, and are more fuel efficient. Yet at the same time, car manufacturers have made and promoted larger cars, which tend to have lower fuel efficiency.

Perhaps the biggest negative factors that alternative car engines have to overcome are not technical, but economic and perceptual. Currently the cost of alternative cars is much higher than standard vehicles. While mass production techniques will bring prices down, it may still be many years before they approach parity with internal combustion engine cars. Operating costs also need to be lowered, and the availability of refuelling stations needs to be increased. Public perception also needs to be changed if alternative engines are to become desirable. Whatever the realities, electric cars are still seen by the majority of car buyers as small, powerless, and operable only over short distances.

There is a seemingly unstoppable trend to new engines in cars to replace the long-serving internal combustion engine, for a variety of compelling reasons. However, when such technology will become commonplace on the roads is still an open question. —DAVID TULLOCH

Viewpoint:

Yes, battery-powered, fuel-cell electric, and hybrid vehicles are technologically viable alternatives to the internal combustion engine now, and they are likely to be economically viable within the decade.

Not only will a viable alternative to the internal combustion engine exist within the next decade, it exists today, although the term "viable" may be cause for disagreement. Applying *Webster's* definition of viable as "workable and likely to survive or have real meaning" raises new questions: viable technologically or viable economically? Technologically, the answer is "yes." The alternative to the internal combustion engine is the electric powered vehicle. Small numbers of battery–powered, fuel–cell electric, and hybrid electric vehicles are in use today, and there is a worldwide race to get more of the new technology vehicles on the road. One factor slowing that race is vehicle cost, so economically, the answer is a qualified "yes."

There are many uncertainties regarding what is economical. In the year 2001, there is no contest. The internal combustion vehicle is more economical than any alternative. The internal combustion engine revolutionized transportation and the economy, as the twentieth century evolved. By the start of the twenty–first century, there were more than 200 million vehicles on the road in the United States. Fueling their internal combustion engines has forced the country to rely on imported oil from the Middle East and to scramble for reserves with the uncertainty of supplies.

Within the next decade, the cost of fuel and the cost of protecting the environment may shift the balance. Environmental issues may be so compelling that health and safety concerns may force a change regardless of cost. There is also the pressing question of global warming and where the internal combustion vehicle emissions fit into the equation.

Although great improvements have been made in reducing emissions from internal combustion engines, the ever–increasing numbers of vehicles, the size/style of the vehicles, and the increasing number of miles driven have combined to negate any progress in emission control. After over 100 years of improvements, the technology of the internal combustion engine is reaching its limits for improvement. The United States Environmental Protection Agency (EPA) estimates that motor vehicles in the United States still account for 78% of all carbon–monoxide emissions, 45% of nitrogen–oxide emissions, and 37% of volatile organic compounds in the atmosphere.

Background for the Shift to Alternatives
How do the three types of electric vehicles that are in use—battery powered, fuel cell, and hybrid electric—compare? The technology for batteries is very different from the technology for fuel cells, although neither has any moving parts, which makes them both very quiet and reliable. Batteries store electricity, and when they run out, they have to be recharged. Battery–powered vehicles are limited for the most part to the local utility types because presently there is no battery that can go the distance that an internal combustion–powered vehicle can on a full tank of gas. Some battery–powered vehicles are recharged from photovoltaic cells in California, but most are recharged from the grid. They generally are not viewed as competitors to the internal combustion engine. Fuel cells do not store electricity but rather continuously produce electricity as long as the fuel is available. Fuel–cell technology is well developed. Fuel storage issues are the major focus of current research to give fuel–cell vehicles the performance the public expects of a vehicle. Hybrids are, in some views, the best of both worlds.

Most experts agree that fuel cells are the leading technology as an alternative to the internal combustion–powered vehicle. To maintain a leadership position in energy technologies, major oil companies are joining car and engine manufacturers in research and development to

KEY TERMS

ELECTROCHEMICAL CELL: Where electric energy is produced by a chemical process. Electrons leave the cell at the anode and return to the cell at the cathode. Any device to be run by the cell is attached between the anode and the cathode.
EMISSIONS: Substances discharged into the air.
FOSSIL FUELS: Fuels that are formed in the earth from plant or animal remains. Fossil fuels include coal, oil, and natural gas.
HORSEPOWER: A unit of power equal to 746 watts.
HYDROCARBONS: Organic compounds containing only carbon and hydrogen; often occur in petroleum, natural gas, and coal.
HYDROGEN: The simplest and lightest of the elements; usually colorless and odorless; highly flammable.
INFRASTRUCTURE: The resources (including personnel, buildings, or equipment) required for an activity.
MEMBRANE: A semi permeable surface or thin film. Charged particles and small molecules selectively pass through the membrane separating them from a mixture.
POLYMER: Commonly called plastic. The term literally means many parts. A polymer is produced by many molecules of one or more types repeatedly joining together. The smaller units are called monomers.
SKUNK WORKS: A laboratory where research and development is done usually on proprietary projects or behind–the–scenes.

bring fuel–cell powered vehicles to market. Where hydrogen is the fuel of fuel cells, very probably there will have to be another choice at the pumps before the first decade of the twenty–first century plays out.

The fuel cell is not a new technology, having been first developed in 1839 when William Grove, a British physicist, discovered the principle of the fuel cell. It took over 120 years, however, before NASA (National Aeronautics and Space Administration) found an application for them in the 1960s when fuel cells were considered safer than nuclear power in space flight. Fuel cells were used in both the *Gemini* and *Apollo* missions and continue to be used in space ventures as a source of electricity and water.

With NASA's success, industry became interested, but early research and development efforts were not very encouraging and the technology appeared to be much too expensive. An increase in effort came when the Office of Transportation Technologies of the United States Department of Energy (DOE) started supporting research and development in 1984. In 1990, the United States Clean Air Act

FUEL CELLS IN SPACE

NASA (National Aeronautics and Space Administration) started publishing reports of SPINOFFS annually in the 1970s. The reports feature industry/government collaborations on breakthrough technologies that are being developed as a result of the space program. Had NASA been writing such reports in the 1960s, fuel cell technologies would have been high on the list of early success stories. Fuel cells—cells that generate power through the interaction of oxygen and hydrogen gases—were first invented in the early nineteenth century but not widely used until the early years of the space program. Fuel cells are still being featured in SPINOFF reports; in 1999, for instance, the development of a next–generation PEM fuel cell was featured.

After fuel cells provided on–board power for the *Gemini* and *Apollo* spacecraft, industry became interested. Today, three fuel cell power plants provide the 28–volt direct current needed for the space shuttle. The fuel cell system generates all the electrical power for the vehicle during all mission phases. Cryogenic hydrogen and oxygen are used for the cells. In addition, cryogenic oxygen is supplied to the environmental control and life support system for crew cabin pressurization. The storage temperature for the liquid oxygen is minus 285°F (minus 176°C), and minus 420°F (minus 251°C) for liquid hydrogen.

In addition to providing all the on–board electrical power (there is no backup battery), fuel cells also produce water as a byproduct of the electrochemical reaction. This water is then used as drinking water for the crew as well as for spacecraft cooling. The fuel cells are alkaline fuel cell (AFC) power plants that each contain 96 individual cells. The electrolyte, a solution of potassium hydroxide, gives the fuel cell its name. AFCs have the advantage of a fast reaction and high performance, so they are popular for military and space applications. However, AFCs are also costly. The anode catalyst contains platinum and palladium, and the cathode catalyst contains gold and platinum. The cost factor is no doubt part of the reason for continuing research on fuel cells.

—M. C. Nagel

Amendments along with the National Energy Policy Act of 1992 gave impetus to the development of alternatives to the internal combustion engine for vehicles. The California Air Resources Board recognized in 1990 that it probably would not be possible to meet their first goal of zero emissions with gasoline–powered vehicles by 1998. That goal was unrealistic, so a new timetable was set up. Beginning in 2003, 10% of the new vehicles in California will be required to be zero (or nearly zero) emission vehicles. Similar regulations are now on the books in states on the East Coast.

The Technology of Fuel Cells A fuel cell is an electrochemical device, that is it produces electricity by a chemical reaction. In the fuel cell, hydrogen gas and oxygen from the air are combined to produce electricity, and heat and water are byproducts. Because there is no combustion, fuel cells meet the zero emissions standard. A fuel cell is composed of two electrodes, an anode (positive electrode) and a cathode (negative electrode). The electrodes are separated by a porous medium that serves as an electrolyte. Fuel cells come in five varieties that are distinguished by the electrolyte employed.

The fuel cell that is leading the field for vehicle use is identified as a PEM fuel cell. PEM stands for proton exchange membrane, and also for polymer electrolyte membrane. The PEM is a thin film polymer membrane that is coated with a platinum catalyst. An electrolyte by definition is a substance that dissociates into positively and negatively charged ions in the presence of water. The membrane is moist, and though the membrane does not dissociate, it serves as an electrolyte in the sense that it allows positively charged particles—protons—to pass through it, thus the name proton exchange membrane. In the cell the membrane looks like a piece of thick plastic wrap.

In the fuel cell, hydrogen enters at the anode, where the catalyst on the membrane splits it into a proton and an electron. The proton passes through the membrane; the electron cannot. Instead, the electron moves out of the cell, through the external circuit. The electrons moving in the external circuit are the energy source that drives the vehicle.

On the cathode side of the catalyst–coated membrane, the proton meets oxygen (from the air) and an electron (from the external circuit) on the cathode. As it combines with the oxygen, water is formed. Although this is a heat–producing reaction, and in many cases these reactions are easy to start, this reaction would not happen without the catalyst.

One single cell produces about 0.7 volts. The cells are stacked in series so their voltages add up. PEM stacks were used on the *Gemini* spacecraft, although the early version was not as efficient as present fuel cells, since the former made extensive use of platinum, an expensive element. Researchers at Los Alamos eventually found a way to reduce the platinum by 90%. Additional research at a Canadian–based company, Ballard, further advanced the technology, and the company has about 400 patents on fuel–cell improvements. Ballard's PEM stack has reached well over the power density needed for today's vehicles. International Fuel Cells (IFC), a United Technologies Company (best known for its jet engines), also has made advances in the technology, including reducing size, weight, and cost—important to bring economic viability to

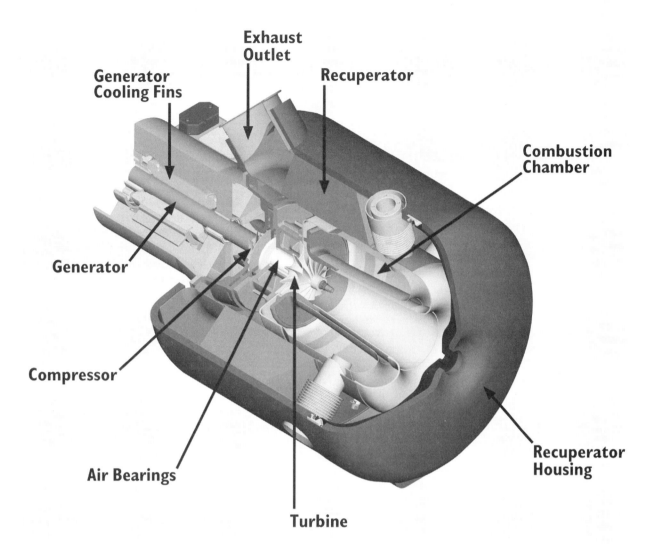

A cross-section of a microturbine.
(Capstone Turbine: www.microturbine.com. Reproduced by permission.)

fuel cells. However, there is still work to be done on streamlining the design and manufacturing processes to produce cheaper cells. Mass production of fuel cells will also bring down the cost.

Some see the key issue in the move toward fuel cells as the fuel itself, now that most of the research and development on the cells is concentrating on fine–tuning to improve the economy of production. Where will the hydrogen come from? All the major car manufacturers are working toward sending a fuel–cell powered vehicle to market in numbers. The oil companies are working with them to develop fuel sources.

There are also technologies related to fuel storage that need to be addressed. Where hydrogen is the fuel, the problems arise of how to safely store enough hydrogen, or how to produce enough hydrogen on board, to give the fuel–cell powered vehicle the same distance a user can get on a fill–up of gasoline with an internal combustion engine. Economic viability is in danger if the fuel–cell driven vehicle cannot provide the same convenience as its internal combustion–driven competitor. Storage of compressed gas on board is being considered. Liquefied gas is also being investigated but that involves very cold temperatures for storage. Cryogenic technologies are involved that may be too costly. There is also the fear of hydrogen stemming from the infamous *Hindenburg* disaster of 1937, when the hydrogen–gas powered airship exploded mid–air, killing thirty–six people.

Allowing that the storage on board problems are solved, oil companies are looking at producing hydrogen to be pumped into a vehicle's storage system. Texaco has over 150 patents on a technology they call gasification that uses otherwise undesirable heavy oil, petroleum coke, and wastes and converts them into hydrogen. The company could market hydrogen at a pump with this system.

Another approach is to store a source of hydrogen in the vehicle. One of the dominant technologies to produce hydrogen on board is the fuel processor, a mechanical device that uses heat and a catalyst to change the chemical composition of a hydrocarbon to free the hydrogen in a system integrated with the fuel cell in the vehicle. IFC has a prototype ready. The hydrocarbon could be methanol, methane from natural gas or other sources, or even gasoline that is stored in the vehicle.

The Technology of Hybrid Electric Systems

The hybrid electric vehicle runs on batteries that are recharged as the vehicle drives. This eliminates the down time at the recharge station for battery electric vehicles and makes the vehicle as useful as any internal combustion–driven counterpart. There are a number of systems employed to recharge the batteries. One that stands out is the microturbine, which meets California's tough emission standards. Capstone, a California company, has a microturbine that was put on a Chattanooga, Tennessee, bus in 1997 as an on–board battery charger, and it logged more than 30,000 miles with no breakdowns. The bus generates less than 1/25th the emissions of a diesel bus. Microturbines are now being used in cities as far away as Christchurch, New Zealand, and Tokyo, Japan. A microturbine is lighter in weight than an equivalent diesel engine. In Christchurch, diesels were replaced with microturbines in hybrid electric buses. A microturbine has only one moving part and is very environmentally friendly. The Capstone microturbine can run on just about any liquid fuel from natural gas to landfill methane to diesel.

In Los Angeles, ISER, a bus manufacturer, has a turbine–hybrid drive system that employs lead–acid batteries with Capstone microturbines recharging them. An onboard computer network continuously monitors vehicle power and battery charge levels, and adjusts the turbine power output to provide just the right amount of power to operate the vehicle. When the generators are not needed, they are automatically turned off and the bus runs on battery power. At that time the bus is a zero emissions electric vehicle. The batteries provide surge power for acceleration and recapture energy during braking. The microturbine is fueled with propane.

Presently all the major auto manufacturers are producing hybrid electric test vehicles under the DOE Office of Transportation Technologies Hybrid Electric Vehicle Program. One of the recent models is a Dodge Durango Hybrid sports utility vehicle. Federal legislation to create up to $3,000 in tax incentives for purchasers of hybrid vehicles could make them competitive with internal combustion engines. The down side is that many of them are still using petroleum–based fuels. The good news is that fuel cell and hybrid electric vehicle technologies have moved out of the skunk works and are now on the market to provide a technologically viable alternative to the internal combustion engine vehicle. Within a decade these alternatives might even be economically viable. —M. C. NAGEL

Viewpoint:

No, current internal combustion engine technology has many advantages over its potential competitors—lower costs of production and operation, a longer driving range before refueling, and better overall performance—that will ensure its dominance for several decades to come.

Although concerns over the environment and fuel supplies have fostered an abundance of research to find alternatives to powering cars and other vehicles, each new prototype of the "engine–of–the–future" has demonstrated that replacing the efficient and dependable internal combustion engine will be a difficult task. Many informed observers, including officials within the Energy Information Administration at the United States Department of Energy, predict that several decades will pass before any of the new technologies under development—including electric, fuel cell, and hybrid technology—will begin to have an impact on the market supremacy of the internal combustion engine. This versatile engine design, which currently powers more than 200 million vehicles in the United States, has been the backbone of the transportation industry for more than a century and for good reason.

From the very beginning the internal combustion engine has had competitors. For example, the first electric–powered car dates back to the 1860s, and in 1899, an electric car set the world record for speed by going faster than 62 miles per hour. But the internal combustion engine became the technology of choice because of its dependability and convenience.

Over the years, research has continued to improve the internal combustion engine, and it has yet to reach its full potential, including its potential for fuel efficiency and producing cleaner emissions that are less harmful to the environment. During the 1990s, the internal combustion engine was improved significantly in terms of its environmental impact. Increased miles per gallon of gas, for example, has resulted in better fuel efficiency and less pollution.

Lower polluting emissions have also resulted from improvements in gasoline, such as the addition of oxygenates, which has lowered carbon–monoxide emissions by 18%. Furthermore, additional improvements such as progressively reducing the use of sulfur in gasoline will further reduce polluting emissions created by the internal combustion engine.

Overall, because of technologies such as the advanced catalytic converter and electronic combustion control, automobile emissions already are lower by 95% compared to the 1960s, despite the fact that many more cars are on the road today. In addition, most of the cars produced in the United States in 2001 and beyond emit 99% less hydrocarbons than cars made in the 1960s. Even if no further improvements were made in the internal combustion engine, its impact on the environment would continue to decrease solely due to newer cars with far lower emissions replacing older cars as they wear out and are sent to the scrap heap.

With a vast amount of experience in designing internal combustion engines and huge facilities for producing them, vehicle manufacturers are also not about to relegate these engines to obscurity in the near future. Most important, however, current internal combustion engine technology has many advantages over its potential competitors—advantages that the consumer demands. These include lower costs of production and operation, which results in lower costs for the consumer; a longer driving range before refueling; and better overall performance.

The Internal Combustion Engine versus the Electric Car At the beginning of the twentieth century, more than 100 electric car manufacturers in the United States were vying to produce the vehicle of choice in this then young but rapidly developing industry. In less than two decades, nearly all of them had closed up shop. Many of the factors that led to the demise of the electric vehicle, including the large disparity in cost and convenience, still remain valid reasons why such vehicles are unlikely to replace the internal combustion engine in the next decade.

Despite a decade or more of intensive research and testing, electric cars are only able to travel approximately 100 miles (160.9 km) before they need to be recharged. Furthermore, this recharging takes a considerable amount of time compared to the relatively quick refueling that takes place at the local gas station. In comparison, a standard internal combustion engine can take a vehicle about 345 miles (155.2 km) before refueling is required. Electric cars can also only match the internal combustion engine in terms of horsepower for a short period of time (approximately one hour) before their power starts to diminish due to such factors as speed and cold weather. The electric motor's shorter range and slower overall speed might have been acceptable early in the twentieth century when families and businesses usually were condensed into smaller geographic regions. However, in today's society people routinely travel much farther distances, and a consumer public that places a high premium on its "time" has not shown a propensity to accept electric cars that are slower and require more stops and recharging.

In a society that is growing more and more environmentally conscientious, a much touted advantage of electric cars is that they are much cleaner in terms of environmental impact than cars run by internal combustion engines. Although electric cars produce nearly zero emissions from the car itself (perhaps as much as 97% cleaner than an internal combustion engine), this advantage is greatly negated by how electric engines are charged. For example, fossil fuels such as coal and oil are often used to generate electricity, which also produces its own pollutants. Even some noted environmentalists have conceded that this fact offsets any of the environmental advantages electric cars have over the internal combustion engine. Gasoline is now also lead free, but battery wastes still pose significant environmental problems, including the disposal of lead and cadmium.

The Internal Combustion Engine versus Fuel Cells Another technology proposed as the wave of the future are fuel cells, with most of the focus on hydrogen fuel cells. A fuel cell is an electrochemical device that uses hydrogen and oxygen gases to produce electricity. However, like the electric car, fuel–cell cars would have a limited range in comparison to the internal combustion engine–driven cars, probably around 200 miles (321.8 km) or so. Although this range will increase with improvements such as reducing the weight of the car, reduced car weight would also improve mileage in cars powered by the internal combustion engine. Comparably, the fuel cell still would only achieve one–third of the range achieved with an internal combustion engine. In addition, there is the issue of producing the energy stored in the fuel cells. This energy would be created by fossil fuels or the generation of electricity to isolate hydrogen from the air, issues that, like the electric car, would result in environmental pollutants. Hydrogen is also volatile and an extremely small molecule that can filter through the smallest of holes, which increases safety concerns over leaks and pressurized tanks that could burst.

Fuel cells are also extremely expensive to manufacture. The cost of $500,000 per kilowatt of power associated with the first fuel cells used to provide power to space capsules in the early 1960s has been lowered to approximately $500.

Nevertheless, a fuel-cell engine and drive train costs as much as 10 times the cost of producing an internal combustion engine. As a result, on average the cost of fuel-cell technology is approximately $25,000 to $30,000 (and perhaps as much as $45,000 to $65,000) compared to the average $2,500 to $3,000 cost for the standard internal combustion engine in many cars. Few consumers are going to readily accept this significant added expense.

Another factor to consider is the local gas station, which represents an already existing nationwide infrastructure for providing fuel to motorists. No such infrastructure exists for fuel-cell vehicles. While the current infrastructure could be adapted to accommodate fuel cells that use on-board reformers to isolate hydrogen from gasoline, diesel, or methanol, the on-board technology would further increase already higher costs. In addition, it would take up even more space than the large tank currently needed for fuel cells to provide adequate driving distances. Fuel-cell technology also still requires fossil fuels just like the internal combustion engine. It would also likely take more than a decade for energy companies to create the number of new or overhauled manufacturing facilities needed to produce enough hydrogen to meet consumer demands. Furthermore, some estimates indicate that only 2% of stations would be able to offer fuel-cell car refueling by the year 2011 and only 3.5% by the year 2020.

The Internal Combustion Engine versus the Hybrid In a sense, the hybrid electric car is a concession that the internal combustion engine will be around for many years to come. Like the electric car, the concept of the internal combustion engine–electric hybrid goes back to the early days of automobiles, with the first United States patent filed in 1905. However, they were never developed as fully as the electric car. In essence, the hybrid uses a battery-powered electric motor for riding around town but also has an internal combustion engine that could be used for traveling longer distances and, in some models, to help recharge the battery-powered electric motor.

Although this alternative to the internal combustion engine as the primary source of power seems attractive, the drawbacks of the current technology used in this approach are substantial. For example, an internal combustion engine combined with the need for many batteries and a large electric motor to power the car requires extra space and more weight, which decreases the vehicle's overall fuel efficiency. Both the National Academy of Sciences and the National Academy of Engineering have stated that this technology is not cost-effective in terms of being environmentally friendly, especially since the use of gasoline or diesel reduces the environmental benefits that are essential to any new motor technology. Hybrids would also cost approximately $10,000 to $15,000 more than current car technology.

No Imminent Alternatives Several reports and studies have stated that current technology will not provide a viable alternative to the internal combustion engine within the next decade and, perhaps, not for many more years following. In its June 2000 policy study *The Increasing Sustainability of Cars, Trucks, and the Internal Combustion Engine,* the Heartland Institute predicted that "it will be 30 years before even a modest 10% of all cars and trucks on the nation's roads are powered by something other than internal combustion engines."

While the technology does exists to make alternatives to the internal combustion engine, it has yet to advance to a level that makes it competitive or viable in terms of consumer needs and wants. For example, although some electric vehicles have been marketed aggressively in places like California, they still make up only a small percentage of the market. And companies such as Honda and General Motors have quit producing them. Even in Europe and Japan, where gas costs two to three times more than in the United States and where significant government and manufacturing subsidies are in place to support consumers in buying electric vehicles, they make up only about 1% of the market.

With the increasing popularity of sports utility vehicles (SUVs) in the Unites States throughout the 1990s, consumers obviously have shown their attraction to vehicles with more horsepower and greater size. To date, alternatives to the internal combustion engine have resulted in smaller cars with less power and less passenger and luggage room. Furthermore, to make up for the technology's additional weight and to increase driving distances before refueling or recharging, these cars are integrating more lightweight materials in their construction. The result are cars that are less safe in the case of accidents, another fact that consumers are not likely to ignore.

Even if the internal combustion engine were never improved upon, it is not likely that alternative technologies can overcome factors such as size, safety, cost, convenience, and power within the next decade. But the race is not against a technology that is standing still. While new technologies will continue to improve, car manufacturers also continue to invest billions in improving the internal combustion engine. Although currently cost-prohibitive, a standard internal combustion engine may one day get 80 miles per gallon and be able to travel 545 miles before refueling. Furthermore, advances in other technologies, such as computerized explorations for gas and horizontal drillings, are increasing the long-term production outputs of oil and gas fields in the United States and other western countries, thus increasing the prospect of a continuous, long–term, and relatively inexpensive supply of fuel.

To become a "viable" alternative to the internal combustion engine, alternative technologies must be as efficient, dependable, and powerful. Furthermore, they must be as competitive in terms of cost to the consumer. To reach these goals within a decade is not possible given the internal combustion engine's current vast superiority in these areas. —DAVID PETECHUK

Further Reading

Anderson, Roger N. "Oil Production in the Twenty–First Century." *Scientific American* (March 1998): 86–91.

Bast, Joseph L., and Jay Lehr. *The Increasing Sustainability of Cars, Trucks, and the Internal Combustion Engine.* Heartland Policy Study No. 95. The Heartland Institute, June 22, 2000.

Bradly, Robert L., Jr. "Electric and Fuel–Cell Vehicles Are a Mirage." *USA Today Magazine* (March 1, 2000): 26.

California Air Resources Board (website). <http://www.arb.ca.gov>.

Fuel Cells Green Power (website). <http://education.lanl.gov/resources/fuel cells>.

Hoffman, Peter. *The Forever Fuel: The Story of Hydrogen.* Boulder, CO: Westview Press, 1981.

Koppel, Tom. *A Fuel Cell Primer.* 2001. <http://www.MightyWords.com>.

Larminie, James. *Fuel Systems Explained.* New York: John Wiley & Sons, 2000.

Lave, Lester B., et al. "Environmental Implications of Electric Cars." *Science* 268 (May 19, 1995): 993–95.

Motavalli, Jim. *Forward Drive: The Race to Build the Car of the Future.* San Francisco, CA: Sierra Club Books, 2000.

Office of Transportation Technologies (website). <http:www.ott.doe.gov>.

Partnership for a New Generation of Vehicles (website). <http://www.ta.doc.gov/ pngv>.

Propulsion Technologies (website). <www.insight-central.net/compare–propulsion.html>.

Wouk, Victor. "Hybrid Electric Vehicles." *Scientific American* (October 1997).

Will wind farms ever become an efficient, large-scale source of energy?

Viewpoint: Yes, wind power is already the fastest-growing source of renewable energy in the world, and economic trends, technological advances, and environmental concerns will eventually transform it into a large-scale contributor of energy.

Viewpoint: No, wind power will not be a large-scale contributor of energy because wind doesn't blow sufficiently everywhere and doesn't blow all of the time.

Wind exists because Earth rotates and has an irregular surface—variable terrain, bodies of water, plants, and trees—and because the Sun heats the atmosphere unevenly. With the right equipment, energy produced by wind flow can be harvested and used to generate electricity.

As early as 5000 B.C., people used wind energy to move boats up and down the Nile. In 200 B.C., people used simple windmills to pump water in China, and vertical-axis windmills to grind grain in Persia and the Middle East. By A.D. 1000, Mid-Easterners used windmills extensively to produce food.

Over time the Dutch refined windmill technology and used more powerful wind turbines to drain lakes and marshes. Wind turbines convert kinetic wind energy to mechanical power that can be used to grind grain or pump water, or be converted to electricity.

By 1910, wind turbine generators produced electricity in many European countries. American settlers used windmills to pump water and generate electricity until the 1930s, when the Rural Electrification Administration brought electric power to the rural United States. Industrialization reduced the numbers of small windmills, but contributed to the development of wind turbines.

In the 1940s the world's largest wind turbine, operating on a Vermont hilltop called Grandpa's Knob, fed electric power to a local utility network for months during World War II. In the 1970s, oil embargoes prompted research and development that introduced new ways of turning wind energy into useful power.

Today, modern wind technology utilizes advances in materials, engineering, electronics, and aerodynamics. Wind farms—large groups of turbines—feed electricity into local utility grids in the United States and Europe. Wind energy is the world's fastest-growing source of electricity generation. But the debate continues about whether wind farms will ever become an efficient, large-scale energy source.

Those who believe in the large-scale potential of wind farms base their convictions on an increasing demand for power and the need for forms of energy other than those provided by fossil fuels and natural gas. Supplies of coal, oil, and natural gas are finite and most often controlled by countries, governments, and private enterprises that can restrict supplies or increase prices, ultimately increasing energy costs to the consumer.

Some experts say the ultimate success of wind power depends on its cost compared to other electricity sources, and on the value to society of reducing pollution. When the 1970s energy crisis caused oil prices to skyrocket, people started considering wind energy as a serious future power source. Wind energy costs are slightly higher today than the cost of energy from natural gas or coal-fired power plants, but wind-energy costs continue to decrease while costs for energy from other sources rise.

Wind-energy proponents say that refining raw materials and producing nuclear power cause serious environmental hazards. In contrast, wind power is a clean, nonpolluting, renewable energy source. Modern wind turbines have a 98% or higher reliability rating, and their larger capacity means fewer turbines generate more kilowatt-hours of energy.

New wind farms added more than 3,800 megawatts (MW) to world wind-energy production in 2000, and four of the largest wind farms ever built were scheduled for completion by year-end 2001, enough power to provide electricity to 1.7 million homes. With strong legislative and policy support, the capacity of wind farms in the United States could increase to 30,000 MW by 2010, providing electricity for 10 million homes and reducing excess utility sector emissions by 18%.

Those who doubt that wind farms will ever become an efficient, large-scale source of energy base their beliefs on the inconsistency of wind as an energy supply, on environmental issues, and on competition from other renewable energy sources.

The big challenge to using wind as a source of power is that it doesn't always blow, it never blows on command, and not all winds can be harnessed to meet the timing of electricity demands. Good wind areas cover only 6% of the contiguous U.S. land area.

Wind resources are characterized by wind-power density classes that range from class 1 (lowest) to class 7 (highest). Good wind resources (class 3 and above) with an average annual wind speed of at least 13 mph (21 kph) are found along the east coast, the Appalachian Mountain chain, the Great Plains, the Pacific Northwest, and others. Because many of these good wind sites are remote areas far from places of high electric power demand, the cost and logistics of transmitting and distributing power to where it's needed have to be considered when siting wind resources.

Environmental issues include noise pollution, hazards to birds, and aesthetic considerations. And wind resource development may compete with other uses for the land that may be more highly valued than for electricity generation.

Competition from other renewable resources will keep wind energy from becoming a large-scale resource. Significant energy alternatives in the renewables category are biomass power, concentrating solar power, geothermal, hydropower, photovoltaics, and solar thermal.

Today wind is one of many clean, renewable sources of electric power. Wind farms will be efficient contributors to the energy mix, but some believe that wind power will never be a large-scale contributor. —CHERYL PELLERIN

Viewpoint:

Yes, wind power is already the fastest-growing source of renewable energy in the world, and economic trends, technological advances, and environmental concerns will eventually transform it into a large-scale contributor of energy.

Overview By the year 2001 wind power was the fastest growing renewable energy source in the world, and its market was expanding rapidly. An article in a 1999 issue of the *Environmental Business Journal* declared that wind power in the United States surged in 1998 and accelerated in 1999 after a slump of several years. Randall Swisher, executive director of the American Wind Energy Association (AWEA), Washington, D.C., was quoted in that article as saying: "We are celebrating the revitalization of the American wind market."

Europe is the center of this US$16.5 billion industry, generating 14,200 megawatts (MW) of the almost 20,000 MW produced annually worldwide. Germany (6,900 MW), the United States (2,600 MW), Spain (2,582 MW), and Denmark (2,346 MW) were country leaders, with Denmark producing 7% of its entire electricity supply from wind farms at a cost of 4 cents (US) per kilowatt-hour (kWh). That percentage is projected to increase to 40% by the year 2030.

What Are Wind Farms? Wind farms are areas either on land or offshore where wind turbines (often as many as 500) are clustered together to capture energy from the wind to generate electricity. Wind turbines are tall towers atop which propeller-like blades spin in the wind, transforming the wind's kinetic energy into mechanical power, which, in turn, is generated as electricity.

> ### KEY TERMS
>
> **GRID:** Short term for utility grid, the commercial network of wires that carry electricity from a source to the users.
> **MEGAWATT (MW):** An electrical power unit. Electrical power is measured in watts, kilowatts (kW—1,000 watts), and megawatts (MW—1,000,000 watts). Village power is measured in kW whereas grid-connected sources are measured in MW.
> **PHOTOVOLTAIC (PV) CELL:** A device that uses semiconductor materials to convert sunlight into electricity directly. One PV cell is about 4 in on a side (10 cm). A cell produces about one watt. When 40 cells are connected together it is called a PV module. When 10 modules are connected it is a PV array.
> **WIND FARM:** A network of closely spaced windmills in one location used to produce electric energy to supply a utility grid.

More than 30 years of experience, accompanied by developments in the aerospace industry and a greater understanding of aerodynamics, have led to major advances in materials and methods used to build turbines and strategically locate wind farms. New-generation turbines are manufactured from resilient composite materials, with blades capable of withstanding constant winds, salt spray, snow, and ice over a 20-year or longer life span. The largest units have a rotor diameter of more than 65.6 yd (60 m) with a maximum output of 1.5 MW. Modern units have a 98% or higher reliability rating, and their larger capacity means fewer turbines generate more kilowatt-hours of energy.

New wind farms added more than 3,800 MW to world wind-energy production in 2000 alone. In the United States four of the largest wind farms ever built were scheduled for completion by year-end 2001, increasing installed capacity in that country by 2,000 MW, market growth by more than 50%, and total production to 4,600 MW—enough power to provide electricity to 1.7 million homes.

In the Beginning When did humans begin using wind to their advantage? As far back as earliest recorded history boats caught wind in their sails to propel them through the waters, and windmills blades spun in the breeze turning stones that ground grain. Simple windmills in China pumped water centuries before the birth of Christ and windmills dotted rural landscapes in the United States and other countries as late as the end of the nineteenth century. Windmills still spin in some places, pumping water from wells or streams into cattle troughs and irrigation ditches.

By the beginning of the 1900s small wind systems generated electricity in isolated locations but were superceded in the 1930s as large-scale grid power systems wired their way into rural areas. However, modern wind turbines are an energy alternative for individual homes and small industry.

Revival of Wind Power When the energy crisis of the early 1970s sent oil prices skyrocketing, wind energy began to be considered seriously as a future source of power. In the 1980s California purchased large amounts of wind power at about 38 cents/kWh. The most expensive form of power in 1996 was nuclear power, at between 11.1 and 14.5 cents/kWh. Taking into consideration the inflation factor over 20 years, 38 cents/kWh was a sizeable investment in 1980. However, that investment paid off as experience, knowledge, and new technologies consistently reduced the cost of wind power. In fact, the U.S. Department of Energy (DOE) anticipates wind energy production costs will fall from the 3–6 cents/kWh in 2001, to 2 cents/kWh by the year 2005. If trends continue conventional forms of energy will continue to escalate in cost, making wind power the most economical form of energy.

As with most new technologies wind power ran into its share of snags during its experimental period, causing a temporary slowdown in the industry; however, interest in wind-generated power is bolstered by:

- new and more economical technology that significantly increases wind energy production and decreases mechanical breakdowns and failures
- scientific investigations into, and computerized models of, wind patterns and aerodynamics that allow placement of farms where wind capacity is highest and harvest is most consistent
- public recognition world-wide of the need to reduce greenhouse gas emissions
- domestic and international policies that require emission reductions by a specified date
- governmental subsidy and incentive programs for green energy development
- legislation that requires utility providers to offer consumers the choice of purchasing green energy

Into the Future—Of What Value Is the Planet? As the demand for power increases, forms of energy other than those provided by fossil fuels and natural gas will become increasingly essential. Supplies of coal, oil, and natural gas are not only finite, they are most often controlled by countries, governments, and private enterprise that can, at any given time, restrict supplies or increase prices, ultimately increasing energy costs to the general public.

Also, refining raw materials and producing nuclear power causes serious environmental hazards: fossil fuels spew billions of tons of polluting gasses into the atmosphere every year, damaging human and other life forms, creating greenhouse gasses that cause global warming. Global warming is predicted ultimately to lead to serious and even catastrophic climate changes, melting of the polar ice caps, and subsequent rise in sea level. Generating nuclear energy creates waste that is deadly to all living things and that science has not yet found a way to neutralize. This waste will remain radioactive for hundreds of thousands of years, and no truly safe way has yet been devised to dispose of it.

Environmental pollutants ultimately become huge financial burdens. Cleanup of hazardous waste costs untold billions of dollars annually, while health care costs due to these pollutants are incurred by individuals, corporations, and governments alike. These costs are not reflected on the itemized utility bill that arrives in mail but include lost work hours, increased utility prices to help utility companies pay for medical insurance, and increased taxation to pay for federally funded insurance programs and cleanup of waste sites. The biggest cost, however, is to human, animal, and plant life. Accidental pollution, such as the Chernobyl nuclear power plant malfunction in Russia and the oil spill after the tanker Valdez ran aground off the coast of Alaska, leave in their wake death, destruction, and suffering to countless numbers of people and animals. These factors are all apart from the monetary costs required in the cleanup process. In the case of nuclear pollution effects continue for generations.

In contrast, wind power is a clean, nonpolluting, renewable energy source. Wind is actually a form of solar energy created by the uneven heating of the Earth's atmosphere by the Sun. As long as there is a sun there will be wind. Windpower potential is therefore infinite, cannot be controlled by one particular country nation, and is healthy for the planet and its people. Every kilowatt of electricity produced by wind power displaces one kilowatt of power generated by traditional methods. For 51,000 homes powered by wind-generated electricity, annual environmental pollution is reduced by 94,000 metric tons of carbon dioxide, 830 metric tons of sulfur dioxide, and 280 metric tons of nitrous oxides.

Legislation, Incentives, and Consumer Options Following the 1992 Kyoto Climate Change Treaty established at the Earth Summit in Kyoto, Japan, 38 industrialized nations, including the United States, must reduce greenhouse gas emissions to below 1990 levels by 2008–2012. Major shortfalls of this objective are expected in the United States—by 2010 carbon dioxide emissions from electricity generation alone will exceed the 1990 levels by more than half a billion metric tons. The AWEA believes that, with strong legislative and policy support, the capacity of U.S. wind farms can be increased to 30,000 MW by 2010, providing enough electricity for 10 million homes and reducing the utility sector's excess emissions by 18%.

Legislation and incentive programs help promote renewable energy and reduce pollution. In the United States the federal tax production credit (PTC) program allows a tax credit of 1.5 cents/kWh for the first 10 years of operation of a new wind farm. The AWEA estimates this credit reduced the levelized cost of wind-generated electricity by 0.7 cents/kWh over a 30-year period.

The United Kingdom aims to increase renewable energy sources to 10% by 2010 from its 2001 level of 2.8%. Under the Non Fossil Fuel Obligation (NFFO), a "fossil fuel levy" is added to the electricity bill of each electricity consumer. Income from this levy helps cover the huge costs of nuclear power and funds a program encouraging development of renewable energy.

In 1999 the World Bank approved a US$100 million dollar loan and a US$35 million General Energy Fund grant to China so that millions of people will receive electricity from sources that do not produce greenhouse gasses. Under this plan the proportion of wind-generated electricity will increase to those already receiving electric power.

Texas legislation mandates the generation of 2,000 new megawatts of renewable energy, much of which will come from wind energy. In July 1999 the largest wind farm in that state began generating power. At the opening of the 2,200-acre (736-hectare) facility atop a 600-ft (183-m) mesa, the president of one utility company involved in its development said: "This project's state-of-the-art, cost-effective technology demonstrates the commercial viability of wind energy." President George Bush, then the governor of Texas, praised the project for "...creating jobs, promoting new technology, and responding to customer calls for increased use of renewable energy sources."

Also in Texas, residents of El Paso can contribute to the development of wind-generated electricity under the Renewable Energy Tariff Program by voluntarily purchasing 100kWh blocks of renewable energy for an additional $1.92 per month.

Economic Trends, Viability, and Importance
Despite its youth, wind energy production con-

A wind farm in Hawaii.
(Reproduced by permission of Miriam Nagel.)

tinues to become less expensive and more economically viable. Apart from its positive economic impact on human health and the environment, wind-powered electricity provides other important economic benefits that include:

- an affordable and inexhaustible source of power
- a domestic resource whose cost is not influenced by other countries. (The high cost of importing fossil fuel contributes to the trade deficit that affects the economy of the nation and, ultimately, every individual in that nation.)
- a reliable, sustainable source of domestic employment that fuels local, state, and national economies without depleting natural resources, polluting the environment, or creating hazardous waste
- a provider of more jobs per dollar invested than any other energy technololgy (five times that of coal or nuclear power)
- becoming more competitive as the cost of traditional energy increases.

Also, wind power is cubed by the wind's speed, i.e., when the wind is twice as strong, its energy content is *eight times* as great. Wind velocity of 26 ft (8 m) per second will produce approximately 80% more electricity than wind velocity of 20 ft (6 m) per second, with no added expense for the increased energy production because producing electricity from wind requires no purchase of raw material. Although running costs include the rental of land on which wind farms are built, local economies are stimulated as property owners receive the rental income. In many cases property owners can still graze animals and farm the land on which the turbines stand.

Importantly, wind power creates a positive "energy balance," which means the time it takes to "pay back" the amount of energy used to generate electricity is short. In the United Kingdom the average wind farm will generate enough power in three months to replace the energy it took to manufacture its turbines and build the farm. Over its lifetime, the farm will generate 30 times more power than was required in its manufacture. Fossil fuel and nuclear power plants supply just one-third of the energy required in plant construction and supply of raw materials, and never achieve an energy payback.

Falling Costs The AWEA estimates there was an 80% decrease in the cost of producing wind energy between the 1980s (approximately 38 cents/kWh) and the year 2000 (between 3 and 6 cents/kWh). Although still slightly higher than energy produced by natural gas or coal-fired power plants, wind energy costs continue to decrease steadily while energy from other sources continues to increase. For example, in January 2001, electric power generated from natural gas power plants reached between 15 and 20 cents/kWh in some areas of the United States.

In 1996, comparative levelized costs per kilowatt-hour were:

- wind (with PTC): 3.3–3.5 cents
- wind (without PTC): 4.0–6.0 cents
- gas: 3.9–4.4 cents
- coal: 4.8–5.5 cents
- hydro: 5.1–11.3 cents
- biomass: 5.8–11.6 cents
- nuclear: 11.1–14.5 cents

Potential Power—Potential Growth Although wind power is the fastest-growing renewable energy source in the world, it is still the smallest percentage of America's renewable energy, estimated by The Energy Information Agency of the DOE to be only 1% of all renewable energy sources in the United States. Worldwide, total kinetic wind energy is estimated at more than 80 times greater than the amount of all energy consumed by the planet's population. Variability of wind patterns and viability of locating farms in areas with the highest wind-power potential make harvesting only a small portion of that energy feasible. However, conservative estimates suggest the United States, the United Kingdom, Denmark, and the Netherlands alone could generate 20%–40% of their entire electricity supply from this cost-effective, energy-efficient, and environmentally friendly renewable resource.

Wind potential is plentiful. If, as some experts speculate, the ultimate success of wind power depends upon its cost compared to other sources of electricity, accompanied by the value society places on reducing pollution, wind power has a bright future. —MARIE L. THOMPSON

Viewpoint:
No, wind power will not be a large-scale contributor of energy because wind doesn't blow sufficiently everywhere and doesn't blow all of the time.

Ever is a strong word. In the case of wind farms, what is holding them from *ever* becoming an efficient, large-scale source of energy? The answer is the wind: it does not always blow and it does not blow enough in many locations. Add to that, the competition. There are also some environmental factors. Wind power does not produce toxic emissions, a fact which makes it popular with green energy enthusiasts. However, there are other environmental issues like noise pollution, hazards to birds, and aesthetic considerations. Wind farms are a tourist attraction in Palm Springs, California, in 2001. Rows and rows of humming turbines towering 200 ft (60 m) and higher, with rotors of 144 ft (44 m) diameter for a 600 kW generator, may be spectacular to some. But, would those tourists want a wind farm in their back yard? If the tourists do not live where there is good wind, they will not have to worry.

Wind Power: Not a Large Part of the Energy Mix The Energy Information Administration of the U.S. Department of Energy (DOE) published in its Annual Energy Review 2000 information on energy consumption by source. Not surprising, petroleum is the main source at 39% with natural gas at 24% coming in second. Coal is used for 23% of the energy used in the the United States, and nuclear electric power makes up 8% of the supply. That leaves only 7% for all renewable energy sources. These sources include conventional hydroelectric power which is 46% of the 7%. Wood is 38%, and "waste" is 8%. Waste includes some waste wood and other biomass materials, in a catchall category of municipal solid wastes. Geothermal is 5% of the renewable energy sources, alcohol fuels 2%, and solar and wind are each 1% of the 7% renewable energy consumption by source. (The percents have been rounded in the government data available, and so appear to be more than 100%) This indicates that wind is a very small-scale source in the big picture of all energy sources in the U.S. in 2000.

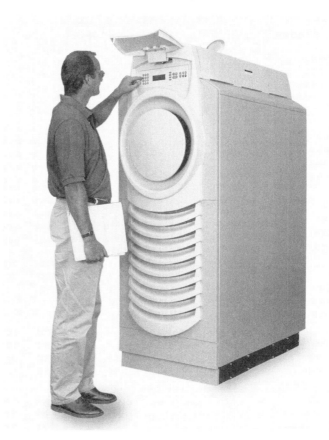

A free-standing microturbine.

(Capstone Turbine: www.microturbine.com. Reproduced by permission.)

According to the DOE Office of Energy Efficiency and Renewable Energy (EREN), wind energy has been the fastest growing source of electricity generation in the 1990s. Most of that growth has been in Europe, although the DOE recently announced the *Wind Powering America* initiative that has a goal to provide at least 5% of the nation's electric power from wind by 2020. Whereas renewables in total were only 7% in 2000, that is a very debatable goal and should it come even close to reality, 5% would still not be *large-scale*.

Although not always dependable, and not everywhere available in sufficient amounts, wind is free. (Of course the technology is not free.) Using wind to generate electricity does not produce any toxic emissions. Why then is wind not a larger part of the energy mix? There are reasons even the 5% goal will be difficult to meet. DOE describes wind resources in the United States as *plentiful*, but adds that good wind areas cover only 6% of the contiguous U.S. land area. Most of the useful wind resources are in the Great Plains from Montana east to Minnesota and south through Texas, according to the DOE. North Dakota, not a populous state, has about the best wind resources. Transmission and distribution of the power produced to where it is needed has to be part of the equation when siting wind resources. California has some good sites, mostly along the coast. California also has the energy crises which gives the state incentive to find any and all sources, including utility-scale wind farms, also called wind plants.

In Northern Europe, Denmark and Germany are leaders in developing wind as a source of electric energy. As a largely flat, small peninsular country with a long North Sea coast, Denmark is well suited to set up windmills. The Danish government has a goal to provide 50% of the country's energy from wind. The topography and geography of the United States does not lend itself to that kind of development. Northwestern Germany has a coastline abutting to Denmark, and so does the Netherlands. With the Netherlands practically synonymous with windmills, it is not surprising wind energy is big there. Even with the favorable landscape for wind to be plentiful, Denmark and Germany are building wind plants offshore now to expand where there is a more reliable high wind resource.

Competition from Other Sources

Competition from other renewable resources will also prevent wind energy from ever becoming a *large-scale* resource. EREN lists the technologies that are significant energy alternatives in the renewables category as biomass power, concentrating solar power, geothermal, hydropower, photovoltaics, solar thermal, and wind. For argument's sake here, the solar sources will be lumped together. Although they are quite different technologies, they use the same free energy source, the Sun.

Hydropower. EREN documents indicate that hydropower is the nation's leading renewable energy source, providing 81% of the total renewable electricity generation. The United States is the world's second-largest producer of hydropower, second only to Canada. As with wind power, not every location is suited to hydropower. Viable hydropower sites have to have a large volume of water and a significant change in elevation. These can be met by river dams as exist along the Columbia River in Northwestern United States, or a dam-created lake as in Hoover dam on the Arizona-Nevada border. Once a dam is in place (not cheap), hydropower is the least expensive source of electricity in the United States with typical efficiencies of 85%–92%. No energy source is without some problems. There are some issues of fish migration and arguments on land covered by water.

Biomass. Biomass is the second on EREN's list of renewable energy resources in the United States. By biomass EREN means plant matter such as trees, grasses, and agricultural crops. There is a broader term, bioenergy, that includes poultry litter and animal wastes, industrial waste, and the paper component of municipal sold waste which increases the energy resources under the *bio* category. Biomass conversion to electric energy involves combustion so it is not entirely clean, but it is a renewable resource. Biomass can be used directly in solid form as fuel for heat and electricity generation, or it can be converted into liquid or gaseous fuel by gasification, also a combustion process, but one that controls the available oxygen. The fuel produced by gasification has many uses, including microturbines.

Microturbines are small (micro relative to a windmill) generators designed for on-site power production, heating, and cooling. They are sometimes described as a miniature jet engine with a high speed alternator on the same shaft. They have emerged on the electric power scene as small-scale localized power sources for their reliability and high quality power that is always available and closer to the site and scale of actual needs. Such power sources are called distributed generation and are generally used for power generation of less than 1 megawatt (MW), a market well suited to microturbines.

Proliferation of the digital computing and communication devices is driving the use of microturbines, as utility electric power can fluctuate, and sometimes fail. Both are disastrous in some markets. Microturbines are also useful for stand-alone power in remote areas such as oil fields where the electric grid is not available. Windmills are sometimes used in remote areas but microturbines have two distinct advantages. They are modular so they can be combined to

A geothermal plant.

(Zunil Binary Geothermal Power Plant in Guatemala, photograph. Ormat Technologies. Reproduced by permission.)

multiply the power available, and they are reasonably portable. Although the fuel for a microturbine is not generally free, except on an oil field where they burn waste gas, the fuel is in constant supply whereas wind is variable and not always available when and where it is needed.

Wind forecasting is a relatively new, and as yet not perfected, art that is intended to help with the problem of intermittent wind even at good sites. Electricity storage in batteries is a solution only on a very small scale. Power producers using wind in their energy mix have to plan for the intermittent nature of wind energy. Wind is clean and renewable, but this problem could limit the large-scale use of wind as a source of electric energy.

Geothermal Energy. Geothermal energy is another competitor to wind power. The current production of geothermal energy places it third among renewable sources, and EREN states it has hardly been used compared to the enormous potential of geothermal energy in the United States. Geothermal energy is not limited to geysers, although that is what immediately comes to mind to the lay person. Such hydrothermal fluids are a part of the story but *earth energy*, the heat contained in soil and rocks at shallow depths, can also be tapped as an energy source. Earth energy is not directly used to produce electricity but the heating and cooling applications are considered as a contribution to the energy mix. Steam and hot water are used to drive turbines to generate electricity. Like wind, the energy source is free. Also like wind it is not found everywhere, but the hot water is more reliable than the wind.

Solar Power. Solar power, whether from concentrating solar, photovoltaics, or solar thermal, is a renewable energy source that also contributes to the energy mix. Combined, solar electricity sources make up about the same percent of the renewable energy resources as wind in the year 2000 energy portfolio. Also, like wind and geothermal, concentrating solar and solar thermal depend on a resource that is not available in sufficient amounts everywhere. Latitude and average cloud cover determine the viability of solar energy as an electric power source. Because photovoltaics require much less concentrated solar energy, they are useful in a much wider region. Photovoltaic can be used for a great variety of electricity applications from a handheld device to a skyscraper in Times Square, New York.

Concentrating solar power plants produce electric power by collecting heat through one of three technologies that can then be used to generate electricity. One technology, the power tower system, uses mirrors to heat salt to the molten state. The fluid salt is used as a heat source in a steam generator to produce electricity. World record solar-to-electricity conversion efficiencies make concentrating solar power attractive in the Southwest United States and other sunbelt regions. One major advantage to these systems where they are applicable is that they produce electricity during the peak use times. Where the sun shines bright, it is a plentiful source. Enough solar electricity could be produced for the entire United States if just 9% of Nevada were covered with solar collectors.

Photovoltaic. Photovoltaic (PV) electric sources are in a class by themselves. PV cells convert sunlight directly into electricity. The electricity can be used as it is produced, or it can be stored in a battery. Large-scale PV arrays produce electricity that is used as part of the local utility in areas of adequate sunlight. More widespread, and gaining in popularity, are building integrated PV systems. PV cells can be built into shingles, wall panels, and windows to produce electricity without being conspicuous. The systems are not cheap, but they are attracting more attention as a clean, on-site, attractive electric energy source. PV is also gaining a lot of attention for its portability. Small panels of PV cells serve as back-up power for homes, local business establishments, camping sites, and anywhere a reliable immediately available electric energy source is needed. PV is one more reason wind power is going to be a source, but not a large-scale source of electric energy.

Windmills have been around since ancient times—for water pumping and grinding grain before electricity was discovered. It might be said that wind power was the first power tapped by man. Today wind is among many good, clean, and renewable sources of electric power. Wind farms will be efficient contributors to the energy mix, but wind power will not be a large-scale contributor in the big picture, ever. —M. C. NAGEL

Further Reading

"Annual Energy Review 2000, Consumption by Source." <http://www.eia.doe.gov/aer/ep/source.html>.

Asmus, Peter. *Reaping the Wind: How Mechanical Wizards, Visionaries, Profiteers Helped Shape Our energy Future*. Washington, DC: Island Press, 2000.

"Australian Wind Energy Association." <http://www.auswea.com.au> (August 14, 2001).

"A Careful Blend of Modelling and Measurement." *CSIRO Land and Water*. <http://www.cbr.clw.csiro.au/windenergy/home.htm> (August 14, 2001).

"Danish Wind Turbine Manufacturers Association." <http://www.windpower.dk>.

"Distributed Energy Resources." <http://www.eren.doe.gov/der/whatis.html>.

"DOE, Industry Improving Technology, Lowering Costs." U.S. Department of Energy—Wind Energy Program. <http://www.eren.doe.gov/wind/wttr.html> (August 14, 2001).

"The Economics of Wind Energy." GALEFORCE. <http://www.galeforce.nireland.co.uk> (August 14, 2001).

"El Paso Electric Customers Can Now Purchase Wind-Generated Electricity." El Paso Electric Online. <http://www.epelectric.com/internetsite/www_news.nsf/5a54b8301a1246f087256845005a85db/b607a6b2112af4c587256a61005a58be?OpenDocument> (August 14, 2001).

"FPL Energy Dedicates Largest Wind Energy Facility In Texas." U.S. Department of Energy—Wind Energy Program. <http://www.eren.doe.gov/wind/weu.html> (August 14, 2001).

"Global Wind Energy Market Report." American Wind Energy Association. <http://environment.about.com/gi/dynamic/offsite.htm?site=http%3A%2F%2Fwww.awea.org> (August 14, 2001).

"Green Energy." Powergen Renewables. <http://www.powergenrenewables.com/greenenergy.htm> (August 14, 2001).

"Innogy, Greenpeace Blow Wind Power Trumpet." *Science—Reuters*. <http://dailynews.yahoo.com/h/nm/20010801/sc/energy_wind_npower_dc_3.html> (August 14, 2001).

"National Wind Technology Center." <http://www.nrel.gov/wind/database.html>.

Perez, Karen. *Wind Energy Basics: A Guide to Small and Micro Wind Systems*. White River

Junction, VT: Chelsea Green Pub. Co., 1999.

"Project Provides Solar-Powered and Wind Generated Electricity to Rural China." World Bank Group. <http://www.worldbank.org/html/extdr/extme/2236.htm> (August 14, 2001).

"Quick Facts about Wind Energy." U.S. Department of Energy—Wind Energy Program. <http://www.eren.doe.gov/wind/web.html> (August 14, 2001).

"Statistics World Wide." Wind Service Holland. <http://home01.wxs.nl/~windsh/stats.html> (August 14, 2001).

"Summary of Opinion Surveys." GALE-FORCE. <http://www.galeforce.nireland.co.uk> (August 14, 2001).

Watts, Martin. *Water and Wind Power.* Buckinghamshire, UK: Shire Publications, 2001.

"Wind Energy in the United States—Production Trends." *Environmental Issues.* <http://environment.about.com/library/weekly/aa021201a.htm> (August 14, 2001).

"Wind Energy in the United States—Worldwide Growth in Wind Energy Production." *Environmental Issues.* <http://environment.about.com/library/weekly/aa011099.htm?terms=Worldwide+growth+in+wind+energy+production> (August 14, 2001).

"Wind Power—Clean Energy for Colorado." Land and Water Fund of the Rockies. <http://www.cogreenpower.org/Wind.htm> (August 14, 2001).

"Windpower Surges to Record Levels—Expiring Credits, Lower Costs, and Green Power Revitalize Markets." (from EBJ No. 5/6, 1999). Environmental Business International, Inc. <http://shop.store.yahoo.com/ebj/ar5winpowsur.html> (August 14, 2001).

Yago, Jeffrey R. *Achieving Energy Independence—One Step at a Time.* Gum Spring, VA: Dunimis Technology, 1999.

Do the potential dangers of nanotechnology to society outweigh the potential benefits?

Viewpoint: Yes, from unforeseen pollution to deliberately designed weapons of mass destruction, nanotechnology holds significant risks, and the legal, technical, and cultural work that must be done to tip the balance toward safety is not being conducted.

Viewpoint: No, nanotechnology research contributions to advanced electronics and medical progress far outweigh any perceived danger.

Everything we see around us is made of atoms, often combined together in molecules of particular chemical compounds. For example, a molecule of water consists of two hydrogen atoms and an oxygen atom. When we deal with materials in our ordinary lives, we handle many billions of molecules at a time. This limits the precision and versatility with which we can manipulate the matter around us.

Nanotechnology is a word used to describe techniques by which we can work with matter on the atomic or molecular scale. Experts predict that it will have a major impact on our lives by about 2015. In the United States the National Nanotechnology Initiative is funding research and training in the field, and other industrialized nations are establishing programs in nanotechnology as well.

The potential applications of nanotechnology are boundless, although medicine and electronics manufacturing are areas of particular interest. Advocates predict that nanotechnology systems will allow us to cure diseases by directly repairing the intricate biochemical mechanisms and structures in human cells, or to clean up the environment by removing toxins, molecule by molecule. Others warn that nanotechnology could be used by terrorists to construct undetectable weapons, or launch microscopic machines that run amok and process the entire biosphere into "gray goo."

Nanotechnology implies the ability to select and manipulate individual molecules and atoms, to repair a defective structure or build one from scratch. So nanotechnologists are working on molecular-scale tools to go where ordinary tools can not. One promising area of research and development involves carbon nanotubes, cage-like molecules consisting of 60 or more carbon atoms forming a narrow cylinder. The nanotubes can be used as microscopic probes or as tools for manufacturing miniaturized electronic devices. Nanotechnology must also rely on robotics to scale down human movements to the submicroscopic level.

Perhaps the most controversial area of research in nanotechnology is self-replication. If nanotechnology systems could be engineered to build copies of themselves, they would be much less expensive to produce in large numbers. However, allowing the self-replication of tiny robotic devices intended, for example, to modify the environment or the workings of the human body, is a scenario viewed with trepidation by many. The thought of combining this ability with artificial intelligence advances such as machine learning raises fears of an invisible army of devices evolving to pursue their own agenda, out of our control and probably to our detriment.

Like most powerful tools, nanotechnology can be a double-edged sword. It can be used for good or for evil, and may have consequences unintended by those who deploy it. Attempting to legislate against knowledge is, in general, neither helpful nor effective. Therefore, as with other technologies in the past and present, our role must be to understand what we have made, and try our best to ensure that it is used wisely. —SHERRI CHASIN CALVO

Viewpoint:

Yes, from unforeseen pollution to deliberately designed weapons of mass destruction, nanotechnology holds significant risks, and the legal, technical, and cultural work that must be done to tip the balance toward safety is not being conducted.

Against the background of the many promises of nanotechnology, there is a strong and legitimate concern about the potential dangers of this new capability. The true power of nanotechnology is unproven, but with claims ranging from self-assembling nanobots to sky hook elevators into space, the potential for both good and harm exist with this new technology. Can we handle all this power? Do we have the social frameworks and technical understanding and skills to deal with both the direct and indirect effects of nanotechnology? History suggests that we do not.

The power to heal is also the power to harm. Technology has always been a two-edged sword. It may actually be easier to design a nanobot to target healthy cells than to attack cancer cells. One can imagine a new age of weapons, specifically designed to avoid the body's defenses. Similarly, nanodevices, aimed at removing toxins and restoring the environment, could be refashioned to destroy crops or selectively cause environmental damage within an enemy's territory. Unlike biological warfare, which is indiscriminate and can turn on the aggressor, nanodevices could be programmed to work within boundaries and to self-destruct when their jobs were completed.

Monofilaments, if manufacturable, might become essential building blocks for a sky hook and open up space exploration. They could also be used to make nearly invisible, gruesome weapons. Finer than a spider web but as strong as steel, these incredibly thin polymers could slice through almost any materials, making vandalism, breaking and entering, and even murder easier.

Of course, the most celebrated and remarkable promise is the promise of immortality. The flip side of this promise is the risk of extinction itself. Indeed, nightmare scenarios of the entire biosphere being transformed into "gray goo" of nanodevices have become an object of serious discussion.

Dangers of Nanotechnology Recognition of the two-edged nature of nanotechnology is not new. It dates back to at least to 1986, with the publication of *Engines of Creation*, a largely upbeat view of nanotechnology. More recently, Bill Joy, one of the pioneers of computer technology, has raised concerns about the dangers of nanotechnology, most notably in an article in *Wired* magazine. Although the risks of nanotechnology have been expressed in many fashions over the years, they can be viewed in four distinct ways: nanotechnology puts powerful means of destruction into the hands of irresponsible people; the unknowns of nanotechnology threaten us with pollution and other unintended consequences; nanotechnology can take away our humanity; and nanotechnology contains the seeds of mass extinction.

Nanotechnology puts powerful means of destruction into the hands of irresponsible people. Among the promises of nanotechnology is the ability to manufacture sophisticated devices at extremely low costs. Much of this is predicated upon the ability of nanodevices to create nanodevices, either through the use of specialized assemblers or via self-replication. This could put exquisitely designed weapons into the hands of individuals, terrorists, and rogue nations. Both known weaponry, such as chemical weapons, and unimagined nanoweapons could become available. With current weapons of mass destruction, testing, obtaining exotic precursors, and specific evidence, such as radioactivity, make monitoring and control possible in many instances. In contrast, nanoweapons, by definition, very small and made from common materials, would be almost undetectable.

Many people have the ability to create computer viruses, with over 1,000 new ones created each year. In fact, it's even possible for teenagers with low levels of skill (so-called "script kiddies") to copy code and vandalize sites. The combined forces of government and industry have not stopped the creation of new viruses, and there have been numerous instances in our connected world of the quick and destructive spread of these damaging programs. If such capability were extended to the world in general (a not unlikely event given the pervasiveness of networks and smart devices, combined with significant availability of assemblers), the results could be tragic. Even if this technology doesn't become available to the individual, it's likely to be available to many nations. It is not unreason-

KEY TERMS

BIOVOROUS: Means literally life eating. The term was used to describe the possible destructive forces of replicating nanorobots.

GLOBAL ECOPHAGY: From eco—the environment, and phagy—to eat or consume something. The term refers to the possibility of replicating nanorobots running amuck and consuming their, and our, environment.

NANOTECHNOLOGY: The science of manipulating atoms and molecules. Nano comes from the Greek for dwarf. It is a prefix meaning one billionth. A nanometer is about the width of 3–5 "average" atoms, or 10 hydrogen atoms, hydrogen being the smallest of all atoms.

QUANTUM: Literally means how much? The term evolved to describe the motion and interaction (quantum mechanics) of particles in the range of nanometer size where the forces of their interaction make classical laws of motion as described by Newton not relevant.

able to project a significant probability of a costly and risky arms race based on nanotechnology.

Advances in nanotechnology are driven almost entirely by curiosity and economics, with little concern for precedents that might be set and the establishment of both moral guidelines and inherent security. This means that it is left to chance who will become the first mover for significant advances, such as the development of assemblers, and how much safety will be built into these devices.

There is not even good agreement on when the different advances are likely to take place. In fact, discussions tend to smear out options and possibilities. Speculation about using nanobots to clean the blood stream are likely to be made in the same conversation with predictions of sky hooks and advanced computer components.

Going further, there may be even greater challenges if nanotechnology is hybridized in any way with other technologies. No technology exists in isolation, and the combination of nanotechnology and genetic engineering, or nanotechnology and robotics, could result in even more danger.

Unintended Consequences of Nanotechnology The unknowns of nanotechnology threaten us with pollution and other unintended consequences. Physically, nanodevices have the potential for being persistent, like plastics, and invasive (because of their small size). One could easily imagine a worse problem than was seen with DDT in the environment, with nanomaterials lasting for a thousand years after taking up residence in the tiniest niches of living things. For specific nanomaterials, there may be consequences along the lines of what was discovered with "chemically inert" chlorofluorocarbons, which turned out to catalyze the destruction of the ozone layer. Heat pollution is an unavoidable side effect of the activity of nanodevices. In fact, one defense suggested against a gray goo surprise is simply to look for the heat signature of replicating devices.

There's an old phrase in computing, "It's not a bug, it's a feature." Bad design raises the risks of unintended consequences. Often designs are done in piecemeal fashion or without regard to the users and the environment. This sort of engineering approach could have devastating consequences when the actualization of the design lies in the real world rather than in cyberspace. Even a good design could easily be extended beyond the originator's intent. This has been seen in the use of combination drugs for weight loss, with deadly results. In addition, genetic engineering has already witnessed this, when corn not intended for human consumption made its way into the general market, causing allergic reactions among sensitive individuals.

The risk of unintended consequences rises steeply in the face of greater capability of nanodevices. If the devices have the ability to replicate, their effect will be similarly amplified. If they are modular, that is, they can be combined with other nanodevices to form more complex devices, predicting how they might interact with the users and the environment becomes much more difficult. If the devices are purposely evolved rather than designed, a practice that is already in evidence in the world of software, then it will become impossible to understand the details of the nanodevices or to predict their effects on the environment. The likelihood and the danger of evolved nanodevices devices will go up significantly if it is proven that evolved software has an economic advantage.

This scenario, again, only gets worse if nanotechnology is used in combination with other powerful technologies.

Nanotechnology: Danger to Humanity
Nanotechnology can take away our humanity. Inexpensive, invisible, powerful devices threaten freedom and privacy if they fall into the wrong hands. A totalitarian regime could use nanodevices to coerce its citizens. In the short term, it could monitor them pervasively with tiny sensors or threaten them with nanoweapons. It the long-term, modifying the populace with nanosurgery or even using nanobots to transform the genes of future citizens is not inconceivable. With nano-

devices to do the work, large sectors of the population could be selectively destroyed, and intervention by other nations could be discouraged by threats based on expertise in nanotechnology.

A more subtle threat is that represented by the cyborg. Taking nanodevices into our bodies, either to extend our powers or to extend our lives, creates an intimate relationship between our machines and ourselves. Whether the result is viewed as symbiotic or parasitic, at some point, the needs, values, and orientation of these new individuals may become drastically different from what is currently defined as human. To some, this is an opportunity to revel in, but there is by no means a consensus as to whether this is good for our species or not. The social consequences could be profound: Are unmodified humans obsolete (and possibly expendable)? As people become cyborgs, are they still part of our community with all the same rights? How would the accumulation of power and wealth by "immortals" be handled? How are benefits distributed? Who is responsible for the costs of side effects? Philosophically, if all natural parts are replaced by nanodevices, and the resulting individual passes the Turing test, do you still have a human? Is it a good thing for the species if all humans are replaced this way, or is it an empty fantasy that ends humanity?

Nanotechnology Harbors Seeds of Mass Extinction Nanotechnology contains the seeds of mass extinction. In 1974, leading scientists in the field of genetic engineering called for a moratorium on certain areas of research. In 1975, the Asilomar agreement went into effect, and scientists voluntarily suspended their work until the consequences were explored, rules were written, and remedies were found. No such agreement has been made among leading nanotechnologists.

Experience and Funding Is the threat of nanotechnology less obvious or less real than what the genetic engineers faced? This is unlikely. Genetic engineering had its *Andromeda Strain* concerns just as the gray goo haunts nanotechnology (and both had and have their healthy skeptics). In 1975, genetic engineering was still in its infancy, with cloning, embryo research, "Frankenfood," and gene therapy still many years in the future. The primitive state of the art of nanotechnology doesn't explain the lack of attention devoted to its potential dangers. There are, however, two key differences between nanotechnology and earlier technological advances: the experiences of the scientists and the funding of their research.

Asilomar was an offshoot of the Pugwash conferences, where scientists took on the tough issues of living with nuclear weapons. The threat of nuclear annihilation was more vivid in the polarized Cold War years. Testing, proliferation, and safety problems were both urgent and intractable. Scientists were at the center of nuclear debates and were asked to make policy recommendations. Many also felt guilty about developing the Bomb and not opposing its use in Japan. Although the issues of living with nuclear weapons haven't gone away, they have faded—and almost no nanotechnologists have personal experience with building weapons of mass destruction.

While funding for genetic engineering was largely in the hands of government, specifically the National Institutes of Health (NIH), nanotechnology gets much of its funding from industry. Its pressures are market pressures, and competition encourages speed and applicability over deliberation and understanding. A nanotechnologist who advocates a pause in the development of nanotechnology would not only be challenging the current market ethos but taking on the corporation that funds his or her work.

Nanotechnologists are not restrained by their experiences and are driven by their funders to put nanodevices into the real world. In this context, it seems unlikely that precautions like secrecy, testing, and isolation of nanodevices will be taken. It is probable that nanodevices will not be engineered for safety with features such as traceability, self-destruction, and dependency. All this makes abuse more likely and error more probable. The consequences can range from vandalism to the infamous gray goo.

The power of nanotechnology is not intrinsically bad, but its destructive potential is great. The legal, technical, and cultural work that must be done to tip the balance toward safety is not being conducted. Laws are needed to control information, proscribe areas of activity, and control use and distribution. We must, as a species, develop a cultural response to this power that includes education, discussion, and approaches toward agreement and consensus. Timing is everything. Unlike our experience with nuclear weapons, we still have the time to be thoughtful, to prepare. But today, with work on safety still in front of us, we face a dangerous future. If Bill Joy is correct, that future is less than 30 years away. —PETER ANDREWS

Viewpoint:
No, nanotechnology research contributions to advanced electronics and medical progress far outweigh any perceived danger.

Birth of Nanotechnology To make a rational argument, pro or con, one should examine both sides of a question, including the background for the debate. There was not even a word "nanotechnology" in 1959 when sometimes-jester Richard Feynman (Nobel Laureate in 1965) presented his classic lecture at the California Institute of Technology inspiring his audience with a view of a working world so small that all of the world's books could be stored on something the size of a dust speck. In a speech titled "There's Plenty of Room at the Bottom," Feynman said, "A biological system can be extremely small. Many of the cells are very tiny, but they are active; they manufacture substances; they walk around; they wiggle; *and they do all kinds of marvelous things—all on a very small scale. Also, they store information. Consider the possibility that we too can make a thing so small which does what we want—that we can manufacture an object that maneuvers at that level.*"

Feynman said, "I am not inventing antigravity, which is possible someday only if laws are not what we think. I am telling you what could be done if the laws are what we think; we are not doing it simply because we haven't gotten around to it." Feynman's speech is viewed as the beginning of nanotechnology. He stressed better tools were needed. Feynman spoke of using big tools to make smaller tools suitable for making yet smaller tools, and so on, until researchers had tools sized just right for directly manipulating atoms and molecules.

Feynman did not envision the controversy his ideas would produce when researchers began speculating on the prospect that what these tiny devices would manufacture is copies of themselves, over and over and over at tremendous speeds; that these nanorobots could *replicate*. Another scientist, Bill Joy, co-founder of Sun Microsystems, published his essay titled "Why the future doesn't need us" in the April 2000 issue of *Wired*. Bill Joy wrote, "From the moment I became involved in the creation of new technologies, their ethical dimensions have concerned me, but it was only in the autumn of 1998 that I became anxiously aware of how great are the dangers facing us in the 21st century." After considerable rationalization, he added, "Thus we have the possibility not just of weapons of mass destruction but of knowledge-enabled mass destruction, this destructiveness hugely amplified by the power of self replication."

The Gray Goo Problem Joy's essay popularized, and was based on, the 1986 book by K. Eric Drexler, *Engines of Destruction: The Coming Era of Nanotechnology* in which he said, "The early transistorized computers soon beat the most advanced vacuum-tube computers because they were based on superior devices. For the same reason, early assembler-based replicators could beat the most advanced modern organisms. "Plants" with "leaves" no more efficient than today's solar cells could out-compete real plants, crowding the biosphere with an inedible foliage. Tough, omnivorous "bacteria" could out-compete real bacteria: they could spread like blowing pollen, replicate swiftly, and reduce the biosphere to dust in a matter of days. Dangerous replicators could easily be too tough, small, and rapidly spreading to stop—at least if we made no preparation. We have trouble enough controlling viruses and fruit flies.

"Among the cognoscenti of nanotechnology, this threat has become known as the 'gray goo problem.' Though masses of uncontrolled replicators need not be gray or gooey, the term 'gray goo' emphasizes that replicators able to obliterate life might be less inspiring than a single species of crabgrass. They might be 'superior' in an evolutionary sense, but this need not make them valuable." He adds, "We have evolved to love a world rich in living things, ideas, and diversity, so there is no reason to value gray goo merely because it could spread. Indeed, if we prevent it we will thereby prove *our* evolutionary superiority."

The gray goo threat makes one thing perfectly clear: we cannot afford certain kinds of accidents with replicating assemblers. The graygoo part especially continues to attract a lot of attention from science fiction fans, and some from the media.

What happened between Feynman's speech and Joy's essay were many discoveries and technological breakthroughs that have led to several Nobel Prizes for their scientific merit and value to humanity, and new fields like nanomedicine with untold promise for breakthroughs in treatments of cancer, diabetes, and other high profile diseases. Clearly there is much support for the view that any potential dangers of nanotechnology to society *do not* outweigh the potential benefits.

There have also been knowledgeable reports published to refute the gray-goo concept. Robert A. Freitas Jr., research scientist with Zyvex, wrote in 2000 on *Some Limits to Global Ecophagy by Biovorous Nanoreplicators, with Public Policy Recommendations*. He sees a potential danger, but puts it in perspective. Freitas concludes, "All ecophagic scenarios examined appear to permit early detection by vigilant monitoring, thus enabling rapid deployment of effective defensive instrumentation." Although there are some who propose *outlawing* certain research activities they deem dangerous, others point out the lack of success in such political measures throughout history, giving biological warfare agents and nuclear proliferation as examples. Cloning of humans can be added to the list.

Zyvex is a unique startup company. It was specifically founded to build what they call the assembler, the key tool to creating atomically precise materials and machines. As the researchers at Zyvex envision it, the assembler would start with a generic feedstock such as methanol and manufacture any precisely defined object which can be built from stable arrangements of the feedstock atoms. The assembler would start with the atoms and build up, an ultimate goal of many researches in nanotechnology. One team at Zyvex believes that they will eventually build a devices that can build smaller versions of themselves, which will then build smaller versions, and so on down to the nano world, but all this is as yet in the future.

Feynman was very positive. He said, "I am not afraid to consider the final question as to whether, ultimately—in the great future—we can arrange the atoms the way we want; the very atoms, all the way down!" The future got closer in 1981 when Gerard Binning and Heinrich Rohrer at IBM's Research Center in Switzerland invented a scanning tunneling microscope (STM), a tool that has become a workhorse in nanotechnology. The inventors of the STM received the Nobel Prize in 1986 for making it possible for researchers to finally *see* atomic and molecular surfaces. Advances have come steadily since. Richard Smalley, and collegues, at Rice University received the Chemistry Nobel Prize in 1996 for discovering fullerenes, which led to the discovery of carbon nanotubes by Japanese scientist Sumio Iijima in 1991, soon after the Smalley discovery was first announced.

Stronger than steel, more conductive than gold, and with electrical characteristics of either metals or semiconductors, carbon nanotubes have been at the center of a great deal of nanotechnology research at IBM's T. J. Watson Research Center, and other laboratories worldwide. The electronics industry is pushing the dimensions of transistors in commercial chips down into the nano dimensions. In other applications, carbon nanotubes are being used as reinforcing materials in composites today, making stronger parts for lighter vehicles possible. DNA molecules, about 2.5 nanometers wide, are also being used in the development of new materials both in electronics and in medicine using nanotechnology.

Nanotechnology and Moore's Law Part of the momentum for the rise in nanotechnology comes from the realization by the semiconductor industry as it has had to come to grips with Moore's law, a driving force of the digital era. Moore's law is not a law of nature or government but an observation made in 1965 by Intel co-founder Gordon Moore when he plotted the growth in memory chip performance versus time. That observation described the trend that has continued to be remarkably accurate. He observed that the number of transistors that can be fabricated on a silicon integrated circuit is doubling every 18 to 24 months. Moore's observation held true for over 4 decades. But the Law is in trouble. It is expected that the industry will be making components down to the 100 nanometer range by 2005. At that point the quantum mechanical world of atoms and molecules takes over and the physics of the larger world no longer applies.

In response to the fundamental problems with Moore's law, Stan Williams, director of the quantum structures research initiative at Hewlett-Packard Labs, says over the next 10

NANO DIP-PEN

The nano dip-pen could be the key to making computer nanocircuits, and possibly needle-tip sized medical instruments. It can now be used to modify existing circuits especially for sensors. Chemistry Professor Chad Mirkin and his colleagues at Northwestern University's Center for Nanofabrication and Molecular Self Assembly call their discovery *dip-pen nanolithography.*

The new writing tool is able to use an organic oily ink to make nano-scale letters with near-perfect alignment. The area surrounding them can be filled with a second type of ink. Some of the technology that makes the device work is remarkably similar in principle to dip-pens that Mirkin estimates have been used for 4,000 years (although not too much in recent times). The pen is another story.

Mirkin's new dip-pen nanolithography uses an atomic force microscope (AFM) with an extremely fine pen tip made of silicon nitride. The "paper" is gold-plated silicon and the "ink" is a single layer of molecules called ODT that were selected because they are attracted to the writing surface. The ink is first allowed to dry on the tip, then the tip is brought in near contact with the surface. A water meniscus forms between the surface and the coated tip. The size of the meniscus is controlled by relative humidity. The appearance of water on the surface had been considered an obstacle to the AFM operation by earlier researchers, but Mirkin's team found that it was the key to success for nano dip-pen lithography. The team turned a major problem into a solution.

To demonstrate the success of nano dip-pen lithography, the Northwestern University researchers printed a nano-sized excerpt from physicist Richard Feynman's written version of his famous December 1959 speech, "There's Plenty of Room at the Bottom." Feynman's speech is considered the beginning of nanotechnology.

—M. C. Nagel

years researchers believe a hybrid molecular/silicon technology will evolve that utilizes the best of both worlds. Some see quantum computers as the future. Williams believes they are out there, possibly more than 10 years away. He also predicts by 2020 electronics will be 10,000 times as capable as they are today due to nanotechnology research.

Although the challenges of Moore's Law will dominate nanotechnology research in the next decade, as it has driven it in the past decade, there are other important areas of research benefiting from it. With an increased demand for drugs to solve all ills, there are a number of remarkable nanodevices that could make a difference in the not-too-distant future. A team under the direction of Jean M. Frechet at the University of California, Berkeley, is working on drug delivery with dendrimers, chemical nanodevices. iMEDD, a startup drug delivery company, has a patented process using nanomembranes to produce a controllable rate of drug release.

One product of nanotechnology that has made the news for its novelty has been described as "smart bombs" for the connection to its funding source. The development of the novel technology from the Baker Laboratory at the University of Michigan Medical School was funded by the U.S. government Defense Advanced Research Project agency (DARPA). The nanobombs are molecular-size droplets that proved to be 100% effective in a test as a potential defense against anthrax attacks. The application, it is hoped, will never be needed. But, the formulation can be adjusted to attack such disease organisms as those causing influenza and herpes.

DNA nanoparticle-based probes and detection and screening assays are being developed at Nanosphere, Inc., a startup company incorporated in 2000, and founded by Chad Mirkin of Northwestern University and a colleague, both pioneers in the field of DNA and nanoparticle research. Chad Mirkin is the Director of the Institute for Nanotechnology and Center for Nanofabrication and Molecular Self-Assembly at Northwestern. (It might be an exaggeration to say that every university is doing research in nanotechnology, but it's not a large exaggeration.)

Mirkin and his colleague found that single-stranded DNA can link metal nanoparticles in a way that controls the structure's optical properties. The method relies on a specific color change that occurs when the functionalized nanoparticles, which look red in solution, are induced to form nanoparticle aggregates, which are blue in solution. A conspicuous color change makes them useful in diagnostic testing for many forms of cancer. The test procedure can be modified to be used for virtually any sequenced organism or genetic disease. The procedure can also be used for food testing.

With the proliferation of nanotechnology research, the driving forces of a worldwide economy that is tied to the advance of electronics, and a society that demands medical progress, the question of whether the potential dangers of nanotechnology to society outweigh the potential benefits may be an entirely academic debate. The genie is out. Society has tasted the successes of nanotechnology and they are many. Even with the *science fictionese* suggestion of replicating robots, gray goo and all, nanotechnology is here to stay. —M. C. NAGEL

Further Reading

Baringa, Marcia. "Asilomar Revisited: Lessons for Today?" <http://www.usc.edu/dept/law/Pacific_Center/Main_Links/Asilomar/Science_article.html>.

Barlow, John Perry. "Is Technology a Threat to Humanity's Future?" <http://seattletimes.nwsource.com/news/business/html98/techweb_20000319.html>.

Crandall, B. C. *Nanotechnology: Molecular Speculations on Global Abundance*. Cambridge, Mass.: MIT Press, 1996.

Drexler, K. Eric. *Engines of Creation*. New York: Anchor, 1987.

———. *Nanosystems: Molecular Machinery, Manufacturing, and Computation*. New York: John Wiley & Sons, 1992.

Gross, Michael. *Travels to the Nanoworld: Miniature Machinery in Nature and Technology*. New York: Perseus Books Group, 1999.

Joy, Bill. "Why the Future Doesn't Need Us." *Wired* 8.04 (April, 2000) <http://www.wired.com/wired/archive/8.04/joy.html>.

Kahn, Herman. *On Thermonuclear War*. Westport, Conn.: Greenwood Publishing Group, 1978.

Krieger, Lisa M. "Genetic Researchers Cite Specter of Profits Asilomar: Scientists Say Demands for Financial Gain Threaten Public Health by Concealing Dangers." <http://www.biotech-info.net/asilomar_mercury_news.html>.

"Nanoelectromechanical Systems Face the Future." <http://www.physicsweb.org/article/world/14/2/8>.

"Nanotechnology: The Coming Revolution in Molecular Manufacturing." <http://www.foresight.org>.

"Preparing for Nanotechnology." <http://www.foresight.org/index.html>.

"U.S. Nanotechnology Initiative." <http://www.nano.gov>.

Have technological advances in sports such as tennis, golf, and track and field supplanted the athletic achievements of the participants in those sports?

Viewpoint: Yes, advanced technology in sports has significantly narrowed the gap between world record holders and the way the athletic achievement is perceived.

Viewpoint: No, technological advances in sports such as tennis, golf, and track and field have not supplanted the athletic achievements of the participants in those sports.

In the twentieth century, the entire face of sports changed drastically with the advent of new technologies. Today advertisements for new types of running shoes, golf clubs, tennis rackets, and hundreds of other sports accessories bombard us. The level of scientific research into something as simple as a golf ball can be mind-boggling. With new materials and computer engineering, improvements are being made on sporting equipment faster than marketers can publicize them.

This advance in technology has broadened the spectrum of athletes. Improvements in safety standards, cost, and accessibility have allowed more people to take advantage of formerly exclusive sporting events. Self-improvement and exercise are thriving businesses, and sporting manufacturers have capitalized on this interest, promising better performance as the result of their equipment or product.

Sports technology has found its greatest proponents among professional athletes. Athletes like Pete Sampras, Tiger Woods, and Nick Hysong now have access to far better equipment than ever before. Advances in sports equipment have undoubtedly played a role in the achievement of these athletes in their respective fields.

Athletes are often elevated to the status of superheroes, revered by the public for their successes. Many consider the players even more important than the game. This, in turn, has made competition fiercer than ever. With enormous amounts of money, and sponsorship, riding on an athlete's performance, the old adage "it isn't whether you win or lose" has fallen to the wayside. Some would argue that today's athletes are willing to take any advantage they can get. They want to hit further, jump higher, or run faster and further than their opponent. Could it be argued that winning for the sake of winning has begun to replace the wonder of athletic achievement? If so, how large a role does technology play in that success? What truly makes these athletes exceptional? Is it personal achievement in and dedication to their sport, or has technology made them into the athletes they are today? Did Venus Williams or her racket win Wimbledon in 2001? Did David Duval become the 2001 U.S. Open golf champion because of practice or because he used a certain type of club? Was it the shoes Marion Jones wore or her tireless training that captured the 2000 Olympic gold medals in the Women's 100M and 200M races?

While some would argue that there is no replacement for raw athletic talent, many maintain that the technological advances in sporting equipment have added significantly to athletic performance. For example, there is disagreement about whether current-day athletic achievements should be viewed with the same regard as records established in the past—when athletes were performing without the benefit of graphite tennis rackets, fiberglass poles, ultra-light running shoes, and titanium golf clubs. Even the athletes themselves are divided on the issue. Despite the ongoing debate, it is certain that athletes will continue to utilize advances in sports technology to enhance and better their performance. —LEE ANN PARADISE

Viewpoint:
Yes, advanced technology in sports has significantly narrowed the gap between world record holders and the way the athletic achievement is perceived.

When people talk about the prowess of Tiger Woods, Michael Chang, or Stacy Dragila, the 2000 Olympic gold medalist in the women's pole vault, it's impossible to ignore how improvements in sports equipment contributed to their success. While technological advances do not negate the achievement of the athletes or fully explain their accomplishments, better golf balls, tennis rackets with bigger "sweet spots," and springier poles have enabled athletes to perform better in this century than ever before. No competitive professional would think of returning to the wooden tennis racquet or smooth golf ball. Today, the alternatives are just that much better.

Technological Improvements of Sports Equipment In 1963 with the advent of aluminum, fiberglass, and graphite poles, the pole vault record shot up 2 ft (0.61 m) in three years and now stands at over 20 ft (6.09 m). Prior to that technological innovation, the pole vault record increased only about 2 in (5 cm) to 16 ft (4.88 m) between 1942 and 1960. At the Atlanta Olympics, Michael Johnson became the first male athlete to win both the 200- and 400-meter sprint. He also set the new world record in the 200-meter sprint—wearing specially designed ultra-light Nike running shoes weighing just 3.4 oz (96.39 g). Now, many sprinters also wear full-body suits to reduce wind resistance, which could make the critical hundreth of a second difference in a race.

In the future, the advanced technology of the vaulting pole, running shoes, or golf ball will make more of a difference in the future than it has in the past. In some sports it will be because the time differences between first and second place will continue to shrink, allowing more room for technological improvements in equipment to give athletes an edge. Also, sports technology is not just limited to improvements in equipment. The modern-day athlete can now depend on computerized training systems to analyze their swing, stride, and follow-through.

When discussing technological improvements to sports equipment a distinction must be drawn between legitimate improvements and improvements that give athletes an unfair advantage—the equipment equivalent of performance enhancing drugs. In almost every sport there's been debate about how this distinction should be drawn. After a German inventor devised a "spaghetti racquet" with a standard frame and double strings fitted with plastic tubes designed to give the ball more topspin, the International Tennis Federation jumped in to prevent potential uproar. A tennis racquet could be made out of any material, be any weight, size, and shape, the Federation said, but it cannot alter the flight of the ball and the strings must also be evenly spaced.

Sometimes the distinction is not so clear. A new golf club, the ERC II, does not conform to United States Golf Association specifications because of a "trampoline effect" that gives the club more spring. However, the USGA does allow some spring in golf clubs—as long as it's below a set "spring-back number" of 83%. Similarly, with the introduction of fiberglass poles for use in the pole vault there are now over 200 poles of varying stiffnesses from different makers. In 1998, a special United States Track and Field Pole Vault Equipment Task Force was created to determine if there should be a standard "flex number" for maufacturers to follow.

Sports Technology—Optimizing Athletic Performance The recent push of athletic associations to standardize rather than outlaw new innovations points to the growing importance of sports technology for optimizing athletic performance. Although some current changes are more apparent than others—like form-fitting body suits that some world-class swimmers and sprinters now wear—there has been a steady evolution of even seemingly mundane sports equipment for decades.

One of the best examples is the golf ball. The first generation of golf balls resembled hacky sacks, covered and leather and filled with feathers. The ball, called a "feathery," could be struck up to 200 yd (182.88 m) but slowed when it became damp. Around the same time, ballmakers started experimenting with the gutty ball, a golf ball made out of a rubbery substance called "gutta-percha" from India. When softened in hot water, gutta-percha could be rolled into the shape of a ball with the hands.

Soon after, ballmakers made a sports-changing discovery: The newly made smooth balls did not fly as well as older balls. The reason? As golfers hit the balls, the surface of the balls became more nicked, and the bumpy surface is what made the difference in performance. The bumps or dimples on a golf ball trap a layer of air that extends the flight of the ball. Since the golf balls travel forward with a backspin, the layer of air on top of the ball moves faster than the layer of air at the bottom. This creates more air pressure underneath the ball, resulting in better lift. Years later, the dimpled golf ball still survived but in a different form. The new golf ball consisted of rubber strips wound around a rubber core—signaling the death knell of the gutty ball.

In the past century, carbon fiberglass—lighter than metal or wood—revolutionized both tennis and the pole vault. When pole vaulting began as a competitive sport, athletes used bamboo poles with a sharp point at the bottom to plant in the grass. Today, modern poles are made out of fiberglass and are much lighter than their bamboo or metal counterparts. The lighter poles allow athletes to run faster and gain the momentum they need to vault higher.

The new poles also have more spring. The fiberglass pole absorbs more of the vaulter's energy when it bends and as it straightens. Along with rubberized track surfaces, special boxes for planting the pole, and padded landing pits, the new fiberglass poles help athletes attain records that would be impossible to obtain using the traditional equipment.

Today, tennis rackets that used to be made out of wood have also turned composite. This in spite of former Wimbledon champion Bjorn Borg's stubborn insistence on using his wooden racket even after the introduction of the new lighter metal and graphite rackets. In the 1960s, manufacturers started introducing metal frames of steel and aluminum, later turning to an array of materials from titanium to graphite shells with plastic foam cores. Modern-day tennis rackets also feature larger "sweet spots" that minimize vibration.

Aside from using new materials, manufacturers also started experimenting with the physical design of the tennis racket. When Head increased the size of the face of its racket 20%, it resulted in a 300% increase in the size of the

> **KEY TERMS**
>
> **FIBERGLASS:** A vaulting pole made of fiberglass has long, stiff filaments of glass fiber that is combined with a more flexible polymer. The stiffness-to-weight ratio can be engineered accordingly to different specifications.
>
> **GRAPHITE:** A carbon-based material commonly used for the frames of tennis rackets. It's lighter than wood and 10 times stiffer.

sweet spot. Later, Prince introduced its Long-Body tennis racket designed to have a greater length and give the player more control. New lighter materials like graphite helped maufacturers to lengthen rackets, giving shorter players like Michael Chang a longer reach without a lot of additional weight.

As for golf, the traditional wooden golf club has changed as much as the ball. Golfers now use drivers with the weight distributed around the edge of the clubhead, specially designed to maximize the efficiency of each hit. The titanium-based shafts and club heads have also helped professionals and weekend golfers improve their game.

It's difficult to assess precisely how much of a difference improved technology makes in the competitiveness of an athlete. But it's undeniable that the equipment—whether it's the shoes, ball, or pole—is receiving more attention than in the past. The evolution of the pole vault is a prime example. In their coverage of the 2000 Olympics, Newsday.com ran an infographic about the switch from fiberglass to metal poles. Another Web site, NBCOlympics.com, hosted a chat with Stacy Dragila, the American gold medalist of the first ever women's Olympic pole vault. Participants asked her about her training regime, mindset, injuries, and the improvements seen in pole technology. To the questioner who asked, "How much of the sport depends on the pole? How much depends on the athlete?" Dragila gave this telling response: "That's a tough question. I think it's a 50/50 take right there. You have to be able to maximize your pole selection as well as your athleticism. If you're not comfortable with your pole, I think it's very hard for the vaulter to maximize her strength on the pole, so it's a 50/50 toss."

On the golf course, as the technological revolution of the golf ball continues, it appears that the wound ball with its core of tightly wrapped rubber bands will now go the way of the feathery and the gutty ball. In the 2000

Masters, 59 out of 95 players hit wound balls. A year later, only four competitors chose to use wound balls. The solid-core ball is quickly becoming the ball of choice. When Tiger Woods used wound balls, he averaged a 288.9 yd (264.17 m) drive. With the solid-core, he averages 305.4 yd (279.26 m). But the innovation doesn't stop at the golf ball core. When Jesper Parnevik won the Honda Classic, he cited the new Titleist Pro VI as contributing factor in his win. The new ball featuring an innovative dimple design and "ionamer casing" results in higher ball speed with lower spin.

In addition to revolutionizing equipment, improved sports technology has also shrunk the difference between winners and losers, allowing thousandths of a second to decide who finishes first and who finishes second. Modern-day timing systems start with the firing of the starting gun and stop when a light-based sensor detects the winner across the finish line. Now that there are fewer improvements that can be made to the equipment—namely limited to slight adjustments to the surface of the track and weight of the shoes—the difference in winning times of the 100-meter sprint has shrunk. Improvements are now made in about 0.006 second increments rather than the 0.015 second improvements seen in the early 1900s. Although improvements in time, at least in the shorter running events, will be largely due to the training regime of the athletes, the technology that measures their performance will be distinguishing a "win" that might have been a "tie" a century ago.

Computerized measuring devices might even change the judging of other sports like the pole vault. Historically, the pole vault has been judged strictly on the height of the bar that the athlete scaled. Now, poles with light-emitting diode (LEDs) can measure the actual altitude a pole vaulter has achieved.

Aside from altering the way we conceive of winning and losing, in sports like tennis and golf, computer-based systems are now helping players analyze their performance over weeks, months, or even during the course of a season. Since 1997, both PGA and LPGA Tour players have used a tracking system called SportsTrac to monitor their performance. Similarly, IBM computers at Wimbledon receive an input of shots played and points won, the umpires record information about the match in personal digital assistants, and a radar measures the speed of players' serves. At the end of a match, players and coaches receive a 40-page report of the collected data.

As the equipment and training regimes of athletes become more sophisticated, technological innovation will play a larger role in dictating winners and losers. Races that would have been ties when measured with older technology will now be definitively decided. Access to the latest golf balls that fly farther than their predecessors will not replace hours-long practice sessions, but the technology has become so advanced that it could make the one or two stroke difference at the end of the tournament. In the future, records will be broken in vanishingly small increments, not because there are no more great athletes, but because technology is helping to optimize the performance of all. —EILEEN IMADA

Viewpoint:
No, technological advances in sports such as tennis, golf, and track and field have not supplanted the athletic achievements of the participants in those sports.

Sports have an intimate quality that makes the use of technology seem too intrusive. However, the use of technology in sports is not new and has led to many benefits for mankind. In fact, in some ways, these advances in technology help to justify the expense of time and money that we invest in sports. Probably the greatest and most persistent concern about technology and sports is that it diminishes the prestige of athletes by blurring the distinction between their accomplishments and those that are made possible because of the technology.

It certainly is the case that in the hands of a major leaguer an aluminum bat becomes a lethal weapon. Not only is an aluminum bat a more consistent tool than a wooden bat, but it also creates more force, allowing balls to be driven further and harder. In fact aluminum bat, although allowed in college play, is banned in most professional leagues, including the major leagues. Similar concerns have been raised about the differences created in tennis when larger metal rackets totally replaced wooden rackets at the professional level. And professional golf has put specific limits upon the materials that can be used for golf clubs.

It is unlikely that anyone would insist on professional football players going back to the days when they wore leather helmets on the field. Similarly, there is no move to prevent athletes from improving their nutrition or engaging in exercise, often with the most advance resistance equipment and monitoring instruments. So where does the controversy lie?

Objections to Sports Technology There are two situations in which the move to new technology is most opposed. First, when it is believed that the "integrity" of the game has been put in jeopardy. Often the aesthetics and fun of the competition are keenly dependent upon such things as the dimensions of the field, the bounce of the ball, the height of the net, and other factors that developed over time and have proven their value. The second objection to new technologies is based primarily upon concerns over the health of the athlete. For instance, performance drugs, particularly street level performance drugs, have already demonstrated that they can cause illness and shorten the life spans of athletes. However, there is no doubt that some performance drugs can provide a margin of victory. This use of chemicals is perhaps the most difficult technological question that sports faces today. Looking at the Olympics, we see that there has been a continual struggle to set standards of fairness and safety, yet there is strong evidence that some athletes continue to seek an edge in this way, and, in fact, may be a step ahead of those trying to test them.

There is no question that integrity of the game argument holds in many cases. The same sort of caution should be shown in the introduction of new technology as when modifications are made, say, to the size of the strike zone in baseball or in the limits to roughness allowed in football. In cases where it could damage the aesthetics of the game or too abruptly separate today's athletes from past players, it should be rejected.

And while it is reasonable for athletes to assume some risk in their professions (in fact, it is expected as part of the game in many events), there are levels of unreasonable risk, and there are occasions when risk can be easily prevented. In the cases of performance drugs as they exist today, it is reasonable to protect athletes from the economic pressures that might force them to submit their bodies to unreasonable risk. On the other hand, we have many instances where technology can and does reduce the probability of injury to athletes. For instance, the chances of injuring one's arm in tennis go up significantly if one uses an old wooden racket. And it's obvious that the protective equipment we have today is of great benefit to athletes. No one would want to go back to playing baseball without batting helmets, or playing football without proper padding.

Injuries can be further minimized by the use of ergonomics, sports psychology, and sports medicine. For instance, a better understanding of ergonomics can help the pitcher move more fluidly and spare his body from injuries, even crippling ones. Sports psychology can help athletes perform better, and sports medicine has given us new approaches to treating injuries and to maintaining physical health.

Value of Sports and the Role of the Athlete

To show that technology has supplanted athletes would require proving that the value provided by athletes has shifted in some way to technologies or away from the sport. So what value does sports promise? While different people might appreciate sports for different reasons, clearly sports provide entertainment, celebrity, instruction and development, commerce, aesthetics, and stories. Is the athlete's role in any of these degraded by technology?

Entertainment. Overall, the public's attention to sports, as measured by their investments in time and money, has never been higher. The status of individual sports may wax and wane depending upon a number of factors. One of these, of course, is how much fun the sport is, and this may be related to how closely it connects with daily life. For instance, it has been observed that in times of peace Americans tend to gravitate toward baseball, while in times of war they tend to gravitate toward football. Probably the most dependable predictor of whether a specific sport will be of interest is how well it keeps up with the pace of daily life (one reason given for the growing popularity of soccer over baseball among the young). Life's pace has continued to increase, and technology has helped to increase the pace of sports with faster serves, better runs, and quicker race cars. Technology also helps athletes demonstrate their full potential, so all manner of records are continually broken. Judging from the press attention to such feats, it would appear that these are appreciated by the public, and that they add to the excitement of sports.

Celebrity. Sports provides us with people to identify with, cheer for, and boo. Often our greatest attraction is to people that we know, people that we essentially share a history with. Technology can often displace these people if they do not respond well to the changes in the game. (One crisis in golf came when a younger generation quickly overcame the older, familiar players, and gate receipts declined.) However, it can also lengthen careers as better conditioning, new surgical treatments, and better medication make it possible for highly skilled athletes to have long careers. With effects on both sides of the ledger, it might be difficult to determine whether technology, overall, degrades the star power of athletes.

Instruction and Development. Sports often provides instruction in many areas. People often come to a better understanding of the possibilities for improving their health by observing sports and hearing details of sports treatments and training. Similarly, sports psychology has helped people to appreciate the connection between mental health and achievement, if not happiness. Strategists, particularly in business, have often relied on sports to provide analogies. And much of the original intent of sports was as

Olympic gold medalist Marion Jones in a full-body running suit.
(© Duomo/Corbis. Reproduced by permission.)

practice for warfare and other conflicts. This may, in some cases, still have value today.

Given the pressure for success in sports, it is not surprising that advances in medicine, nutrition, and therapy have their origins in the world of sports. Since sports can accelerate the toll that life takes on the human body—wearing down joints and demanding more of muscles and ligaments—sports medicine has found particular application among the large and growing population of older people.

Sports also provides a market that supports the development of new materials and the study of the physics of athletic equipment. Helmets have been improved in football, and safety equipment is honed to perfection in auto racing.

Commerce. Sports, even amateur sports today, is big business. The sales opportunities for athletes who have been on the scene for a long time and who can sell to his or her peers are enhanced. In addition, the updating of equipment, which can be recommended by athletes, can make sports more lucrative for athletes. This has been seen especially in the case of athletic shoes. Technology here can support the extension of careers, and, in a world where wealth is often admired, raise the respect for athletes. In can also damage the images of athletes who might be seen as having "sold out" or who have extravagant lifestyles that seem wholly unlike those of their fans, making them difficult to identify with. However, the role of technology here would seem to be minimal. Ticket prices and strikes are the main targets of fan disapprobation when commerce lays too heavy a hand on the game.

Aesthetics. At its best, sports is an artistic endeavor. The motions of athletes challenge those of the best dancers. The sounds of the crowd or the crack of a bat or even the gasp of the player surging over the goal line can rival music. The interaction between players on the field can be both intricate and intellectually satisfying. It is difficult to quantify the effects of technology on the aesthetics of sports. Certainly, there is no artistic merit to a performance that has been compromised by equipment failure.

There is a large overlap between aesthetics and the "integrity" of the game. We want to know that the achievements are real, not faked. Elegance should not come from fancy camera work or hidden wires. It must be about the body and the mind, and it must be connected to the traditions of the game. Replace baseballs with over-sized golf balls, and you'll get more home runs. But one would miss the timing, rhythm, and judgment of a "real" home run. And the game itself will no longer be the one that Babe Ruth played. Have sports ever stepped over the line? The argument can certainly be made that they have. But not far and not often. As much as synthetic grass fields are disparaged in baseball, they have never been as controversial as a non-technical change, the introduction of the designated hitter. Free substitution made more significant changes to football than any amount of body armor. The best evidence that the integrity of sports has, overall, been maintained is that history is an important point of comparison in the world of sports, and fans continue to cheer for the achievements of the best.

Stories. There may be no greater contribution of sports than the stories it tells about itself and about us. The triumphs and tragedies of our heroes in sports provide us with lessons on how we should or should not lead our lives. They also inspire us and sometimes provide guidance for our lives. Just by virtue of the careers it lengthens, technology changes the nature of these stories: there are more novels, fewer short stories.

Against these positives is, again, the question of the integrity of the game. Is this drama true? Or is the game "fixed?" The 1919 Black Sox scandal stills lives in memory because the public wants, even needs, an honest game. A technical marvel, such as an advanced rudder, might take the America's Cup to Australia, but a resulting movie script, if honest, would more likely deal with the engineers than the sailors. It is possible to take the athletes out of the story with technology that actually changes the game, and it is the responsibility of the commissioners, referees, and judges to keep that from happening. Based on the reactions of fans, they are doing a pretty fair job, and the appreciation of athletes' accomplishments seems to be secure.

There are many ways that a sport can lose its currency with fans: It can be replaced by another sport that better reflects the spirit of the times. It can lose the fans' trust and affection through corruption, the attitude of athletes, or disaster (such as the death of a popular athlete). It can even be outlawed, as has happened with cockfighting and has been threatened with contact sports from time to time. But the tools athletes use and the advances they take advantage of in medicine, training, nutrition, and psychology have not presented any serious threats to sports. As long as the integrity of the game and the safety of athletes are not seriously violated by technological advances, athletes will still be the heroes of the games. —PETER ANDREWS

Further Reading

Akins, Anthony. "Golfers Tee Into the Future." *The Futurist* (March/April 1994): 39-42.

Canfield, Jack, ed. *Chicken Soup for the Sports Fan's Soul: 101 Stories of Insight, Inspiration and Laughter from the World of Sports.* Deefield Beach, Fla.: Health Communications, 2000.

Coakley, Jay J. *Sport in Society: Issues and Controversies.* New York: McGraw-Hill College Division, 1997.

Cox, Richard H. *Sport Psychology: Concepts and Applications.* New York: WCB/McGraw Hill, 2000.

Froes, F. H. "Is the Use of Advanced Materials in Sports Equipment Unethical?" *JOM* no. 2 (1997): 15-19.

Haake, Steve. "Physics, Technology and the Olympics." *PhysicsWorld* no. 9 (September 2000).

Micheli, Lyle J., and Mark D. Jenkins. *The Sports Medicine Bible: Prevent, Detect, and Treat Your Sports Injuries Through the Latest Medical Techniques.* New York: HarperCollins, 1995.

Schrier, Eric W., and William F. Allman. *Newton at the Bat: The Science in Sports.* New York: Charles Scribner's Sons, 1984.

"Sports Technology Hotlist." <http://www-white.media.mit.edu/~intille/sports-technology.html>.

LIFE SCIENCE

Should the threat of foot-and-mouth disease be met by the destruction of all animals that might have been exposed to the virus?

Viewpoint: Yes, all animals that might have been exposed to the foot-and-mouth disease virus should be destroyed.

Viewpoint: No, animals suspected of being exposed to the foot-and-mouth disease virus should not be destroyed because the threat of the disease to humans is minimal.

Foot-and-mouth disease (FMD) is a highly contagious viral disease that affects cloven-hoofed livestock animals like pigs, cattle, sheep, and goats. The foot-and-mouth disease virus (FMDV) is a member of the picornavirus family, which includes important human pathogens, such as poliovirus, hepatitis A virus, and rhinovirus. These viruses are characterized by a small RNA genome. Usually, the virus is transmitted by contaminated passive vectors, such as people and farming equipment and implements that have come in contact with infected animals. Because the virus is found in the upper respiratory tract of infected animals, aerosols may also play a role in the transmission of the disease. The movement of livestock, or livestock products, however, seems to be the most important factor in the management and control of the disease.

During the nineteenth century microbiologists used the term "virus" in a general sense to describe pathogenic agents with infectious properties. Eventually, the term was used in reference to submicroscopic infectious agents. Because the agents associated with certain diseases could be separated from visible microorganisms by special filters, they were often called "filterable, invisible viruses." Several important diseases of humans and animals were thought to be caused by these mysterious "filterable" and "invisible" microbes. Studies of a disease of the tobacco plant led to the hypothesis that certain disease were caused by entities that needed the living tissues of their hosts in order to reproduce.

Foot-and-mouth disease provided the first experimental demonstration of a filterable viral disease of animals. Friedrich Loeffler (1862–1915) and Paul Frosch (1860–1928) were able to culture bacteria from the fluid in lesions in the mouths and udders of sick animals. In 1897 they discovered that the filtered fluid could transmit the disease to experimental animals and that experimentally infected animals were capable of transmitting the disease to other animals. These experiments suggested that a living agent, rather than a toxin, must have been in the filtrate.

FMD was introduced into the Americas in 1870. FMD was reported in certain parts of the United States, Argentina, Chile, Uruguay, and Brazil. By the beginning of the twentieth century FMD had spread to the rest of Brazil, Bolivia, Paraguay, and Peru. During the 1950s the disease was introduced into Venezuela, Colombia, and Ecuador. Since an outbreak of FMD in 1929, the United States has not had a confirmed case. The last outbreak in Canada took place in 1952. Outbreaks of FMD occurred in Mexico in 1947, but no cases have been reported since 1954. The establishment of the Pan-American Foot-and-Mouth Disease Center (PANAFTOSA) in 1951 has been an important factor in the battle against FMD in the Americas. PANAFTOSA developed a plan for FMD eradication, which included vaccination, and established a highly successful surveillance system. Epidemiologists believe that this model could be used to control and eradicate other diseases as well.

The magnitude of the 2001 FMD epidemic was quite different from previous modern outbreaks, including an outbreak in Britain in 1967 and another in 1981. Many authorities thought that the disease had been effectively controlled, or perhaps even eliminated, by the end of the twentieth century. The outbreak of 2001 began in February with reports of FMD in pigs at slaughterhouses in England. The first outbreak was confirmed on February 20, and livestock movement restrictions were imposed on February 23. Despite harsh measures instituted to stop the spread of the virus, the disease quickly spread to many locations in Britain and to many European nations. Millions of cattle, sheep, and pigs were destroyed and the export of British meat, milk, and livestock was banned. These expensive measures devastated the British farming industry, which was already depressed because of damage caused by outbreaks of swine fever and international fear of mad cow disease. Other measures affected people who were not directly involved in agriculture, as the government banned parades, fox-hunting, horse racing, visits to farms, fishing in streams, and hiking through the countryside, forests, and bird sanctuaries. Whether or not the wholesale destruction of animals that might have been exposed to the virus is warranted remains a subject of active debate.

Although vaccines for FMD were available before the 2001 outbreak, European governments have opposed routine vaccination because vaccinated animals test positive for FMD antibodies. Vaccination for FMD, therefore, makes it impossible to qualify for official disease-free status, which results in trade restrictions. Some of the experimental vaccines developed in the 1990s might allow the authorities to distinguish between vaccinated and infected animals.

The danger that FMDV poses to humans has been the subject of considerable debate and uncertainty. During the 2001 outbreak in Europe many reports assured the public that FMD is generally not regarded as a threat to humans who might consume meat or pasteurized milk from affected animals, although it was conceded that humans might contract the disease through close contact with infected animals. Some epidemiologists argued that the threat to humans must be evaluated more cautiously, even though the occurrence of the disease in humans is clearly quite rare. There is some evidence that FMD could be spread to humans through infected milk and dairy products. Proven cases of FMD in humans have been reported in Europe, Africa, and South America, and virologists have isolated and typed the virus in more than 40 human cases.

Scientists are developing novel vaccines and more rapid diagnostic tests for FMD. New diagnostic tests could cut the waiting time for the diagnosis of FMD from two days to 90 minutes. Better diagnostic tests might limit the need for destruction of whole herds of animals. New vaccines might make it possible to distinguish between infected animals and vaccinated animals. Reacting to the serious and widely disseminated outbreak of 2001, epidemiologists emphasize the importance of recognizing the fact that the human residents of our "global village" must understand the dynamics of infectious diseases in the twenty-first century "global farm." —LOIS N. MAGNER

Viewpoint:
Yes, all animals that might have been exposed to the foot-and-mouth disease virus should be destroyed.

Foot-and-mouth disease is so contagious and such a serious threat that all animals believed to have been exposed to the virus should be destroyed. It was the only way to eradicate the 2001 outbreak in the United Kingdom (UK), which had spread to France, Ireland, and the Netherlands before being stopped.

Foot-and-mouth disease (FMD), a highly contagious illness that has been likened to a biblical plague, is devastating to livestock herds. The disease gets its name from its symptoms: the virus causes blisters to form on the animals' tongues, lips, and nostrils and between their hoofs. The animals go lame, experience a tremendous amount of pain, stop eating, produce less milk, and are unable to fatten. Most could survive the disease but would be left debilitated and could not thrive. They would remain contagious and be an economic drain to their owners.

Outbreak in the UK In February 2001 the worst outbreak in history devastated the livestock industry in the UK and threatened the livelihood of farmers in Europe. Outbreaks were reported in 300 British locations and in France, Ireland, and the Netherlands. Outbreaks also occurred on the continents of South America, Africa, and Asia. To prevent the further spread of FMD and in an attempt to eradicate the disease and preserve its meat export business, over six million cattle and sheep were destroyed in the UK at a cost of hundreds of millions of dollars.

Because the disease is so contagious, authorities in the UK had no viable alternative but to destroy not only infected animals, but also those believed to have been infected. The authorities had no real choice other than killing all suspected animals, because the FMD virus can be transmitted from animal to animal via contaminated humans or their clothing, through contaminated equipment or feed, or even blown through the wind. Nearly any cloven foot animal—cattle, pigs, sheep, and goats as well as wild animals including deer, antelopes, llamas, camels, giraffes, and elephants, but not horses—exposed to FMD will develop the infection. Once an animal is infected, the virus reproduces and spreads to the animal's mouth, feet, stomach, and teats as well as to the heart muscle. The incubation period can last for up to three weeks, a time when the animal is highly contagious yet demonstrates no symptoms.

How do you fight an invisible viral enemy that might float in the wind or be carried on someone's shoes? After less intensive methods failed to slow the disease, the UK instituted draconian methods, an aggressive policy of control known variously as "ring culling" or "slaughter on suspicion." All cloven hoof animals (except horses) within 3 km (1.9 mi) of a diagnosed case of the disease were sacrificed, and their carcasses were buried or burned.

It was a painful decision for the stakeholders involved, yet even leading farmers' organizations supported the government's mandate. At a March 2001 press conference announcing his support for the government policy, Ben Gill, president of the UK's National Farmers, told news reporters, "The pain and anguish members . . . are going through now with the announcement of all livestock within three kilometers of an outbreak are going to be slaughtered is a death penalty for their businesses that is so devastatingly cruel it is beyond description. I am absolutely heartbroken for them, but . . . we have to get rid of it."

UK prime minister Tony Blair, who ultimately made the decision to rigorously cull the animals, declared, "There is no alternative to this policy." Nick Brown, the UK minister for agriculture, stated repeatedly in the House of Commons that the government's policy was based on "The latest scientific and veterinary advice. Killing healthy sheep and pigs in a three-kilometer zone around infected sites in the worst affected areas is vital. This is necessary for the whole of the livestock sector. It is not being

KEY TERMS

BOVINE SPONGIFORM ENCEPHALOPATHY: Fatal neurological condition in cattle, commonly called "mad cow disease."

CLOVEN FOOT: Divided hoof of cows, sheep, and other ruminants (animals that chew a cud and have a stomach containing four chambers).

CULL AND RING CULLING: To cull is to pick out substandard, weak, or sick members of a group or herd. Ring culling, the practice of destroying even apparently healthy animals in a 3-km ring around the central point where an infected animal was confirmed, was instituted by the British government in March 2001 to combat FMD.

CULLING: Slaughter of a selected group of livestock in an effort to stem the spread of disease.

DIOXINS: Group of highly toxic compounds that can increase the risk of cancer and cause a severe skin disease called chloracne.

DRACONIAN: Harsh or very severe measures. Named after the Athenian lawgiver (621 B.C.) who drew up a code of laws prescribing fixed punishments.

GENERATION INTERVAL: Time between the infection of one animal and its ability to infect other animals.

RING VACCINATION: Vaccination of livestock in a several-mile radius around an infected farm to create a firebreak against disease.

STAMPING OUT: Slaughter of infected animals along with other livestock on the same farm in an effort to stem the spread of disease.

VIRUS: Submicroscopic agent smaller than bacteria. Composed of a core of RNA or DNA and an outer coat of protein. Viruses grow and multiply within living cells and are able to infect almost all types of organisms. Besides FMD, viruses cause mumps, rabies, and a host of other ills including the common cold.

done for any other reason than to follow the explicit veterinary advice on what is necessary to control the disease."

Vaccination Some critics of the slaughter on suspicion policy have argued that vaccination against the disease might be a viable alternate. But critics of the vaccination approach point out that vaccination cannot alone completely eliminate the disease. Furthermore, because the vaccine only confers about six months of immunity, an immense mechanism would have had to be established for continual repeat vaccinations. Those against the widespread use of vaccines cautioned that such intensive use of antiviral serums would, in turn, pose an increased risk of the virus escaping from laboratories. They also

A funeral pyre of livestock carcasses in the United Kingdom during the 2001 outbreak.
(Photograph by Michel Spinger. AP/Wide World Photos. Reproduced by permission.)

warned of the danger of further FMD spread if the vaccine was administered improperly.

"Moves to vaccination would be a diversion of resources and could mean the disease was not completely stamped out," stated Dr. Alex Donaldson of the Institute for Animal Health. A computerized study later backed up Donaldson's opinion. Ferguson et al. did a computerized mathematical model of the transmission of FMD that simulated the impact of various control strategies. Writing in *Science,* they endorsed the preemptive slaughter of the animals, saying, "Culling is predicted to be more effective than vaccination."

Indeed, objective evidence and recent experience indicates that the vaccine approach falls short, whereas "slaughter on suspicion," although painful, works. For example, both Japan and South Korea suffered FMD epidemics in the year before the British outbreak. Japan instituted and stuck with a vigorous culling program and regained its FMD disease-free status by 2001. South Korea, however, chose to vaccinate. By mid-2001 South Korea still was not disease free, according to Donaldson.

The experience of Greece, in contrast, supports vigorous destruction of all animals believed to be exposed. Greece, the last European Union country to suffer an FMD outbreak before the UK was hit, chose not to vaccinate when FMD struck there in 2000. By 2001, after a vigorous slaughter on suspicion program, Greece was certified disease free.

That is the ultimate goal of the ring culling program. The United States and other countries do not allow the import of meat or animals from countries not certified as disease free. Billions of dollars are at stake.

"FMD is the most infectious animal disease and its eradication is of crucial importance, . . . not only for the farming industry in terms of resumption of normal trade, including experts, but for all parts of the rural economy," emphasized UK Secretary of State Margaret Beckett in a May 2001 notice to the public calling for continued cooperation to prevent the spread of the disease.

Possible Threat to Humans Animal FMD should not be confused with hand, foot, and mouth disease, a human infection fairly common in children caused by an entirely different virus. A telling argument for the slaughter on suspicion program, however, is the fact that animal FMD has jumped the species barrier and infected humans. Although health authorities maintain that FMD causes no real danger to humans, a review of the medical literature conducted by Paul Greger, a physician in Jamaica Plain, Massachusetts, revealed over 400 reported cases of FMD in humans during the twentieth century. A 1923 report told of one patient who developed a serious heart infection, but the vast majority of the cases were considered mild. Most people recovered within two weeks. However, the results of the disease must have been uncomfortable. There were reports that the "skin of [the] soles peeled off like sandals, in one piece, in some of those people," Greger said.

Conclusion In summary, all animals believed to have been exposed to the virus should be destroyed. The policy puts the animals out of their misery and has the highest odds of eradicating even large numbers of localized FMD outbreaks. In the long run, the policy may also be protective of human health, because the disease has crossed the species barrier. —MAURY M. BREECHER

Viewpoint:
No, animals suspected of being exposed to the foot-and-mouth disease virus should not be destroyed because the threat of the disease to humans is minimal.

The 2001 epidemic of foot-and-mouth disease (FMD) in the United Kingdom (UK) began in late February. By early July, nearly 3.5 million animals had been destroyed in an effort to halt the disease: 2.8 million sheep, 548,000 cattle, 129,000 pigs, and 2,000 goats, according to statistics from the UK Department for Environment, Food and Rural Affairs (DEFRA).

Few scientists dispute that the limited slaughter of infected and exposed animals is a vital first line of defense against the disease. For many, however, destruction on this massive scale is deeply upsetting. More than that, it may be ineffective at stemming widespread epidemics. Over four months after the mass slaughter started in the UK, for example, new confirmed cases were still cropping up at an average rate of three per day. Large-scale slaughter also may be economically counterproductive. By some estimates, the economic benefits of the UK policy may be far exceeded by the costs to the government for compensating farmers and disposing of carcasses and the costs to society at large for lost revenues from farming, farm-related businesses, and tourism in the affected regions. In addition, the disposal of such a large number of carcasses presents its own problems. Burning them can release highly toxic compounds called dioxins into the air, and burying them could lead to the pollution of water supplies. Finally, and perhaps most importantly, the disturbing specter of animals being killed by the tens of thousands and burned in massive pyres may be unnecessary. Today, many scientists believe a combination of limited slaughter and mass vaccination is a better way of fighting FMD.

The Culling Fields Some animal lovers might question whether any animals at all need to be destroyed to combat a disease that is itself rarely fatal. FMD usually runs its course in two or three weeks, and most animals recover. Even after recovery, though, the animals may grow more slowly, making it more expensive to bring them to market for meat. In dairy cattle, the ability to produce milk may be markedly decreased. Because the disease is highly contagious, infecting nearly 100% of exposed animals, the result of an uncontrolled outbreak could be a devastating loss of valuable meat and milk. As a result, most scientists concede that it is necessary to destroy infected animals along with other livestock on the same farm, a process known as stamping out. The real debate is over whether to move beyond the farm to destroy apparently healthy animals from nearby areas that *may* have been exposed to the FMD virus—but then again, may not.

In the UK, the slaughter of a selected group of livestock, known as culling, was applied to well animals on the same farm as infected ones as well as those on neighboring farms. In some of the hardest hit areas, such as the county of Cumbria in northwest England, the cull was extended to all livestock within a 1.9-mi (3-km) radius of infected farms. Proponents say a wider cull was needed in these hot spots because the FMD virus travels so easily. It can be spread not only directly by contact with an infected animal, but also indirectly by contact with people, clothing, vehicles,

FMD AND THE UNITED STATES

The United States has been free of foot-and-mouth disease (FMD) since 1929, and efforts are under way to try to keep it that way. In 2001, when the epidemic of FMD first broke out in the UK, then spread to a few other European nations, the United States quickly banned imports of livestock and related products from the entire European Union. As of July, that ban had been lifted for countries where no cases of FMD had been reported, but it was still in place for infected countries.

Import bans are just the start, however. Although FMD is not a health threat to humans, people can carry the virus from place to place on their clothing, shoes, personal items, and even in their throats and noses. As a result, travelers entering the United States from an infected country are warned to take special precautions:

- Launder or dry-clean all clothing before traveling, and remove all dirt and soil from shoes. Any soiled luggage or personal items, such as watches, cameras, mobile phones, and laptop computers, also should be wiped clean.
- Avoid visiting farms, stockyards, animal laboratories, packing houses, zoos, and other animal facilities for at least five days before leaving the infected country. Also, avoid contact with livestock and wildlife for five days after arriving in the United States.
- Be honest with customs officials. Travelers arriving from an infected country are asked whether they have visited a farm, been near or around animals, or are carrying any live animals, meat, meat products, fruits, vegetables, plants, or other items that might harbor pests or disease. If the answer to any of these questions is yes, the person's baggage is x-rayed or physically inspected, and any prohibited items are taken and destroyed. Shoes worn around a farm are checked for dirt or manure, and dirty shoes are cleaned and disinfected before being returned to the person.
- For more information, travelers can call the U.S. Department of Agriculture's information line at (866) SAFGUARD.

—Linda Wasmer Andrews

---

equipment, dogs, and vermin that happen to be carrying the virus. This explains why movement restrictions also were imposed around infected farms. In addition, though, the virus can spread through the air. Critics say this is one reason that the 1.9-mi cull was doomed to failure. The FMD virus has been known to travel up to 37 mi (60 km) over land and as far as 186 mi (300 km) over water. Some of the longest airborne transmissions on record occurred under weather conditions very similar to those found during the 2001 UK epidemic.

Another problem with a slaughter-only policy is the sheer number of animals that must be killed in modern farming regions very densely populated with livestock. The time between the infection of one animal and its ability to infect other animals, called the generation interval, is quite short for FMD. Studies have shown that sheep with the type O strain of the disease, the type involved in the 2001 outbreak, are able to infect other sheep and pigs in as little as two days. For culling to work, therefore, it must be started immediately and finished quickly. Unfortunately, the size of the UK cull soon overwhelmed authorities, causing them to fall well behind the government's own timetable for slaughtering animals and disposing of the carcasses. Two months into the crisis, the nation's agriculture minister admitted that the goal of destroying livestock on neighboring farms within 48 hours still was being met only 70% of the time. This time lag was another important reason that the cull alone was unlikely to stop FMD.

Money Matters It is no surprise, then, that the UK crisis soon turned into a catastrophe. The World Organization for Animal Health—also known as the OIE for its French name, Office International des Epizooties—is the international group charged with collecting data about FMD worldwide. It also decides which countries can claim its coveted FMD-free status. Most other countries will not import animals or meat from places without this designation. By early July 2001 more than 1,800 cases of FMD had been confirmed in the UK. The disease also had spread elsewhere in Europe, with a small number of cases reported in France, Ireland, and the Netherlands. As a result, these four countries all had their FMD-free status suspended, temporarily shutting off exports.

The potential economic losses from such a situation are staggering. Thus, once FMD occurs, the strategy used to get it under control is dictated as much by economic concerns as by animal health needs. The main alternative to mass culling is mass vaccination combined with smaller scale slaughter. European countries have resisted this option, however, because standard tests for FMD cannot tell the difference between antibodies from a vaccinated animal and those from an infected one. To be safe, other countries also ban the import of animals and untreated products from places that vaccinate. The chief argument against large-scale vaccination, then, is the cost in additional lost exports. In the UK, the best estimate of the value of threatened exports, adjusting for predicted changes in the domestic market, is $423 million (£300 million). The cost of a vaccination program itself would run about $21 million (£15 million), bringing the total cost of a vaccination-based strategy to about $444 million (£315 million).

Yet the economic toll of mass culling was probably much higher, according to Peter Midmore, an agricultural economist at the University of Wales in Aberystwyth, who made these calculations. Much of the income in rural areas of the UK comes not from farming but from tourism. This tourist trade was sharply curtailed by the travel restrictions and bad press that went along with the mass cull. By Midmore's estimates, the cost of lost tourism alone amounted to $846 million (£600 million). Other costs, including payments to compensate farmers, expenses for carcass disposal, and uncompensated losses from farming and farm-related businesses, brought the total cost of mass culling to about $1.1 billion (£781 million)—more than double what a vaccination-based plan might have cost.

The Dioxin Dilemma Clearly, mass culling is a far from perfect solution to FMD. It also creates a brand-new problem of its own: what to do with all those dead bodies. In the UK, many thousands of carcasses were burned in giant pyres. Residents were forced in some areas to live for weeks with the sight of smoke-filled skies and the stench of burning animals. Their greatest worry, however, was about the possible health effects of dangerous chemicals released into the air by the fires. Among these chemicals are dioxins, a group of highly toxic compounds that can increase the risk of cancer and cause a severe skin disease called chloracne. Some evidence indicates that dioxins may also affect reproductive health, weaken the immune response, and cause behavioral problems. Manufactured sources of dioxin in the air include fuel burning, waste incineration, the manufacture and use of certain herbicides, and the chlorine bleaching of pulp and paper. In the case of the FMD pyres, dioxins and other harmful chemicals were emitted by objects, such as wooden railway sleepers, coal, and old tires, used to light the fires. According to the UK government's own statistics, pyres lit during the first six weeks of the FMD crisis released 63 g of dioxins into the air—18% of the country's average annual emissions. It is still unclear whether this level of exposure to dioxins poses a public health threat.

Burial is another option for disposing of carcasses, but it carries its own pollution risks. The UK Environment Agency, the government agency that oversees environmental protection in England and Wales, says the main concern is the possibility that breakdown products of carcass decomposition might leach into groundwater over time. Potential pollutants include ammonia, chlorides, phosphates, fatty acids, and bacteria. Metal concentrations in the water might increase, and the taste and smell of the water might be affected. Once groundwater has become contaminated this way, it can be difficult or impossible to fix the problem. In addition, the carcasses of older cattle cannot be buried, because they might be infected with bovine spongiform encephalopathy (BSE), a fatal neurological condition in cattle commonly called "mad cow disease." BSE is thought to be caused by feeding cattle material derived from diseased sheep. People, in turn, seem to get a rare human brain condition, a form of Creutzfeldt-Jakob disease (CJD), from eating certain parts of meat from BSE-infected cows. Scientists say the danger of anyone getting CJD from drinking water contaminated by buried cattle carcasses is very low. For caution's sake, though, the UK government has prohibited the burial of cattle born before 1996, when a ban on risky material in cattle feed went into effect.

A third option for carcass disposal, and the one most favored by the Environment Agency, is processing the bodies in carefully monitored rendering plants. Scientists say rendering is likely to be very effective at destroying the FMD virus while causing less damage to the environment. During the 2001 outbreak, however, burning and burial still were practiced widely, partly because there were not enough rendering plants in some places to handle the huge number of carcasses produced by mass culling.

Vaccination Facts If mass culling is not the answer to FMD, what is? A growing number of scientists favor a program of mass vaccination combined with more limited slaughter. Although livestock from infected farms might still have to be killed, vaccination rather than culling could be used on animals in a several-mile radius around the farm to form a kind of firebreak against the disease. This strategy,

A poster from the U.S. Department of Agriculture warning about foot-and-mouth disease.
(Created by the United States Department of Agriculture. Reproduced by permission.)

often called ring vaccination, would greatly reduce the number of animals needing slaughter and the carcasses needing disposal. It would decrease the spread of FMD virus both from direct and indirect contact and through the air. Some people argue that, although vaccination might buy time, vaccinated livestock would still have to be slaughtered to protect a country's ability to export its animals and animal products. However, even if the vaccinated animals were not killed, European Union rules say an emergency vaccination program would mean a 12-month ban on exports, not a permanent one. Because exports are banned as long as the disease lasts anyway, an effective policy to end it quickly would make both scientific and economic sense.

Compelling evidence indicates that a vaccination-based program can work. When FMD broke out in Macedonia in 1996, for example, at least 18 villages in two districts were involved, and 120,000 cattle had to be vaccinated. Yet only 4,500 had to be destroyed, and the outbreak was controlled within three weeks. The main drawbacks to vaccination are the limitations of the vaccine itself. The vaccine used in this kind of emergency is potent, but it still can take up to a week to work. However, critics have noted that it took longer than this for authorities in the UK to mobilize a mass cull. Also, the vaccine is only effective for a period of months, and it only works against one particular strain of FMD. Nevertheless, this may be all that is needed to halt an epidemic.

The strongest argument against vaccination is that approved tests cannot distinguish between vaccinated animals and those with the disease. Yet that soon may change. The current FMD vaccine is made of deactivated live virus. Animals given the vaccine make antibodies to the virus's protein coat, but they are never exposed to other viral proteins produced when the virus infects a cell. In 1999 a scientific commission of the European Union concluded that experimental tests for noncoat proteins can tell the difference between vaccinated and infected animals 90% of the time. Although not perfect, a test with 90% accuracy is probably good enough to declare a herd uninfected. Also under development are experimental synthetic vaccines that likewise might make it easier to tell vaccinated animals from infected ones. With better diagnostic tests and vaccines, vaccination may become an even more attractive option in the near future. —LINDA WASMER ANDREWS

Further Reading

Barnhart, Robert K., and Sol Steinmetz. *Hammond Barnhart Dictionary of Science*. Maplewood, N.J.: Hammond, 1986.

BBC News. <news.bbc.co.uk>. A good source of current and past news about the UK outbreak of FMD.

"Current Prevention and Detection Efforts Examined." *Vaccine Weekly,* April 11, 2001, p. 13. Also via <NewsRx.com> and <NewsRx.net>. Item No. 4352473 (April 11, 2001).

Elm Farm Research Centre. <www.wfrc.com>. A vocal opponent of the slaughter-only policy in the UK.

"The English Patient." *Science Now,* April 26, 2001, p. 2.

Enserink, Martin. "Intensified Battle against Foot and Mouth Appears to Pay Off." *Science* 292 (April 2001): 410 (in News of the Week). (Available online; published April 12, 2001).

European Union animal health. <www.europa.eu.int/comm/food/fs/ah_pcad/ah_pcad_index_en.html>. The latest news about European outbreaks and the European Union's stance on vaccination.

Ferguson, Neill M., Christi A. Donnelly, and Roy M. Anderson. "The Foot-and-Mouth Epidemic in Great Britain: Pattern of Spread and Impact of Interventions." *Science* 292 (April 2001): 1155–60. (Available online; published April 12, 2001).

"FMD Shatters Argentine and Uruguayan Hopes for Increased Beef Exports." *International Agricultural Trade Report.* Washington, D.C.: U.S. Department of Agriculture, June 8, 2001.

"Foot-and-Mouth Compels Britons to Pursue Plan B." *Washington Times,* April 16, 2001.

"The Foot and Mouth Disease Epidemic: Is It a Threat to Children?" *Child Health Alert* 19 (May 2001): 1.

"Foot and Mouth—The Work Continues." News release by the UK Department for Environment, Food & Rural Affairs (DEFRA) issued June 11, 2001, and revised June 12, 2001. Available at <www.maff.gov.uk/animalh/diseases/fmd/new2s/statement.asp>.

MacKenzie, Debora. "Annihilate or Vaccinate." *New Scientist* 169, no. 2284 (March 2001): 16.

Midmore, Peter. "The 2001 Foot and Mouth Outbreak: Economic Arguments against an Extended Cull." <www.efrc.com/fmd/fmdtext/fmdecon.pdf>.

Mina, Mark. "Attachment to FSIS Constituent Update." Food Safety and Inspection Service. Washington, D.C.: U.S. Department of Agriculture, April 6, 2001, at <www.fsis.usda.gov/oa/update/040601att1.htm.>

The New Lexicon Webster's Dictionary of the English Language. New York: Lexicon, 1988.

"*Science* Paper Calls for Rapid, Pre-Emptive Slaughter Called 'Ring Culling' to Combat Outbreak." *Tuberculosis & Outbreaks Week* via NewsRx Network, April 24, 2001.

Sieff, Martin. "Epidemic Plagues Britain, Spares Blair." UPI, April 4, 2001.

"Summary of Current Policy on Foot and Mouth Disease." News release by the UK Department for Environment, Food & Rural Affairs (DEFRA) issued March 30, 2001, and revised June 15, 2001, available at <www.maff.gov.uk/animalh/diseases/fmd/disease/strategy/current.asp>.

Sumption, Keith. "Foot-and-Mouth Disease in the United Kingdom: Problems with the Current Control Policy and the Feasibility of Alternatives." <www.efrc.com/fmd/fmdtext/fmdvet.pdf>.

U.K. Department for Environment, Food, and Rural Affairs. <www.defra.gov.uk>. Government agency that coordinated the mass cull in the UK.

U.K. Environment Agency. <www.environment-agency.gov.uk>. Government agency that monitored the environmental effects of carcass disposal in England and Wales.

U.S. Department of Agriculture foot-and-mouth disease. <www.aphis.usda.gov/oa/fmd>. The latest information on import restrictions and other efforts to keep FMD out of the United States.

Are XYY males more prone to aggressive behavior than XY males?

Viewpoint: Yes, the best studies of XYY males indicate that they are more prone to aggressive behavior than XY males.

Viewpoint: No, the presence of the extra Y chromosome in XYY males does not in and of itself produce aggressive behavior in those affected; dealing with aspects of the condition during adolescence is a more likely explanation for any later social difficulties experienced by XYY males.

The debate about the XYY karyotype can be seen as part of the old debate about nature and nurture. The belief that nature, or biological determinism and inheritance, is more important than environment and education was the basis of the field that Francis Galton (1822–1911) called "eugenics." According to Galton, talent, character, intellect, disposition, and other aspects of "natural ability," as well as physical features, such as height and eye color, are governed by heredity. Similarly, a tendency to vice, alcoholism, feeble-mindedness, and criminality are inherited. Although eugenics became a disreputable concept in the first half of the twentieth century because of its association with involuntary sterilization laws and Nazi barbarities, the essence of Galton's premises were incorporated into human sociobiology after E. O. Wilson published *Sociobiology* in 1975. Those who supported the hypothesis of a direct link between the XYY karyotype and aggressive, even criminal, behavior tended to favor the old concept that "biology is destiny." In this case, "biology" was manifest in the chromosome number of XYY males.

Since the 1950s clinical cytogeneticists have discovered several major syndromes in which the number of chromosomes per cell nucleus differs from 46, the normal human chromosome number. The human chromosome complement consists of 23 pairs of varying size and shape. Normally, 23 chromosomes are inherited from each parent. In each set of chromosomes, 2 are known as the sex chromosomes (X and Y), and the other 22 pairs are known as the autosomes. Human females have two X chromosomes, and males have one X and one Y chromosome. Errors in chromosome patterns can occur during the formation of the egg or sperm or during embryological development. As a group, chromosomal abnormalities contribute to reproductive loss, infertility, stillbirths, congenital malformations, mental retardation, abnormal sexual development, mental retardation, and cancers. Chromosomal abnormalities appear to cause a significant fraction of all spontaneous abortions. At least 60 syndromes of varying severity are associated with specific chromosomal abnormalities. Abnormalities in chromosome number (a condition known as aneuploidy) in humans are usually attributed to maternal origin and increase with advancing maternal age.

Segregation of the chromosomes during meiosis, the mechanism that produces sperm or egg cells, is a complex process that appears to be quite error prone in humans. About 20% of eggs and a small percentage of sperm may be aneuploid; that is, they have a numerically unbalanced set of chromosomes. Most chromosome imbalances and abnormalities are so detrimental to development that affected embryos die during the first trimester of pregnancy. The exceptions include embryos with abnormal numbers of the sex chromosomes. Apparently sex chromosome anomalies are less likely to be spontaneously aborted than aneuploidy involving the autosomes.

Abnormal numbers of the sex chromosomes seem to produce less severe clinical manifestations than aneuploidy involving the autosome. Geneticists suggest this is due to the fact that one X chromosome is normally inactivated and the Y chromosome has very few genes. Although the X and Y chromosomes are very different in size and do not exchange regions during meiosis, they do share two small regions called "pseudoautosomal regions" at which they pair during meiosis. In addition to a considerable amount of so-called junk DNA, the Y chromosome contains a sex-determining factor. Aneuploidy of the sex chromosomes is fairly common. When considered as a group, the most common aberrations of the sex chromosomes (XXX, XXY, XYY) are found in about 1 in every 500 live births.

When clinical patterns associated with abnormal numbers of sex chromosomes were discovered in the 1930s and 1940s, these conditions were named after the physicians who first described them. Thus an extra X chromosome (47,XXY) results in Klinefelter's syndrome, and the lack of one of the X chromosomes (45,X) is known as Turner's syndrome. Since the 1960s, cytogeneticists have used the chromosome constitution to name chromosomal aberrations. When the chromosome constitution 47,XYY was discovered in 1961, individuals with this chromosome pattern were called XYY males.

About 1 in 1,000 newborn males has two Y chromosomes rather than one; this is known as the 47,XYY chromosome pattern (other names: 47,XYY karyotype, 47,XYY syndrome, diplo-Y syndrome, polysomy Y, YY syndrome). The 47,XYY chromosome karyotype is the result of errors that occur in the formation of sperm cells (paternal meiotic nondisjunction) or errors that occur when the fertilized egg begins to divide (postfertilization errors). Therefore, unlike many autosomal chromosome abnormalities, the XYY type is not associated with advancing maternal age. Advancing paternal age has little or no effect either. About 80% of males with two Y chromosomes have the 47,XYY chromosome constitution. Some men diagnosed as XYY actually have a "mosaic" chromosome constitution; that is, some of their cells are 46,XY and others are 47,XYY. Sex chromosome abnormalities can be diagnosed before birth by amniocentesis and chorionic villi sampling. During the 1970s and 1980s, reports associating the XYY chromosome pattern with criminal behavior led many parents to abort fetuses diagnosed with XYY during prenatal chromosome examinations. The frequency of induced abortions of XYY fetuses decreased as further studies cast doubt on the original hypothesis.

At birth, XYY babies are usually of normal birth weight and length. Their development is generally normal during early childhood, and they enter puberty at about the same age as their XY counterparts. Some studies indicate that the IQ of XYY males is generally within the normal range, but usually lower than that of their normal siblings. However, XYY males seem to have delayed speech development, and as children and young adults they may exhibit a tendency toward behavioral and learning problems caused by distractibility and hyperactivity. XYY males do not have an increased risk of diseases as boys or adults, except for severe acne, and life expectancy is normal. As adults, XYY males are generally taller and thinner than their brothers or fathers. In other respects, their appearance and fertility are essentially normal. Indeed, the level of testosterone (male hormone) is not significantly elevated in XYY males. Sexual development, secondary sex characteristics, and fertility are essentially normal, despite somewhat lower sperm quality. Studies of the sperm produced by XYY males suggest the extra Y is usually eliminated during the formation of sperm. Thus XYY fathers rarely have sons with two Y chromosomes.

When the XYY type was discovered, some researchers speculated that the presence of an extra Y chromosome might make a male more aggressive and prone to criminal behavior. The popular press referred to this condition as the "supermale" syndrome. Some early studies of prison populations and mental institutions seemed to confirm this hypothesis. But further research raised questions about the initial concept and methodology. Later studies of the general population, and close studies of the maturation of XYY individuals, cast doubt on any direct and simple linkage between the XYY type and criminal behavior. Initial reports of a slight deficit in mental abilities eventually gave way to findings of a wide spectrum of IQ scores. Researchers have attempted to define a specific psychological profile of XYY males, but some researchers argue that no specific behavioral characteristics can be related to the XYY chromosome pattern.

Early studies of XYY males suggested they were 10 times more likely than XY men to be found in criminal populations. XXY men and XXX women were said to be more commonly found among mentally retarded or psychotic patients than XY men and XX women. Popular science writers promoted the idea that men with an extra Y chromosome were more aggressive than XY males. Some reports claimed that the prevalence of XYY men in prison was at least 25 to 60 times as high as the prevalence of XYY males in the general population. This led to the belief that XYY males are more likely to commit acts of criminal violence. Richard Speck, the killer of eight student nurses, falsely claimed he was a "victim" of XYY syndrome. Later researchers argued that sampling errors caused by conducting such studies in mental hospitals and prisons had misrepresented the effect of anomalous chromosome patterns.

The medical literature of the 1990s suggested that although 1 male in 1,000 live births is an XYY male, most go through life undiagnosed. Apparently, most do not look or behave in a way that results in testing for chromosomal abnormalities. Carefully controlled longitudinal studies of individuals with various sex chromosome anomalies conducted during the 1990s suggested that previous reports of antisocial behavior and mental disorders were misleading because of bias and sampling errors.

As early as 1974, prominent geneticists Jon Beckwith and Jonathan King called the notion of a dangerous XYY supermale syndrome a dangerous myth. This idea was primarily based on assumptions about the tendency of males to be more aggressive than females and early studies of XYY males in prisons. Beckwith and King argued that the assumption a male with an extra Y chromosome would develop antisocial or even criminal behavior was social and medical folklore, not science. Nevertheless, although researchers have generally rejected the erroneous impressions created by early studies of XYY males in prison, researchers continue to suggest that XYY males as a group exhibit an increased tendency toward aggressive behaviors. Scientists continue to debate the possibility that further research will uncover and explain subtle effects of the extra Y chromosome. —LOIS N. MAGNER

Viewpoint:
Yes, the best studies of XYY males indicate that they are more prone to aggressive behavior than XY males.

The XYY syndrome is a genetic irregularity that gives the male an extra Y chromosome. All males inherit two "sex chromosomes" from their parents—an X chromosome from their mother and a Y chromosome from their father. The Y chromosome determines sex. In 1 out of about 1,000 males (XYY males), an extra Y chromosome is inherited.

XYY men (sometimes called supermales) have been of interest to psychologists and criminologists because of the suggestion that these males may be more aggressive and more prone to violence than males with a single Y chromosome. When evaluated by a battery of psychological tests, XYY males do exhibit more aggressive behaviors than XY males.

Flawed Studies Although some studies suggest (and many researchers imply) that XYY are more violent and less intelligent than normal males and, as a result, are more often institutionalized in mental hospitals and prisons, the best exploration of XYY male aggressive behavior looks only at measurable aggressive tendencies, not at end points such as criminal acts of violence and incarceration rates.

Studies of XYY men and criminal tendencies aim at examining violent and aggressive behavior in XYY males, and most start their research in prisons and mental institutions—places where they are bound to find such behaviors, whether exhibited by XYY males or XY males. These researchers are less interested in examining aggression and more interested in understanding the roots of criminal behavior. Generally, researchers do not find a greater number of XYY males in prison, and they use those findings to suggest the XYY male is *not* more aggressive than normal males. This thinking confuses cause and effect with correlation.

For example, renowned criminal behaviorist S. A. Mednick concluded after research on the criminal records of Danish men that XYY men are no more likely to commit crimes of violence than XY men. Although his conclusion might be valid, many researchers have used it to suggest XYY males are no more aggressive than XY males. The data, however, should not be extended to this conclusion. Incarceration rates and types of crimes committed are data that reflect incarceration rates and type of crimes committed. They do not reflect XYY men's aggressive psychological attributes and tendencies, many of which may be noncriminal in nature.

One measure of aggressive behavior outcomes may be violence and subsequent incarceration, but aggressive behavior (if it is a trait of XYY males) may not always have violence and prison as an end point. Thus the many studies that show the percentage of XYY males incarcerated for violent crimes is not significantly greater than XY males misses the point. Many factors can intrude on aggressive behaviors so that crimes are not committed and XYY men not imprisoned.

KEY TERMS

ANEUPLOIDY: Condition where a copy or copies of a chromosome are lost or gained so the total number is more or less than the normal 46.

AUTOSOMES: The chromosomes that are not sex chromosomes (i.e., those that are the same in both sexes).

CHROMOSOME: Structure composed of DNA (deoxyribonucleic acid) that conveys genetic information.

KARYOTYPE: Chromosomal constitution of an individual as seen in the nucleus of a somatic cell. For a normal human male the karyotype would be 46, XY (i.e., 46 chromosomes in total, including one X and one Y chromosome).

PHENOTYPE: Physical characteristics of an organism. Often not identical to the genotype because organisms can carry genes that are not expressed.

RORSCHACH: Psychological test in which participants explain what they see in an inkblot.

TAT (THEMATIC APPERCEPTION TEST): Written tests, verbal stories, and picture interpretation.

Testing Validity A better, more valid measure of XYY tendencies toward aggressive behavior can be examined with appropriate psychological testing employing outcomes suitable for valid testing. Validity in scientific testing is important. Validity does not mean accuracy. Validity means that a study or experiment tests what it is meant to test. Therefore, if we seek to find out if XYY men are more aggressive than XY men, aggressive behavioral tendencies—not acts of criminal violence and subsequent incarceration rates—should be the measure.

Thus when appropriate and comprehensive psychological testing is implemented, men with XYY syndrome are shown more likely to be aggressive and exhibit more aggressive behaviors than normal XY men.

Theilgaard Perhaps the best study of the psychology of XYY men and aggression was carried out by Danish researcher Alice Theilgaard and published in 1984 in *Acta Psychiatrica Scandinavica*. Theilgaard studied the personality traits of XYY men compared to normal men (XY) and men with another genetic variation, XXY, with an extra X chromosome inherited from their mothers. XXY men are reputed to be more female-like and much less aggressive than either XY males or XYY males. The most valuable aspect of Theilgaard's study with respect to the question at hand is comparing aggressiveness in XYY men with the XY male participants, called the control group here.

According to Theilgaard, the frequency of men with the XYY karyotype (the number, types, and forms of chromosomes in a cell) is about 1 in 1,000. She reported in her literature review that most studies of XYY men found them to be taller than average males. In medical terms, Theilgaard added that "no single characteristic except height . . . has been associated with the XYY condition." She noted that because XYY males are generally taller than average, they may exhibit behavior that is more aggressive than average.

Theilgaard also noted that studies of testosterone (the male hormone) levels have been contradictory. Some studies find that XYY testosterone levels are average; others have found them slightly higher than normal. Testosterone is thought to be associated with some typical male behaviors, and aggressiveness is one such behavior.

Some researchers, according to Theilgaard, have found that not only are XYY men taller, but they are more apt to be "clumsy" and slow witted, perhaps adding to their social frustration levels. Also at question has been if XYY men test lower on intelligence tests and whether lower intelligence has a correlation with aggressive, and perhaps violent, behaviors. Some studies, reported Theilgaard, showed that XYY men had "poor performance scores" on mental tests and showed a "lack of persistence and weak concentration power." Yet other studies showed "no overwhelming evidence to suggest that the IQ of XYY individuals is lower than average."

Personality profiles of XYY men in comparison to XY men are an important source of information. In her literature review, Theilgaard noted a 1970 study suggesting that the XYY male is "a difficult child, has problems in school, shows excessive daydreaming, is a loner, a drifter, presents unrealistic future expectations, and manifests impulsiveness with sporadic, sudden violence and aggression." Note that Theilgaard's literature review of relevant studies did not find research suggesting that XYY males were more often in prison than other males. Thus she sought to answer whether XYY men might be more aggressive than XY men. Consequently, in her study she attempted to document psychological characteristics to see if aggressiveness was a common trait among XYY men. To do so, she employed standard and appropriate psychological tests.

Among the tests she used were the TAT (Thematic Apperception Test), in which study participants write stories that reveal inner psychological states; tests that use word associations to indicate psychology; and other psychological profiling examinations to evaluate XYY men's potential for behaviors and ways of thinking.

In her study, Theilgaard tested both XYY and XXY men and XY controls. Her results after examining XYY males had many implications regarding the tendency toward aggressiveness and—more importantly—many of the psychological tendencies that may lead to aggressive behaviors.

For example, Theilgaard found that XYY men "give more aggressive and less anti-aggressive content in their TAT stories compared to their controls." She also found that XYY men are more "rigid" in their thinking. Generally, she found that on all tests XYY men score higher on "aggression" than their controls. Too, the "habitual mood" of XYY men is "judged to be more low and pessimistic" than that of controls. Also, XYY men show more "evasiveness" and report more "negative relations" with people. They live lives that are "disharmonious" with more "unstable feelings" than controls.

In addition, Theilgaard found that the quality of a XYY man's childhood was judged "less positive" by the test interviewers as well as by the XYY men themselves. She also found that XYY men get along less well with their "partners" and act more aggressively toward them. Theilgaard found that XYY men were more restless and impulsive than XY men, that they were more prone to "sporadic outbursts" with more immature, intense feelings. She says the typical XYY male appears "disharmoniously" integrated with an "unfinished personality." He seems "insecure, emotionally frustrated and perhaps a little blunt in his reactions toward others," concluded Theilgaard. Once more, the XYY male is more "rigid" in thinking and may find it harder to control himself. It may be that all of these tendencies result in more aggressive behaviors.

In terms of criminal behavior, she found that XYY men are arrested more frequently than controls, but their conviction rates were not higher. However, in the test of Aggression Against a Person (AAP), she found that XYY men show less "antiaggression" behavior and more often report being a "bully." The XYY group showed a "shorter reaction time on words with aggressive content," said Theilgaard, "and provided more aggressive content in the Rorschach (inkblot) test," a test where they are asked to describe their reaction to a shape made by an inkblot. "Positive correlations are seen (in the XYY group) between reported aggression against persons and aggressive content in TAT," wrote Theilgaard.

Why XYY males might have higher levels of aggression is unclear, however. Theilgaard reported that her study found higher levels of testosterone, the male hormone, in the XYY men she studied. Interestingly, she found that testosterone levels were also high among XY men who had been convicted of crimes.

Conclusions When looking at her conclusions with regard to aggression and comparing the literature of other researchers, Theilgaard wrote, "In the literature the richly faceted concept of aggression (applied to XYY men) has often been represented as a unitary idea, without attention being paid to behavioral, emotional, motivational or arousal aspects, and the emphasis has been on the destructive features while the positive side has been ignored."

She added that the word "aggression" comes from the Greek word *ad-gerdios*, meaning "approaching" and suggested the word be evaluated in its positive and negative aspects, which means adding achievement, leadership, dominance, and assertiveness to the more negative attributes of hostility and violence. She concludes that "consensus" on what aggression means has been hard to reach.

"This study," she concluded, "sought to let the unitarian concepts give way to more elaborated behavior constructs." What Theilgaard meant was that although she looked at XYY males and concluded they were, in fact, more prone to aggression than normal XY males, their aggressive tendencies did not necessarily end in violence, prison, or other forms of institutionalization.

"The complexity of the matter requires that the question of aggression is not viewed in isolation," wrote Theilgaard. That men with the XYY chromosome are more aggressive than normal XY men is supported by Theilgaard's psychological testing. But, in keeping with Theilgaard's request, we must recognize that aggression has many forms, with both positive and negative outcomes.

Although many studies have concluded that XYY men are *not* more aggressive than XY men, most of these studies have looked at prison and mental institution populations and have identified XYY men in these settings at only a slightly greater rate than would be expected. These studies may be flawed because they are self-fulfilling: they look at prison populations for aggressive and violent men and—not surprisingly—find them. That there are a few more XYY men among these populations is not too surprising, given some of the XYY psychological characteristics Theilgaard identified.

Data that reveals not greatly inflated rates of XYY men in prison can be used to suggest that XYY men are not more aggressive than XY men, but this conclusion does not account for the great variety of aggressive behaviors in XYY men that may incarcerate them. As Theilgaard noted, "aggressiveness" does not necessarily translate into violence and criminal behavior.

When she tested personality rather than incarceration rates, Theilgaard found that yes, XYY males have aggressive tendencies, that the

XYY male is more aggressive than XY males, but their aggressive tendencies do not necessarily manifest in violent behavior toward other people.
—RANDOLPH FILLMORE

Viewpoint:
No, the presence of the extra Y chromosome in XYY males does not in and of itself produce aggressive behavior in those affected; dealing with aspects of the condition during adolescence is a more likely explanation for any later social difficulties experienced by XYY males.

Reports of a biological basis for antisocial or criminal behavior always raise considerable interest among both the public and the scientific community. There are several reasons for this. First, criminals and their defenders hope to gain acquittal or reduced sentences by pleading mitigating circumstances, suggesting the defendant was not responsible for his or her actions. Second, there may be a wish to absolve society in general of the responsibility for creating criminals or a strong need to believe that humans are basically good and antisocial or criminal behavior is caused by influences beyond our control. And finally we may also hope that if a biological or physical basis for criminal behavior is found, there may one day be a cure for it. Although the idea of a biological basis for criminal behavior is not new, stretching back at least to Francis Galton's theory of eugenics in the mid-1800s, the suggestion of a link between the XYY genotype and aggressive behavior was the first to provide a chromosomal demonstration.

XYY Syndrome Also known as polysomy Y, XYY syndrome occurs in males who, instead of the usual 46 chromosomes (including one X and one Y chromosome), have 47, due to the presence of an extra Y chromosome. This type of chromosomal anomaly is known as aneuploidy of the sex chromosomes. Such aneuploidies, which can involve either the X or Y chromosome and a varying number of extra chromosomes, are the most common type of chromosomal disorders, occurring in around 1 in 500 live births. Aneuploidies of the sex chromosomes, especially those involving the Y chromosome, generally show less serious clinical symptoms than do abnormalities of autosomes, because of the restricted and specialized functions of the sex chromosomes. However, some studies suggest that the loss of balance of the gene products of the sex chromosomes can cause an increased susceptibility to other genetic or environmental factors. This susceptibility can lead to behavioral problems or the beginning of mental disorders in childhood.

XYY syndrome is not always visible in the phenotype. Affected individuals may show no signs of the condition and never be detected, unless their chromosomes are assessed or karyotyped for another, unrelated reason. However, there are some characteristics that are often found in XYY males. They are frequently very tall, compared with both the general population and other people in their own family. Over 40% suffer from a very severe form of acne in adolescence. Skeletal abnormalities are sometimes found in XYY males, but these are not normally visible and have little or no impact on the affected person. Intelligence is often lower than average, although this is not inevitable and some XYY males have a normal to very high IQ. Some affected individuals have learning problems, such as delayed speech and difficulty with articulation. EEG traces show significantly slower background activity than those of similar aged nonaffected males. This is normally a characteristic of earlier stages of development, suggesting a developmental lag in XYY individuals.

In the general population, the incidence of XYY males is around 1 in 1,000. Among tall males, this rises to 1 in 325 and among tall males in prison as high as 1 in 5, according to some studies.

History The first recorded case of an XYY male was reported by Avery Sandford and colleagues in 1961 at the Roswell Park Memorial Institute in Buffalo, New York. The individual concerned was karyotyped because his wife had given birth to a Down syndrome child, and investigators were looking for any chromosomal abnormalities.

Initially, it was suggested that, since the Y chromosome defines the male, and males are more aggressive than females, a double complement of Y would increase aggressiveness. To test this idea, in 1965 Patricia Jacobs and colleagues conducted a study of inmates of a Scottish maximum security hospital. The subjects of this study were defined as "mentally subnormal male patients with dangerous, violent, or criminal propensities." Jacobs found that 1 in 28 of these subjects were XYY. This high rate, compared with the incidence in the general population, seemed to suggest that the XYY condition could indeed be linked to increased violent or antisocial behavior. Studies of other similar populations confirmed Jacobs's findings. These results sparked fierce debate, especially in the United States, fueled by sensationalist reports in the media. A large number of professional people, committed to the idea that the environment is responsible for human development, were unable to accept the idea of a genetic component to human behavior, to the extent that individual researchers were harassed and, ultimately, research in the field was curtailed.

A later study by D. A. Price and A. J. Whatmore (1967) found a significant difference in the age of first criminal conviction for XYY males (age 13, compared with age 18 for controls). The XYY group had a much higher number of convictions for crimes against property (an average of 9.0 for XYY males as opposed to 2.6 for normal XY males) but a much lower number of convictions for crimes against the person (0.9 for XYY males and 7.7 for XY males). Criminality was much higher among the siblings of the XY males than the XYY males, suggesting the XYY condition overrides familial or environmental influences. The families of the XYY males also reported a history of behavioral problems from early childhood. However, these reports were given with hindsight and may not be entirely reliable.

A number of issues cast doubt on the reliability of such studies. Many used very small sample groups, with inadequate or mismatched control groups. Some were even based on a single case of extreme antisocial behavior. Many investigators drew their samples from groups considered likely to have a high proportion of XYY men—for example, tall institutionalized men—rather than a randomized group. Finally, the investigating psychologists who were assessing the personalities of the subjects often knew whether or not they were XYY at the time of assessment, which could have led to bias.

Witkin In 1976 a definitive study was done by Herman A. Witkin, S. A. Mednick, and Fini Schulsinger, and others. Using the comprehensive records of the Danish draft board, which included tests for intellectual function, educational history, socioeconomic background, and criminal history, Witkin was able to screen a cohort of over 4,000 men. The original sample was in excess of 28,000 individuals, but due to the time and expense involved in karyotyping, only the tallest 16% of the population were screened. The sample contained 12 XYY men (1 in 345), a considerably higher proportion than would be expected in the general population.

Witkin tested three possible hypotheses to explain the previously recorded data on increased criminality in XYY males. First was the aggression hypothesis, which proposed that an extra Y chromosome would cause increased aggression and hence increased criminal activity. The second hypothesis was that the increased height of XYY males would cause them to be perceived as more threatening, and so more likely to be suspected of crime and also more successful in aggressive encounters because of their greater size. The final hypothesis, known as the intellectual dysfunction hypothesis, was that the XYY condition causes decreased intelligence, which in turn leads to more antisocial behavior, plus an increased risk of being caught.

The results of careful analysis of the data showed that the XYY males were convicted of significantly more crimes than the controls. In fact, 42% of XYY males had criminal records compared with 9% of normal males. However, these were mainly nonviolent crimes such as vehicular offenses and shoplifting. For violent crimes against people, no difference was found between the XYY males and the controls. This does not support the aggression hypothesis.

Individuals in the XYY group were significantly taller than the control males, but among the controls, noncriminals were taller than criminals. The height hypothesis was therefore disregarded.

Both mean IQ and educational attainment were significantly lower in the XYY group than in the control males. In addition, among the controls, those with criminal records had a significantly lower IQ than those without. Both these findings support the intellectual dysfunction hypothesis.

Allowing for background variables—that is, comparing XYY males with XY males of similar IQ, height, and social status—the rate of criminality in XYY males is significantly higher than for normal males, although the difference is slight. The conclusion is that XYY males are more prone to criminal behavior, but only in crimes against property and not against the person. This elevated crime conviction rate may be mediated through reduced intellectual capacity.

Much other evidence indicates that an XYY constitution does not lead to aggressive behavior. In institutions, XYY males tend to show less aggression and better adjustment than normal males. Some studies have confirmed that XYY criminals commit fewer violent crimes than the mean for the general criminal population.

Much more recently, a study of over 31,000 men found no strong correlation between the presence of an extra Y chromosome and criminality (Mednick, 1988). In 1999 M. J. Gotz, E. C. Johnstone, and S. G. Ratcliffe studied a sample of over 34,000 infants screened at birth, comparing XYY males with matched controls. Gotz also found a significantly higher frequency of antisocial behavior in the XYY males in both adolescence and adulthood. The XYY males also had a higher rate of criminal convictions—again, of crimes against property rather than people. Complex analysis of the data shows that this effect is a result of lower intelligence.

What are needed are long-term studies of the development of XYY males from birth onward, with matched XY controls for comparison. However, such studies can be problematic. In the past, a prenatal diagnosis of XYY syndrome led parents to terminate a pregnancy. Fortunately, this rarely occurs today because the theory of increased aggressiveness has fallen from favor. Another problem lies in whether the parents should be told of the diagnosis. If families are aware of the karyotype of the child, they may treat him differently and confound the study. Not to inform parents of the karyotype raises a moral dilemma, since some researchers have shown that awareness of the diagnosis and sustained counseling and advice for families have been helpful in dealing with behavioral problems in XYY boys.

Conclusions Overall, the evidence suggests that XYY males may have certain behavioral and conduct problems, especially in adolescence. Such problems may be compounded by the individuals' unusual height and, in many cases, severe acne, both of which can cause acute distress to an adolescent. XYY males can show hyperactivity, impulsive behavior, and lack of emotional control. They often have a history of outbursts of temper when frustrated and crave immediate gratification. These characteristics, when combined with a lowered intellectual function and the accompanying lack of educational attainment, can lead to increased criminality. However, the crimes tend to be against property and rarely against people. XYY males can be difficult, but they are not generally dangerous. Awareness and appropriate therapy or counseling for both patients and families can overcome the problems and allow these men to lead normal lives. No sound evidence suggests that affected men are more prone to aggression or violence or that they should be absolved of responsibility for their actions, criminal or otherwise.

A prenatal or childhood diagnosis of XYY syndrome need not be cause for alarm. Many such children develop perfectly normally, with no ill effects from their condition. For some, there may be an increased risk of behavioral problems, but environment and upbringing likely will play a greater role in determining how much the individual is affected. —ANNE K. JAMIESON

Further Reading

Beck, W. S., K. F. Liem, and G. G. Simpson. *Life: An Introduction to Biology.* New York: HarperCollins, 1991.

Ciba Foundation Symposium 194. *Genetics of Criminal and Antisocial Behavior.* New York: John Wiley, 1995.

Court Brown, W. M. "Males with an XYY Sex Chromosome Complement." *Journal of Medical Genetics* 5 (1968): 341–59.

Gotz, M. J., et al. "Criminality and Antisocial Behavior in Unselected Men with Sex Chromosome Abnormalities." *Psychological Medicine* 29 (July 1999): 953–62.

Jacobs, P. A. "The William Allan Memorial Award Address: Human Population Cytogenetics: The First Twenty-Five Years." *American Journal of Human Genetics* 34 (1982): 689–98.

———, et al. "Aggressive Behavior, Mental Sub-Normality and the XYY Male." *Nature* 208 (December 1965): 1351–2.

Mednick, Sarnoff A., and Karl O. Christianson. *Biosocial Bases of Criminal Behavior* New York: Gardner Press, 1979.

———, et al. *Explaining Criminal Behavior.* New York: E. J. Brill Press, 1988.

Pallone, Nathaniel J., and James J. Hennessy. *Tinder-Box Criminal Aggression.* New Brunswick, N.J.: Transaction, 1998.

Price, W. H., and P. B. Whatmore. "Behavior Disorders and Patterns of Crime among XYY Males Identified at a Maximum-Security Hospital." *British Medical Bulletin* 1 (1967): 533–6.

Science 289 (28 July 2000). Special edition on violence.

Theilgaard, Alice. "A Psychological Study of the Personalities of XYY and XXY Men." *Acta Psychiatrica Scandinavica* 69(Suppl. 315) (1984).

Walzer, S., et al. "The XYY Genotype." *Annual Review of Medicine* 29 (1978): 563–70.

Witkin, H. A., et al. "Criminality in XYY and XXY Men." *Science* 193 (August 1976): 547–55.

Were dinosaurs hot-blooded animals?

Viewpoint: Yes, the most recent fossil evidence indicates that at least some dinosaurs were hot-blooded animals.

Viewpoint: No, the long-standing view of dinosaurs as cold-blooded animals is still the most compelling; recent evidence may suggest, however, that some dinosaurs had hybrid metabolisms, or aspects of both hot- and cold-bloodedness.

Naturalists have speculated about the nature of fossils since the time of Aristotle (384–322 B.C.), but it was not until the seventeenth century that these remarkable objects were generally recognized as the remains of ancient plants and animals. The work of the great French comparative anatomist and geologist Georges Cuvier (1769–1832) established the foundations of paleontology, the study of fossilized remains, and stimulated interest in the systematic study of fossils. Richard Owen's (1804–1892) investigations of reptilian fossils led to the discovery that some of these specimens were quite distinct from modern lizards. In 1842 Owen named the new group of large, extinct, reptile-like animals "Dinosauria," meaning "terrible lizard." During the second half of the nineteenth century, many new dinosaur fossils were discovered. Indeed, the 1870s were the beginning of a period known to paleontologists as the "Great Dinosaur Rush." The remains of the great dinosaurs were put on display and drew great public attention.

Since their discovery in the nineteenth century, dinosaurs have been portrayed as gigantic, stupid, sluggish, cold-blooded, lizard-like beasts that flourished in the hothouse world of the Mesozoic period before inevitably passing away and leaving the world to superior warm-blooded mammals. Actually, mammals and dinosaurs coexisted during the Mesozoic era, and, despite modern assumptions about the superiority of warm-blooded mammals, dinosaur predators and herbivores (animals that primarily feed on plant food) dominated the earth as long as they existed. Although the terms "hot-blooded" and "cold-blooded" are often applied to the controversy surrounding the nature of dinosaurs, these terms are not really part of the scientific vocabulary. The proper term for "hot-blooded" is endothermic; this refers to the ability to generate internal heat in order to regulate body temperature. In a cool environment, endothermic animals can maintain an average body temperature that is higher than that of their surroundings. Modern birds and mammals are endotherms. On the other hand, ectothermic animals, such as reptiles, rely on the environment and behavioral adaptations to regulate their body temperature.

A controversial new theory about the nature of dinosaurs was brought to public attention primarily through the work of Robert T. Bakker. Like many paleontologists, as a child, Bakker became fascinated by dinosaurs and their place in geological history after seeing dinosaurs in museums and magazine articles. In 1968, just after completing his undergraduate education at Yale, Bakker published "The Superiority of Dinosaurs," his first paper on dinosaur endothermy. Looking at the history of ideas about the evolution of mammals and dinosaurs, Bakker concluded that conventional notions of dinosaur physiology were wrong. Instead of slow-moving reptilian beasts, Bakker envisioned dinosaurs as "fast, agile, energetic creatures" that could run faster than humans. Although critics have labeled one of Bakker's major texts "Dinosaur Hearsay," his articles, lectures, and books, including *Dinosaur Heresies* and *The Great Dinosaur Debate*, have been instrumental in stimulating the "Dinosaur Renaissance" proclaimed by Bakker in a 1975 *Scientific American* article. Since the 1970s, the popular image of dinosaurs has been shifting from cold-blooded reptilian creatures to agile, dynamic, smart, hot-blooded animals. The new image of dinosaurs was reflected in the immensely successful movie *Jurassic Park*.

John H. Ostrom, Armand de Ricqlès, and other scientists less well known than the flamboyant Robert Bakker have also provided insights into dinosaur ancestry, metabolism, growth characteristics, and behavior. In 1969, Ostrom described a newly discovered theropod dinosaur, *Deinonychus antirrhopus*, as "an active and very agile predator." Ostrom, who was Bakker's advisor at Yale, was one of the founders of both the theory that birds evolved from dinosaurs and the hypothesis that some, if not all, dinosaurs were warm-blooded. Unlike the continuing controversy about whether dinosaurs were endothermic, the evolutionary relationship between dinosaurs and birds is now generally accepted. The question of which scientist was the first to propose the idea of dinosaur endothermy remains quite controversial.

According to Bakker, evidence for endothermy in dinosaurs appeared in the nineteenth century, but it was ignored because of prevailing assumptions about the reptilian nature of dinosaurs. In 1836 American geologist Edward Hitchcock (1793–1864) studied footprints (subsequently identified as those of dinosaurs) in Massachusetts and suggested that hawk-like creatures that ran made them. Although Richard Owen coined the term "terrible lizards," he seemed to suspect that the dinosaurs might be more warm-blooded than typical reptiles.

Since the debate about hot-blooded dinosaurs began, observers have noted that the case for ectothermy in dinosaurs has generally been taken for granted, rather than demonstrated on the basis of specific evidence. That is, since the ancestors of the dinosaurs were cold-blooded reptiles, the "terrible lizards" were assumed to be ectotherms that functioned quite well because of the warm, mild climate of the Mesozoic era. More recent studies of the climate of the Mesozoic period suggests, however, that the climate was more varied that previously expected and that dinosaurs lived in areas that were fairly cold, such as polar regions. Proponents of the warm-blooded dinosaur hypothesis argue that the discovery of dinosaur fossils at high latitudes supports the theory because ectotherms generally do not live in such cold climates. Nevertheless, modern snakes, lizards, turtles, and crocodiles are quite widely distributed. Moreover, areas in Alaska, Antarctica, and so forth, though cold now, were probably much warmer during the Mesozoic Era. Dinosaur fossils, including hatchlings and juveniles, from the Late Cretaceous (the latest period of the Mesozoic Era) have been found along the North Slope of Alaska, an area that was presumably very cold during the winter season. Thus, the ability of dinosaurs to adapt to cold climates or to seasonal migrations remains a very interesting question.

Some scientists say that the debate about ectothermic and endothermic might be irrelevant. Large dinosaurs could have been "lukewarm," they suggest, rather than "hot" or "cold." Some scientists have proposed an alternative metabolic model for dinosaurs called "Gigantothermy." Because of their immense size, dinosaurs could have been "inertial homeotherms," warming up and cooling down very slowly despite fluctuations in environmental conditions. However, adult dinosaurs ranged in size from as small as 10 pounds to as large as 80 tons, and all dinosaurs would have been quite small as hatchlings and juveniles.

Dinosaur anatomy and posture have been cited as evidence that they were built to move quickly and, therefore, must have had a high metabolic rate. The weight-bearing limbs of dinosaurs gave them a more erect posture than typical reptiles. Critics counter that the evidence does not really tell us how fast dinosaurs moved. Studies of modern ecological patterns suggest that endotherms like mammals and birds usually outcompete and replace ectotherms in any given area. Proponents of dinosaurian endothermy argue that it would have been impossible for mammals and dinosaurs to coexist for about 170 years if dinosaurs had been ectotherms. However, the outcome of competition between modern endotherms and ectotherms has not been established for all possible combinations of species and environments. Thus, the conclusion that endothermic primitive mammals would necessarily have outcompeted otherwise highly successful ectothermic dinosaurs should be considered speculative.

Modern ecological studies indicate that the predator/prey ratio that can support mammalian predators is lower than the predator/prey ratios that are needed to sustain ectothermal predators. Based largely on studies of modern mammalian predators and evidence that predator/prey ratios for dinosaurs were apparently quite low, some scientists have argued that dinosaurs must have been endotherms. These studies are interesting, but the calculations depend on many assumptions. In particular, the small numbers of surviving dinosaur fossils mean that sampling errors for both predator and prey species could be very large.

Some advocates of the warm-blooded hypothesis insist that comparison of the structure of dinosaur bones with the bones of mammals, birds, and reptiles provides convincing evidence for dinosaurian endothermy. In cross-section, the bones of dinosaurs more closely resemble those of mammals and birds than reptiles. Because bone morphology within a given species varies considerably depending on age, size, health, and growth conditions, evidence from fossilized bones can be ambiguous. However, sophisticated analysis of dinosaur bones does suggest the rapid growth patterns generally associated with modern endotherms.

Despite increasingly sophisticated studies of dinosaur fossils, paleontologists remain divided about the question of endothermy. More cautious biologists point out that many questions about ectothermy and endothermy in modern animals remain unanswered. Therefore, conclusions made on the basis of fragmentary collections of dinosaur materials should be regarded as highly speculative. During the millions of years of dinosaur evolution, perhaps other alternatives to endothermy and ectothermy appeared. Dinosaurs as a group were very diverse in size, form, habits, and habitat preferences. Presumably, the physiological adaptations of small and large dinosaurs might have been quite different. Dinosaurs dominated the world of terrestrial vertebrates for a period of about 170 million years. Given the long course of dinosaur history, it is also possible that some lines of dinosaurs remained ectotherms, like their reptilian ancestors, while others became endotherms. On the other hand, their physiological adaptations might have been quite different from those of modern reptiles and mammals.

Although several lines of evidence seem to support the hypothesis that dinosaurs were endothermic, critics argue that the evidence is not compelling. Debates about dinosaur physiology are based on a combination of evidence, induction, deduction, and speculation. Dinosaurs cannot be directly compared to any living reptiles, birds, or mammals. Thus, the uncertainties about dinosaur physiology are so great that competing hypotheses are likely to proliferate for many years.
—LOIS N. MAGNER

Viewpoint:
Yes, the most recent fossil evidence indicates that at least some dinosaurs were hot-blooded animals.

Because extinct animals—like dead people—tell no tales, scientists determine what their lives and physiology were like by analyzing clues found in fossilized remains and comparing these findings with attributes in living organisms. Over the last few decades, paleontologists have unearthed dinosaur fossils containing well-preserved tissues and organs which indicate that at least some dinosaurs were warm-blooded animals.

In 1968 Robert Bakker, a Yale undergraduate, wrote "The Superiority of Dinosaurs," showing that dinosaurs were "fast, agile, energetic creatures" with an advanced physiology strongly suggesting warm-bloodedness. Bakker showed that 10-ton ceratopsian dinosaurs could outrun a human—an impossible feat for a cold-blooded reptile.

In 1969 Yale professor John H. Ostrom published a paper describing *Deinonychus*, a theropod dinosaur, as an "active and very agile predator" that had to have a high metabolic rate to sustain active hunting. Bakker's and Ostrom's findings kindled the warm-blooded versus cold-blooded debate.

Metabolism The terms "cold blooded" or "warm blooded" refer to an animal's average body temperature in relation to its environment. Body temperature is determined by metabolism, the rate at which fuel (food) burns to maintain life.

The technical term for "cold blooded" is *ectothermic* (*ecto*, Greek for "outside"). An ectothermic animal relies on the outside air temperature to regulate its metabolism and body temperature. Ectotherms have low metabolic rates, because food is not used to maintain their internal temperature.

The technical term for "warm blooded" is *endothermic* (*endo*, Greek for "inside"). Endothermic animals (mammals and birds) create heat inside their bodies to maintain body temperature within an optimal range. Endotherms consume lots of food to fuel the inner metabolic engine that maintains their body temperature. Because their bodies burn so much fuel, endotherms have high metabolic rates.

A body's metabolism affects all aspects of its growth and physiology; tissues and organs are formed by and in accordance with metabolism. Dinosaurs' endothermy is established by comparing key characteristics of their physiology and lifestyle with the lifestyle and physiological characteristics common to living endotherms. Each of the following sections describes these comparisons to show that dinosaurs were warm blooded.

KEY TERMS

ECTOTHERM: "Cold-blooded" animal; an animal that relies on the outside air temperature to regulate its body temperature; ectotherms (animals that are ectothermic) have a low metabolic rate.

ECTOTHERMY (POIKILOTHERMY): Also called cold-bloodedness; metabolic state by which an animal's internal or body temperature fluctuates depending on the temperature of the surrounding environment.

ENDOTHERM: "Warm-blooded" animal; an animal whose high metabolic rate maintains a fairly constant internal body temperature.

ENDOTHERMY (HOMEOTHERMY): Also called warm-bloodedness; metabolic state by which an animal's internal or body temperature remains relatively stable regardless of the temperature of the surrounding environment.

ISOTOPE: Variation on the most common form of the atom of any element; most isotopes occur because of a different number of neutrons in the atomic nucleus.

LINES OF ARRESTED GROWTH (LAGS): Annual growth rings seen in the bones of ectotherms but not endotherms.

METABOLISM: Rate at which food is "burned" as fuel inside the body.

MORPHOLOGY: The form, or shape, of a body and its parts.

PALEONTOLOGIST: Scientist who studies prehistoric forms of life through its remains in plant and animal fossils; a scientist in the field of paleontology.

PREDATOR: Any animal that hunts and eats other animals.

PREY: Any animal that is hunted and eaten by other animals.

RESPIRATORY TURBINATES: Coiled cartilaginous or bony nasal structures found in endotherms but not ectotherms.

SPECIATE: To create a new species; the process in which new species are created through evolution (speciation).

TAXONOMY: The scientific system of classification, particularly of living things, that are grouped according to common characteristics; taxa (sing. taxon): a particular classification of living things; taxa mentioned in the text include from the broadest to the narrowest classification of organisms: family, genera (sing., genus), and species.

THEROPOD: Any of a group (Theropoda) of bipedal, carnivorous dinosaurs.

Role in Ecosystem *As Top Predators, Dinosaurs Must Have Been Warm Blooded.* All animals metabolize food using the same basic biochemical processes: the same enzymes break down food to yield energy. Metabolic processes are most efficient at an optimal body temperature: for every 10°C drop in body temperature, metabolism becomes twice as slow, producing half as much energy. Thus when it is cold outside, cold-blooded animals are sluggish: their body temperature is too low for efficient metabolism. For this reason, endotherms always outcompete ectotherms.

Evolution teaches that in all cases, long-term competition among organisms in an ecosystem favors endothermic over ectothermic animals. The inability to be active when it is cold or at night is such a disadvantage compared with the endotherm's perpetual internal heat engine that no large cold-blooded predators can dominate mammals.

For more than 140 million years during the Mesozoic, dinosaurs dominated terrestrial ecosystems. The few mammals were tiny rodent-like creatures. Mammals continued to be insignificant, and by the end of the Triassic, dinosaurs occupied nearly every slot in the food chain, from top predator to large herbivore to smaller versions of both. Throughout this period, dinosaurs kept mammals in their insignificant place. At no time could mammals outcompete the dinosaurs. Evolution tells us that this must be because the dinosaurs were endothermic. Had dinosaurs not been endothermic, over these millions of years of evolution the mammals would have outcompeted the dinosaurs and dominated the earth's ecosystems. That did not happen.

Dinosaurs Had Endotherms' Ratio of Hunters to Hunted. Large warm-blooded predators consume about 10 times more food than cold-blooded predators. The larger the predator, the higher its metabolic rate must be to fuel its huge bulk. Because a large predator needs such huge quantities of food, there are relatively few large predators in any ecosystem. Scientists can calculate precisely the ratio of predators to prey based on the numbers and weight of both.

In today's ecosystems (e.g., African savannah), the ratio of lions (predator) to grazing prey is about 1:100. Studies of prehistoric ectotherms (e.g., *Dimetrodon*) show a ratio of 20:100, indicating lower food consumption by cold-blooded animals. Prehistoric mammals (saber-toothed tigers) lived in an ecosystem whose predator-to-prey ratio was about 3 to 5:100. Studies of *Tyrannosaurus rex* and other Cretaceous dinosaurs also yield a ratio of 3 to 5:100. In all studies, dinosaur predator-to-prey ratios coincided with those of warm-blooded predators and prey.

Distribution: Arctic Dinosaurs Most dinosaurs are pictured among tropical ferns and cycads. Yet many dinosaur fossils (especially hadrosaurs and theropods) have been found in the Arctic, a region that in the Cretaceous had annual mean temperatures of between 35° and 46°F (2°–8°C). Ectotherms don't thrive in such cold climates. Consequently, almost no ectotherm fossils (e.g., crocodyliform) live in Alaska or other Arctic regions.

Why were dinosaurs able to survive, and thrive, in this cold region when ectotherms could not? The answer must be that the dinosaurs were endothermic.

Some paleontologists suggest that Arctic dinosaurs were ectotherms that migrated seasonally to and from the Arctic. However, Arctic fossil evidence shows the presence of neonates and juveniles, along with adult dinosaurs; this points to year-round habitation. The "migration" theory is questionable, too, because, for all known ectotherms—modern and extinct—a migratory lifestyle is not viable: Their low metabolism does not give ectotherms enough energy for long-term activity.

Bones *Dinosaurs Had Endotherm Bone Structure and Growth.* Bone is a tissue made of a calcium phosphate mineral that is incorporated into a protein called collagen. Collagen occurs in long bundles or fibers.

Dinosaurs had bones that resembled or were identical to endotherm bones in many ways. In ectothermic animals, bone grows slowly. In endotherms bones grow quickly. Fossil dinosaur bones show the rapid growth typical of endotherms.

In ectotherm bones, the collagen bundles are arranged parallel to each other in layers. The resulting bone is dense, with its crystals of calcium phosphate all lined up in the same direction. In endotherms, bones are not dense with neatly aligned fibers but are described as "woven." The collagen is arranged haphazardly, and the calcium phosphate crystals are also oriented randomly. Hundreds of slices of fossilized dinosaur bone, from many different species, show the woven bones typical of endotherms.

Ectotherm bones have pronounced growth rings, because cold-blooded animals grow quickly when it is warm outside and much more slowly when it is cold. Endotherms have bone growth rings, but they are far less pronounced. Dinosaur bones share the poorly defined growth rings found in endotherms.

Ectotherms have relatively few vascular (blood vessel) canals in their bones. Warm-blooded animals have many. Dinosaur bones have the numerous vascular canals common to warm-blooded animals.

Oxygen and Body Temperature. When bone forms, it incorporates an isotope of oxygen along with calcium phosphate crystals. The amount of oxygen isotope in the bone depends on the body temperature at the time the bone forms.

Analysis of the amount of oxygen isotope in *Tyrannosaurus rex* bones shows that they were formed while the animal's body temperature was stable within 4°C. Maintaining such a uniform temperature is consistent with the high metabolic rate of warm-blooded animals.

Teeth In 2001 research described the oxygen isotopes found in 75-million-year-old fossil teeth of crocodiles and theropod dinosaurs. The oxygen isotopes occur in tooth enamel and are not altered during millions of years of fossilization. Fossils of both kinds of animals were obtained from Alaska (cold climate) and Madagascar (hot climate).

The researchers found that the amount of oxygen isotopes in the crocodile teeth varied, depending on the latitude at which the crocodile lived. The high-latitude (cold) crocodiles had relatively little oxygen isotopes in their tooth enamel; isotope levels were higher in those crocodiles from the warmer climate. This variation occurs because in colder climes the reptiles' metabolism is low, so enamel formation is less than in its warm weather cousins.

The dinosaur teeth, however, showed very little variation in the amount of oxygen isotopes, regardless of latitude. Theropods from both sites showed uniform isotope levels, indicating that they maintained a constant body temperature in both cold and warm climates. Thus the dinosaurs must have been endothermic.

Dinosaurs Had Endothermic Organs and Tissues *Brain.* Endotherms tend to have large brains, which function only with a hefty supply of oxygen and fuel (food) and a constant body temperature. Although some dinosaur species did not have large brains, some—particularly the theropods—had brains equivalent to those of similar-sized mammals. This is strong evidence that the theropods were endothermic.

Heart. In 2001 computed tomography (CT) scans of the innards of 66-million-year-old Thescelosaurus fossils revealed that some soft tissue had been preserved. Imaging of the chest cavity showed a highly evolved heart. The image revealed two adjacent cavities: the lower chambers (ventricles) and a tubelike structure emerging from it (the aorta). Although the thin-walled upper chambers of the heart (the atria) likely collapsed when the dinosaur died, the remaining two chambers and single aorta are identical to the four-chambered heart found in birds and other warm-blooded animals.

Four-chambered hearts evolved in endotherms to fully oxygenate the blood

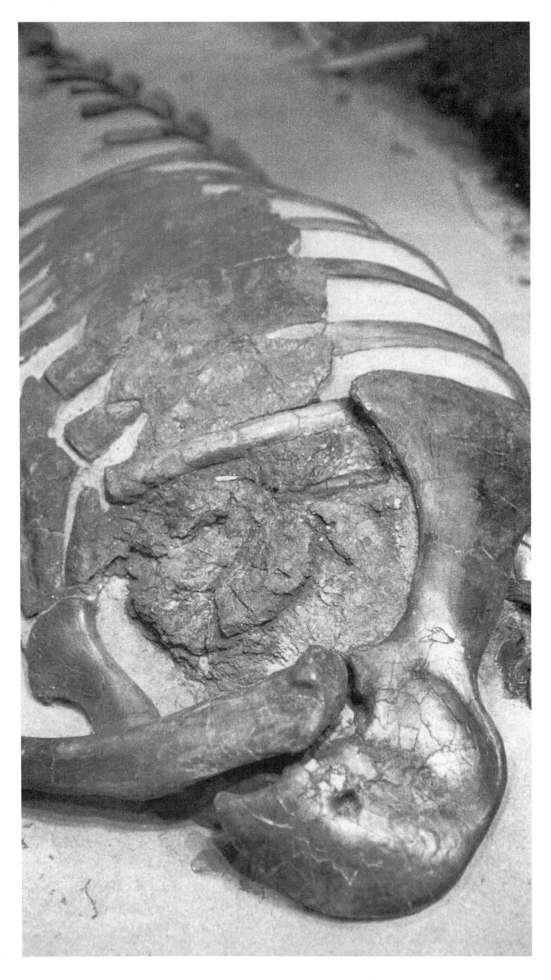

The bones and fossilized heart of a 66-million-year-old dinosaur. Two ventricles and a single aortic stem (lower center) suggest that some dinosaurs were warm-blooded.
(Photograph by Karen Tam. AP/Wide World Photos. Reproduced by permission.)

LIFE SCIENCE

needed to fuel a high metabolic rate. In contrast, ectotherms have relatively little blood oxygen. Reptile hearts typically have only one ventricle, resulting in far less blood oxygenation. Crocodiles' hearts do have two ventricles, but, unlike endotherms, they also have two aortas. No evidence of a second aorta was found in this fossil.

Dinosaur Morphology Supports Endothermy

Bipedalism. Many dinosaurs—and all theropods—were bipedal: they walked on two legs. Bipedalism requires more energy, and thus a higher metabolic rate, than a more sprawling four-legged posture and gait. Most paleontologists insist that some degree of endothermy is a prerequisite for bipedalism.

Running and Speed. Modern mammals have an average walking speed far faster than that of modern reptiles. An animal's speed can be calculated by measuring the distance between its footprints. Measurements of fossil footprints show that dinosaurs' speed was equivalent to that of today's mammals. Analysis of *Thecodont* skeletons, for example, shows that its limbs were built for speed and sustained, rapid running.

To sustain fast and agile running, an animal must have a high metabolic rate. Running is fueled by lungs that efficiently take in large quantities of oxygen and a large heart that pumps this oxygen-rich blood to muscles. Fossils show that most dinosaurs had the wide chest cavity able to accommodate these large organs. Reptiles, and other ectotherms, have much narrower chest cavities incapable of housing large organs. The recently discovered fossilized dinosaur heart (see earlier) confirms this analysis.

Some scientists say that because hadrosaurs and horned dinosaurs do not have wide chest cavities, they were ectotherms. Yet these dinosaurs have air sacs in their vertebrae (backbone) that are connected to their lungs. Their respiratory system is identical to that found in all modern warm-blooded birds.

Dinosaurs' large hips indicate that they had large leg muscles attached to them—leg muscles that needed lots of oxygenated blood and propelled the dinosaurs at considerable speed. Dinosaur hips resemble those of large-hipped birds and mammals with powerful leg muscles, and they are distinctly unlike the small hips and smaller, weaker reptile leg muscles.

Mating Display. Few reptiles have evolved displays to attract mates. This makes sense for animals that depend on the outside temperature to jump-start their metabolism. They cannot waste energy on so-called frills.

Yet many dinosaurs evolved to use a great deal of energy on developing displays either to attract mates or repel territorial competitors. Various species of dinosaurs flaunt ornamental crests on their head, huge "sails" on their back, powerful horns for head-butting intruders, and other remarkable—and energy-intensive—"extras." Such morphological "extravagances" are typical of endotherms.

Insulation. It takes far less energy to heat a well-insulated house than an uninsulated one. Insulation is needed for energy efficiency because the heat is generated inside and must be retained there.

There are absolutely no examples of cold-blooded animals developing body insulation. Insulation is intended to keep heat inside; if the animal obtains heat from outside its body, insulation is counterproductive.

Modern endotherms have various kinds of insulation: blubber, fur, feathers, and down. Some dinosaur fossils reveal the presence of feathers or protofeathers, the forerunner of modern feathers. Recent fossil discoveries in China (e.g., *Beipiaosaurus*) indicate that these were indeed feathered, although flightless, dinosaurs. Paleontologists assert that feathers functioned as insulation for these theropod dinosaurs.

Two 120-million-year-old dinosaur fossils found in China in 1998 were unquestionably feathered species. The *Caudipteryx* had a fan of feathers on its tail and down-like feathers on most of its body. The *Protarchaeopteryx* was covered with feathers. The find is significant because in both cases the wing feathers were arranged symmetrically, indicating that they were not for flight (birds fly with asymmetrical wing feathers). The researchers therefore concluded that the feathers had to be for insulation.

In April 2001 scientists unearthed a 130-million-year-old fossil dinosaur in China that was covered from head to toe with primitive feathers and down. The bipedal theropod (a dromaeosaur, cousin to velociraptor) yields the strongest evidence yet that dinosaurs are the ancestors of endothermic birds and first developed feathers to insulate their warm-blooded bodies.

Scientists agree nearly universally that dinosaurs are the ancestors of birds. Dinosaurs and birds share many characteristics, including erect posture, similar bone structure, similarly efficient heart and respiratory systems, relatively large brains, the presence of insulating feathers, and care for their young (see later). Birds evolved from dinosaurs in the Jurassic, so it is reasonable to assume that warm-blooded dinosaurs gave rise to warm-blooded birds.

Rate of Dinosaur Evolution Equals That of Endotherms Reptiles are among the most ancient species on Earth, with some species having remained unchanged for 50 to 60 million

GIGANTOTHERMY

Some scientists suggest that when animals become extremely large, entirely new processes come into play to maintain an internal body temperature. The argument for gigantothermy can be explained through a simple analogy.

Imagine a bathtub filled to the top with hot water. Then imagine a wide shallow soup bowl also brimming with hot water. In the case of the bathtub, only the water on the surface is exposed to the air that cools it. The vast volume of water in the tub is not exposed to the air, so does not cool as readily. In the case of the soup bowl, the water surface exposed to the cooling air is relatively large compared to the volume of water in the bowl. Thus the relatively small volume of water beneath the surface will cool rather readily. The key difference in these two examples is the surface-to-volume ratio. The lower the surface-to-volume ratio (the case of the bathtub), the more inner heat is retained.

Obviously, dinosaurs had a low surface-to-volume ratio. Their skin covered an enormous gut that was far more voluminous than the skin surface. Thus scientists have argued that the dinosaurs would lose little internal heat to the outside. The sheer bulk of the dinosaur innards was sufficient to maintain a fairly stable internal temperature.

But several problems arise with this argument. First, gigantothermy is generally able to maintain an internal temperature within 6° to 8°C. Fossil studies have shown that some dinosaurs maintained their internal temperature within 4°C. This range of temperature stability, although larger than that seen in modern mammals, is indicative of some degree of endothermy.

Second, as discussed in the text, the dinosaurs outcompeted the mammals with whom they lived for many millions of years. Had the dinosaurs been gigantothermic, evolution tells us that the warm-blooded mammals would have outcompeted them. They did not.

Third, dinosaurs' ancestors were relatively small animals for whom gigantothermy was impossible. Over time, evolution allowed these small ancestors to dominate the small mammals that also lived then. For the small dinosaur ancestors to have outcompeted the mammals, these ancestors must have been endothermic—and endothermy served them well in the competition for dominance. Why then, would the descendants of small endothermic animals abandon dominance-conferring endothermy for gigantothermy? In evolutionary terms, there would be no reason for the dinosaur descendants to give up a winning strategy.

Finally, we must consider the fact that some dinosaurs lived in relatively cold climates. Although adult dinosaurs might have been able to tolerate cold conditions through gigantothermy, there is no way their young could have done so. Dinosaur hatchlings were small—too small for gigantothermy. Had they not been endothermic, the young would have died and no dinosaurs would have been found in these cold regions.

—Natalie Goldstein

---

years. Because they do not generate internal heat and therefore eat less, ectotherms are far less vulnerable to drastic changes in the environment (e.g., drought) than endotherms. Thus catastrophes put relatively little evolutionary pressure on ectotherms. Their ability to survive adverse conditions accounts for the longevity of ectotherm species.

Warm-blooded animals are highly susceptible to the evolutionary pressures of environmental change. They are also vulnerable to the evolutionary pressure that arises from intense competition. In general, endotherms compete fiercely for resources. They reproduce rapidly and evolve quickly to occupy many different ecological niches. Their ability to diversify, or speciate, quickly to take advantage of newly created niches means that warm-blooded species are relatively short lived. On average, mammals create five or six new genera every 10 million years, compared with reptile species that last from 20 to 50 million years or more. Entire endotherm families survive about 25 million years.

Some dinosaurs—particularly the horned dinosaurs—are known to have generated new species every 5 to 6 million years, just like mammals. Like mammals, these dinosaur families lasted about 25 million years.

Taxonomy tells us that mammals produce more species per genus, and more genera per family, than ectotherms. Scientists know that dinosaurs produced these taxa at a rate similar to that of mammals. For example, dinosaurs averaged 3 to 4 species per genus, and 12 genera per family, the same as mammals. Hadrosaurs and horned dinosaurs created nearly 7 and 6 new genera, respectively, over a period of about 10 million years—again identical to mammals. In contrast, the taxonomic group that includes giant turtles has produced only one new genus in the last 5 million years.

Dinosaurs Had Endotherms' Lifestyle

Hunting. Most modern reptiles are rather sedentary hunters, simply waiting for prey to blunder by, then lunging and snaring it. The "cold-blooded" lifestyle does not lend itself to active pursuit and hunting.

In contrast, dinosaurs were active hunters. They pursued prey, running relatively long distances on two legs to catch it. Many predatory dinosaurs also had to attack prey and fight long and hard to subdue a large animal with a well-armored body, sharp horns, and other defenses. This level of predatory activity requires endothermy; no ectotherm has the energy for this degree of active hunting.

It is also worth noting that several dinosaur species, particularly *Deinonychus*, are thought to have hunted at night. Night hunting is almost unknown among cold-blooded animals because nighttime temperatures are generally too low to allow them to be active.

Parental Care. Some dinosaurs nested in large colonies, where parents incubated the eggs until they hatched. Although this behavior is very much like that of nesting birds, hatching eggs is not, in and of itself, proof that dinosaurs were endotherms. Yet although reptile young hatch from eggs, ectotherms cannot afford to expend their limited energy on feeding their offspring once they have hatched. Once born, reptile young are on their own.

Studies of *Maiasaura* nesting areas have yielded evidence that after hatching, the young remained in the nest for a significant period of time. The fossil evidence includes eggshells repeatedly trampled by the young, and fossils of juveniles of different ages found in the nest. From this evidence, scientists stipulate that the dinosaur parents must have gathered food and brought it back to feed their growing young in the nest—behavior common only to endotherms.

Conclusion Overwhelming evidence indicates that at least some dinosaur species were endothermic. There can be no doubt, based on the data and their clearly endothermic characteristics, that some dinosaur species were warm blooded, with some form of endothermic metabolism. —NATALIE GOLDSTEIN

Viewpoint:

No, the long-standing view of dinosaurs as cold-blooded animals is still the most compelling; recent evidence may suggest, however, that some dinosaurs had hybrid metabolisms, or aspects of both hot- and cold-bloodedness.

Despite recent media depictions of dinosaurs as warm-blooded animals, evidence strongly favors the hypothesis that the long extinct creatures of the Mesozoic era were cold-blooded organisms. Although the decades-old notion that dinosaurs were cold blooded is still sound, many scientists are now considering the possibility that dinosaurs may, in fact, have had a so-called hybrid metabolism that would have given them the best of both worlds. They would have had the benefits of a warm-blooded animal, complete with on-demand explosive energy spurts, while enjoying the much lower energy requirements of their cold-blooded lifestyle the rest of the time. Hybrid or not, the evidence for cold-bloodedness is persuasive.

Cold-bloodedness (the scientific terms are *ectothermy* or *poikilothermy*) means that an animal's internal or body temperature fluctuates with the outside temperature. For example, a dragonfly, grasshopper, or other insect is ectothermic, so it is sluggish on a cool summer morning but becomes increasingly active as the day warms. The temperature rises, the insect warms, the chemical reactions that control its metabolism speed up, and the insect can move more quickly. The same process happens in amphibians (like frogs and salamanders) and reptiles, all of which are ectotherms.

Historically, the scientific community has placed dinosaurs in the cold-blooded category because of their close phylogenetic relationship with reptiles, which include turtles, lizards, snakes, crocodiles, and alligators. Based on overall bone structure and other clues from the fossil record, scientists have concluded that dinosaurs evolved from reptiles. Because dinosaurs branched off the reptilian evolutionary tree, it naturally followed that dinosaurs were cold blooded like their ancestors, and that hypothesis has yet to face any conclusive evidence to the contrary.

The greatest hurdle in the debate of warm-bloodedness (also known as *endothermy* or *homeothermy*) versus cold-bloodedness is the lack of information. The fossil record, although growing, is far from complete. Paleontologists frequently find only a few broken pieces of dinosaur bones and then have the daunting task of making inferences about physiology, behavior, and social structure from these tiny remnants. On those rare occasions when scientists unearth an entire skeleton, they often struggle to answer definitively even the basic questions: Was it male or female? What did it look like? How did it live? In light of these most fundamental uncertainties, cold-blooded proponents typically see the warm-blooded argument as little more than a collection of ambiguous findings and ungrounded conjecture.

Hearts and Heat Loss For example, warm-blooded supporters frequently tout the highly publicized discovery of a fossilized dinosaur heart in 1993 as proof of their claim. Fossil collector Michael Hammer of Augustana College in Illinois made the find in South Dakota among the bones of a 660-pound, 13-foot-long, 7-foot-tall *Thescelosaurus*, a plant-eating beast that roamed the earth 66 million years ago.

Soon after he announced his discovery, scientists began questioning whether the grapefruit-sized clump of red rock was actually a heart at all: The chest (thoracic) cavity contained none of the other preserved soft tissues that, they said, should have also been present; and some of the purported heart was actually outside the cavity rather than inside where it should be.

Even if the structure was cardiovascular tissue, the cold-blooded camp rejected the Hammer group's conclusion that it originated in an endothermic animal. Using computed tomography (CT) scans to see inside the find, Hammer's group suggested the heart's structure matched that of warm-blooded animals, like mammals and birds: it had two lower chambers, or ventricles, that separated the oxygenated, inhaled air from the exhaled air, but it had only one aorta, the main blood vessel leading out of the heart.

Reptiles, in contrast, have two aortae, and most have a three-chambered heart with two upper chambers (atria), but just one ventricle. Of Hammer's finding, cold-blooded proponents noted that the fossil was incomplete, lacking the atria as well as carotid and pulmonary vessels. Likewise, they asserted that a second aorta may well have been present, but not preserved.

In addition, the presence of a second ventricle does not exclude ectothermy, because some reptiles have four-chambered hearts. The modern cold-blooded monitor lizard, for example, exhibits two-phased pumping in its ventricle, so its heart functions as a four-chambered organ. Crocodilians (crocodiles, alligators, and their allies) have a completely divided ventricle and thus a four-chambered heart.

With this reasoning, cold-blooded supporters contended that even if the Hammer group's description of the heart is correct, it still comes nowhere near proving that the dinosaur was warm blooded, only that it had a two-ventricle, one-aorta heart.

Some warm-blooded supporters also use the allegation that cold-blooded animals could not have survived in particularly cold climates, because they could not extract enough heat from the environment to support them. However, many present-day reptiles survive quite well in extreme temperatures.

A 1990 study on leatherback turtles showed how dinosaurs could cope with the temperature extremes. For this study, the scientists investigated how the large turtles regulate their body temperatures on their long migrations from the hot tropics to the downright cold North Atlantic waters. They found that the turtles use their size—some tip the scales at more than 1,500 pounds—and insulating tissue to maintain a fairly constant internal temperature, even in highly variable weather conditions.

The researchers, whose work appeared in *Nature*, extrapolated from the turtle's present-day abilities and concluded that dinosaurs could have had a similar capacity to survive in wide-ranging climates. According to Drexel University's James Spotila, one of the authors of the research, "Because of their large body size, dinosaurs were able to use their bulk as insulation and as a countercurrent heat exchanger." He added, "At a certain bulk level, warm-blooded and cold-blooded aren't that much different."

Clues in a Dinosaur's Breath Work published in 1997 and 1999 provided a glimpse of dinosaur metabolism and helped confirm the cold-blooded hypothesis. In these studies, scientists examined fossils from various dinosaurs.

In the 1997 study, the researchers studied the nasal bones of three different species: the ostrich-like *Ornithomimus*, a duck-billed *Hypacrosaurus*, and the *Tyrannosaurus* relative called a *Nanotyrannus*.

Using scanning equipment, they found that the dinosaurs lacked the coiled cartilaginous or bony structures, called respiratory turbinates, that are found in nearly all endotherms, but not in ectotherms. Actually, the scientists looked for the bony ridges that support the turbinates rather than the coiled structures themselves, because turbinates are delicate and rarely survive fossilization. They readily found the telltale ridges in fossilized mammals, but not in dinosaurs. The dinosaurs also exhibited narrow nasal passages similar to those in present-day reptiles. Endotherms, however, require large passages to house the turbinates and also to admit the greater amount of oxygen required by a warm-blooded animal.

Lead researcher John Ruben, a professor of zoology at Oregon State University at Corvallis, said, "The nasal structures we saw in dinosaur fossils simply would not have been able to accommodate the higher lung ventilation rates that dinosaurs would have had if they were warm-blooded."

In the 1999 research, Ruben's group added fuel to the ectothermy supposition following an analysis of a fossil of a baby *Scipionyx samniticus,* a small, velociraptor-like, meat-eating dinosaur. Using ultraviolet light to see the fossil in detail, they found and examined internal organs that had been preserved along with the fossil. Most importantly, they discovered that the lungs and heart were located in one part of the body cavity and the liver in another part. A diaphragm separated the two cavities. This division is significant, because it only occurs in animals that use a diaphragm to improve the lungs' performance.

By studying the liver's structure and the muscles surrounding it, the researchers surmised that dinosaur respiration was similar to that of the crocodilians living on Earth today. In crocodilians, abdominal muscles drive the liver up into the lungs to force air into and out of the lungs, much like a bellows draws in air, then blows it out to stoke a fire. Likewise, the *Scipionyx* had large muscles that turned the liver into a piston-like device for improving respiration when the animal demanded a higher metabolism. Such a system would allow dinosaurs the bursts of high metabolism needed to chase prey, escape from a predator, or engage in other high-energy activities.

Team member and paleobiologist Nicholas Geist remarked, "These theropod dinosaurs were fast, dangerous animals, certainly not slow or sluggish. They could conserve energy much of the time and then go like hell whenever they wanted to." Besides the liver-aided exhalation, the research team also noted that the fossil contained simple lungs that are much more similar in appearance to modern crocodilians than to warm-blooded animals.

The Hybrid Notion Although many scientists are convinced dinosaurs were not warm blooded, they are becoming increasingly intrigued by another notion: the animals may actually fall somewhere between warm and cold blooded. This hybrid metabolism would allow dinosaurs to be as active as endotherms when necessary, but revert to the much less energy-demanding lifestyle of an ectotherm when at rest. In support of this viewpoint, the scientists point to fossil findings and research showing that although dinosaurs are very close to reptiles, they are not identical.

A 1998 study of growth rates is a case in point. Kristina Curry of the State University of New York at Stony Brook analyzed the growth rate of the bones of a sauropod called a *Apatosaurus* (formerly known as a *Brontosaurus*). She concluded that the dinosaurs grew very quickly, attaining their massive 30-ton adult size in 8 to 11 years. This quick growth rate is more similar to endotherms than ectotherms.

She based her conclusion about age, however, on the number of annual growth rings present in cross sections of the animal's bone, and the growth rings themselves are indicative of reptiles, not mammals. Ectotherm bones display their age by these visible annual lines of arrested growth, called LAGs, because colder seasonal temperature slows their metabolism and, therefore, their growth every year. Modern endothermic bone only acquires LAGS if the animal experiences some type of stress, such as starvation. In summary, the simple presence of the growth rings implies the animals were ectothermic, and the swiftness of development only indicates that the sauropods grew to maturity much more quickly than formerly assumed and at a speed similar to endotherms.

Although the media hype of warm-bloodedness has been in full force recently, the bank of scientific evidence has quietly tilted in favor of dinosaurs having a cold-blooded or perhaps a hybrid metabolism. Just because cold-blooded supporters endorse the traditional view of dinosaur metabolism, it does not mean they are not open to new ideas.

A Velociraptor.

(Velociraptor Dinosaurs, replicas, American Museum of Natural History, photograph. AP/Wide World Photos. Reproduced by permission.)

Most, in fact, no longer view dinosaurs as not particularly bright reptiles that slowly plodded through the terrain of the Mesozoic era. Instead, they believe dinosaurs were smarter (although not on the level of the overly intelligent velociraptors in *Jurassic Park*), more active and something other than just big lizards. With their combination of ectothermic metabolism and wide activity range, said paleontologist James Farlow at Indiana University-Purdue University, "It's not surprising that they ruled Earth for over 100 million years." —LESLIE MERTZ

Further Reading

Associated Press and United Press International. "Huge Turtles May Show How Dinosaurs Endured Hot, Cold." *Detroit Free Press* (April 26, 1990): 2A.

Bakker, Robert. *Dinosaur Heresies*. New York: Kensington, 1986.

———. *The Great Dinosaur Debate*. New York: Citadel Press, 2001.

"Dinosauria." University of California at Berkeley, Museum of Paleontology. <www.ucmp.berkeley.edu/diapsids/dinosaur.html>.

"The Evidence for Ectothermy in Dinosaurs." <www.ucmp.berkeley.edu/diapsids/ectothermy.html>.

Gibbs, Ann. "Lung Fossils Suggest Dinos Breathed in Cold Blood." *Science* 278 (November 1997): 1229–30.

May, Julian. *The Warm-Blooded Dinosaurs*. New York: Holiday House, 1978.

Morell, Virginia. "A Cold, Hard Look at Dinosaurs." *Discover* 17 (December 1996): 98–102.

"New Insights on Dinosaur Metabolism." *USA Today* (June 1999): 8–9.

Ostrom, J.H. "Terrestrial Vertebrates as Indicators of Mesozoic Climate." *Proceedings of the North American Paleontological Convention* (1969): 34–76.

———. The Evidence for Endothermy in Dinosaurs. *AAAS Symposium* 28 (1980): 15–52.

Ross-Flanigan, Nancy. "The Diversity of Dinosaurs: The More Scientists Find, the More They Find Differences." *Detroit Free Press* (June 1, 1993): 1D.

Ruben, John. "Pulmonary Function and Metabolic Physiology of Theropod Dinosaurs." *Science* 283 (January 1999): 514–6.

Thomas, Roger D. *Cold Look at Warm-Blooded Dinosaurs*. Boulder, CO: Westview Press (AAAS), 1980.

Weed, William Speed. "What Did Dinosaurs Really Look Like—and Will We Ever Know?" *Discover* 21, no. 9 (September 2000): 74–81.

Wuethrich, Bernice. "Stunning Fossil Shows Breath of a Dinosaur." *Science* 283 (January 1999): 468.

Has DNA testing proved that Thomas Jefferson fathered at least one child with one of his slaves, Sally Hemings?

Viewpoint: Yes, the genetic and historical evidence strongly suggest that Thomas Jefferson fathered at least one child with Sally Hemings.

Viewpoint: No, the DNA testing is inconclusive and, at best, proves only that any of Thomas Jefferson's male relatives could have fathered a child with Sally Hemings.

A perceptive analysis of Thomas Jefferson (1743–1826) by American historian Henry Adams emphasizes the contradictions and tensions between the philosophy, principles, and political practices of the third president of the United States. Although Jefferson's contributions to the early republic are well known, biographers have both praised and criticized his role in American politics. His private life has likewise been subjected to considerable scrutiny. No aspect of his life has been more hotly debated than the suggestion that Jefferson had a long-standing sexual relationship with Sally Hemings (1773–1853), one of his slaves, and was the father of her children.

John Wayles, Thomas Jefferson's father-in-law, may have been the father of both Sally Hemings and Jefferson's wife, Martha Wayles Jefferson (1748–1782). According to Monticello records, Sally Hemings had at least six children, but only four survived to adulthood: Beverly (born 1798); Harriet (born 1801); Madison (born 1805); and Eston (born 1808). Harriet and Beverly were allowed to leave Monticello in 1822. Madison and Eston were freed by Jefferson's 1826 will. Harriet, Beverly, and Eston apparently passed into white society. Jefferson never officially freed Sally Hemings. Jefferson's daughter Martha Jefferson Randolph probably gave Sally a form of unofficial freedom that allowed her to remain in Virginia with her sons Madison and Eston.

Details Emerge in Print Rumors of the Jefferson-Hemings affair began to circulate among his neighbors, who commented on the striking resemblance between Jefferson and some of the slave children at Monticello. These speculations entered the written record in 1802 in a Richmond newspaper story by journalist James T. Callender, a man who undoubtedly harbored considerable enmity towards Jefferson. Callender asserted that Jefferson had for many years "kept, as his concubine, one of his own slaves," a woman named Sally, by whom Jefferson had "several children." Political enemies in the Federalist press continued to spread the Hemings story. Subsequently, both abolitionists and pro-slavery Southern apologists used the Jefferson-Hemings story. An anonymous poem entitled "Jefferson's Daughter," which appeared in antislavery newspapers, charged that Jefferson's daughter by a slave had been sold for one thousand dollars. British aristocrats cited the Jefferson-Hemings story as a way of criticizing American democracy. After the Civil War, members of the Republican Party used the Jefferson-Hemings story as a symbol of the South and the Democratic Party.

Jefferson's defenders argue that it is impossible to imagine that Thomas Jefferson, author of the Declaration of Independence and third president of the United States, would have engaged in a sexual relationship with one of

Thomas Jefferson
("Thomas Jefferson," portrait painting by Rembrandt Peale, photograph. © Bettmann/Corbis. Reproduced by permission.)

his slaves. However, many Americans also probably find it difficult to imagine Jefferson as a full participant in a system that Mary Boykin Chestnut (1823–1886) described in her famous *Diary from Dixie* as a "monstrous system, a wrong and an iniquity!" Mrs. Chestnut depicted Southern slave owners living like "the patriarchs of old . . . in one house with their wives and concubines." It was common knowledge that the mulatto slave children on each plantation resembled the white children of that family. In a very telling insight, Chestnut observed that "any lady is ready to tell you who is the father of all the mulatto children in everybody's household but her own. Those, she seems to think, drop from the clouds."

Interviews published in 1873 in a newspaper edited by Samuel F. Wetmore, a Republican Party activist, described the alleged Jefferson-Hemings relationship. Madison Hemings and Israel Jefferson, another former Monticello slave, asserted that Jefferson was the father of all of Sally's children. On the other hand, Jefferson's daughter Martha Jefferson Randolph and her children, Ellen Randolph Coolidge and Thomas Jefferson Randolph, denied the possibility of a sexual relationship between Jefferson and Hemings on both moral and practical grounds. They insisted that Peter and Samuel Carr, sons of Jefferson's sister Martha, had fathered the children of Sally Hemings. According to biographer Henry S. Randall, when Jefferson's daughter Martha showed him a poem alleging his relationship to Sally Hemings, Jefferson simply laughed. Those who deny the relationship say this proves that Jefferson disavowed any intimate relationship with Hemings. However, remembering what Mrs. Chestnut said, it is possible that Jefferson was laughing at the naïveté of his daughter.

Fawn M. Brodie, author of *Thomas Jefferson, An Intimate History* (1974), accepted Chestnut's insights into the South's "peculiar institution" and made the liaison between Jefferson and Hemings the central theme of her best-selling biography. In addition to her use of psychological concepts, Brodie documented a pattern of correlations between the time of conception of Sally's children and Jefferson's return to Monticello after various absences. Brodie pointed out that during and beyond Jefferson's lifetime, social conventions required white men to engage in public rejection of miscegenation, or sexual relations, with black women, despite the evidence of mulatto children. Brodie's book stimulated much interest among general readers, but many Jefferson scholars rushed to rebut her conclusions about the relationship between Jefferson's public and private life. Speculations about the sexual relationship between Jefferson and Hemings were revived in the 1990s by studies of the Y chromosome of descendants of Hemings and the Jeffersons. Indeed, although DNA (deoxyribonucleic acid) tests have not conclusively established whether or not Thomas Jefferson had a sexual relationship with Sally Hemings, analysis of the Y chromosome of a descendant of Eston Hemings proved that neither of the infamous Carrs could have been his ancestor.

Studies of the Y Chromosome In 1998 the British science journal *Nature* published the results of Dr. Eugene Foster's study of the Y chromosome of descendants of Eston Hemings, Field Jefferson (Thomas Jefferson's paternal uncle), John Carr (paternal grandfather of Samuel and Peter Carr), and Thomas Woodson. Foster hoped that this DNA study, either alone or combined with historical evidence, would resolve the Jefferson-Hemings issue; that is, if the Y chromosome of the Hemings descendants did not match the Jefferson Y chromosome, Thomas Jefferson could be eliminated with a high degree of probability.

The DNA data proved, however, that Thomas Jefferson and a descendant of Eston Hemings had Y chromosomes that were "identical by descent." Historical and contemporary paternity testing can prove that a man *is not* the father, but it cannot absolutely prove that a man *is* the father. A background level of unanticipated and unexplained "non-paternity" is a problem in all paternity investigations, and the degree of uncertainty or "historical degradation" increases in a study involving many generations.

Interpreting the DNA Testing Results The wide spectrum of reactions to the DNA evidence of Jefferson paternity can be seen as an important symbol of the tense history of racism in American society. Reports that the Y chromosome study proved that Thomas Jefferson definitely was the father of Eston Hemings failed to explain the limitations of the study. The study proved that someone with the rare and distinctive Jefferson Y chromosome fathered Sally Hemings' youngest child, and the data eliminated the "usual suspects," i.e., the Carrs. The tests found no match, however, between the Jefferson Y chromosome and that of Thomas Woodson's descendants, who continue to believe that they are descendants of Thomas Jefferson. Although no other Jeffersons had previously been implicated as the fathers of Sally's children, after the DNA tests were published, genealogists noted that at least 25 adult male Jeffersons, including eight who lived within 20 miles of Monticello, could have fathered Eston Hemings.

It is interesting that the *Nature* article was published shortly before the United States House of Representatives voted to impeach President Bill Clinton. The Jefferson-Hemings story was debated in connection with the sexual indiscretions of William Jefferson Clinton. Those who opposed the impeachment and the investigation of Clinton's private life argued that other presidents, including Jefferson, had their own "sexual indiscretions." In an era in which the sex lives of celebrities and politicians are the subject of constant media attention it is perhaps difficult to understand the attitudes prevalent during Jefferson's lifetime, when secrecy and discretion were the norm concerning sexual relationships, especially when racial issues were involved.

After the publication of Foster's DNA study, Daniel P. Jordan, president of the Thomas Jefferson Memorial Foundation, which owns and operates Jefferson's home Monticello, appointed a committee to review all the scientific and historical evidence concerning the relationship between Thomas Jefferson and Sally Hemings. The committee report concluded that the "currently available documentary and statistical evidence indicates a high probability that Thomas Jefferson fathered Eston Hemings, and that he was most likely the father of all six of Sally Hemings children." A dissenting report by one committee member argued that the historical evidence did not provide a definitive answer concerning Thomas Jefferson's paternity of Eston Hemings or Sally's other children.

In his *Notes on Virginia*, Jefferson discussed a plan to free the children of slaves, but his words and deeds concerning the issue of slavery display remarkable ambivalence. On pragmatic grounds rather than on principle, Jefferson's failure to free all his slaves has been attributed to the fact that he had accumulated enormous debts during his life and all his property was mortgaged. Although it was Jefferson who wrote the immortal words "all men are created equal," he found it impossible to visualize a solution to the political and economic problems associated with the institution of slavery. Those who denied the possibility that Thomas Jefferson was the father of any of Sally's children hoped that DNA evidence would settle the question. Instead, the evidence has resulted in renewed controversy and has demonstrated that even in cases where the scientific evidence is clear, its interpretation remains ambiguous. Moreover, attempts to extrapolate from limited scientific evidence can result in exacerbating controversial issues. —LOIS N. MAGNER

Viewpoint:

Yes, the genetic and historical evidence strongly suggest that Thomas Jefferson fathered at least one child with Sally Hemings.

The rumors that Thomas Jefferson fathered children by Sally Hemings (1773–1835), one of his slaves, began during his lifetime. Several families claim to be descendants of the pair. However, many Jefferson biographers and Monticello historians, some no doubt concerned about what they viewed as a blow to a founding father's reputation and reluctant to credit oral histories, continued to discount the rumors for 200 years.

Recently, however, DNA evidence has been presented that strongly links at least one of Hemings's children with the Jefferson family. Most historians now believe that the burden of proof has shifted, and that Thomas Jefferson was likely the father of at least one of her children, and probably all of them.

Genetic Evidence In a 1998 article in the journal *Nature*, researcher Eugene A. Foster presented DNA evidence connecting the descendants of Sally Hemings's last child, Eston, with the Jefferson family. Foster analyzed the XY chromosome, which is passed down only in the male line. Among Thomas Jefferson's six acknowledged children, only Martha (1772–1836) and Mary (1778–1804) lived to maturity. So the Jefferson male line was represented by DNA samples

> ## KEY TERMS
>
> **DNA:** Deoxyribonucleic acid, the double-stranded, helix-shaped molecule whose sequences encode the genetic information of living organisms. DNA carries information that encodes an organism's traits.
> **XY CHROMOSOME:** The chromosome that determines male sex. A father passes his XY chromosome to his sons. Certain genes are located on the XY chromosome.
> **HAPLOTYPE:** The combination of alleles (alternative forms of the same gene) inherited from a parent.

from descendants of Field Jefferson, Thomas Jefferson's paternal uncle.

The DNA pattern of the Jefferson XY-chromosome, or *haplotype,* is considered by geneticists to be quite rare. It was not seen in any of the samples taken from other "old-line Virginia families." Nonetheless, it matched with DNA from male-line descendants of Eston Hemings. Foster's team estimated that the probability that Eston's father was a Jefferson is greater than 99 percent. However, the results do not establish which Jefferson.

The resemblance of Sally Hemings's children to Thomas Jefferson was common knowledge among contemporary observers. Jefferson's grandson, Thomas Jefferson Randolph, was quoted by biographer Henry S. Randall as saying "it was plain that they had his blood in their veins." Randolph told of a Monticello dinner guest startled by the similarity in appearance between the master of the house and the servant who could be seen over his shoulder. However, he denied that Thomas Jefferson was actually the father of Sally's children, calling the suggestion a "calumny."

Given the obvious family resemblance, those seeking to deny a relationship between Sally Hemings and Thomas Jefferson generally attempted to assign responsibility for her children to a close Jefferson relative. The man chosen was usually either Samuel or Peter Carr, the sons of Jefferson's sister, because both were frequent guests at Monticello. However, because Samuel and Peter Carr's father was not a Jefferson, neither could have been the father of Eston Hemings.

Eston Hemings could have been fathered by Thomas Jefferson's brother Randolph, one of his sons, or one of Field Jefferson's grandsons. However, there is no documented record that any of these men were at Monticello when any of Sally Hemings's six known children would have been conceived. Most of them lived more than 100 miles away. Although a few were occasional visitors, certainly none were at Monticello frequently enough to father six children there. Even contemporaries who denied Thomas Jefferson's relationship with Sally Hemings always picked one alternate candidate; none claimed that Sally Hemings was promiscuous or that the Hemings children had different fathers. The assertion that Sally's children were full siblings is supported by the apparent closeness of the family. They are known to have lived together at various times, and named their children after each other.

Tellingly, none of the other Jeffersons were advanced by contemporaries as a possible father of the Hemings children. Neither were they mentioned by historians in connection with the Hemings children until after the DNA results eliminating Samuel and Peter Carr were published.

Thomas Jefferson has been documented to have been at Monticello during the conception window of each of Sally Hemings's known children. Four of her six children were conceived within three weeks after he returned to Monticello from elsewhere. No evidence suggests that Sally Hemings was not at Monticello at these times. She is not known ever to have conceived a child during Thomas Jefferson's frequent absences, even when one of the other candidates for paternity was present.

A statistical study of Thomas Jefferson's pattern of presence and absence at Monticello with respect to her conceptions yielded a high probability that either Thomas Jefferson or someone with an identical itinerary was the father. Because none of the other men who shared the Jefferson haplotype also shared his travel schedule, the evidence points strongly to Thomas Jefferson as the father of Sally Hemings's children.

The Account of Madison Hemings Because we have no word from either Sally Hemings or Thomas Jefferson about their relationship, the most important documentary evidence is probably the 1873 memoir of her son Madison. The account was dictated to an Ohio journalist after Madison had been listed as Jefferson's son in the 1870 census. Madison Hemings told of his brothers Beverly and Eston, and his sister Harriet, saying they were all Thomas Jefferson's children, and that his mother had no children by other men. Sally Hemings is known to have had six children, according to birth records kept in the Monticello Farm Book. These were the first Harriet (1795) who died as an infant, Beverly (1798), another short-lived daughter, unnamed (1799), the second Harriet, who survived to adulthood (1801), Madison (1805) and Eston (1808).

It should be noted that Madison's account makes no mention of a man named Thomas C. Woodson (1790–1879), whom some had believed was the first child of Sally Hemings and Thomas Jefferson. He is unlikely to have concealed the existence of an additional sibling, because he mentions the others who had slipped into the white world and would therefore have been reticent about their background. Madison's account does say that his mother began her affair with Jefferson in France when she was attending his daughters there, and returned to Monticello pregnant in 1789, but that this first child lived only briefly.

The Woodson family, like that of Madison Hemings, has a strong oral tradition of descent from Thomas Jefferson. They maintain that their forebear was the first son of the future president and Sally Hemings, and was sent to the Woodson plantation as a young boy. Accounts by Jefferson's political enemies do mention a son of Sally Hemings who looked like Jefferson and was named Tom. The newspaper *The Colored American* described Woodson in 1840 as "the son of his master." However, male-line descendants of Thomas Woodson do not share the same DNA haplotype as male-line descendants of Field Jefferson and Eston Hemings (or, for that matter, Samuel and Peter Carr). Neither is there any documentary evidence at Monticello that Thomas C. Woodson was in fact Sally Hemings's son, despite the existence of recorded birth dates for her other children. His parentage, therefore, remains a mystery.

Because Eston Hemings left Virginia and passed for white, his family did not preserve the descent from Sally Hemings in their oral history. Family members became aware of this part of their heritage only recently. However, earlier generations did pass down the knowledge that they were somehow related to Thomas Jefferson. In fact, while Eston Hemings was never known to have commented directly on his relationship to Jefferson, he did change his name to Eston Hemings Jefferson. The *Chicago Tribune* 1908 death notice of Eston's son Beverly described him as "a grandson of Thomas Jefferson." Sally Hemings's children Beverly and Harriet also left Monticello and entered the white world; none of their descendants, if any, are known.

Israel Jefferson, another former Monticello slave who took his master's name, corroborated Madison Hemings's story when speaking to a journalist in Ohio. The men were neighbors at the time; Israel Jefferson was about five years older than Madison Hemings.

Other Contemporary Accounts Many stories of Thomas Jefferson's relationship with Sally Hemings were spread by his political enemies in the Federalist Party. In 1802, the journalist James Thomson Callender wrote in the newspaper *The Richmond Recorder*, "The PRESIDENT AGAIN. It is well known that the man, whom it delighteth the people to honor, keeps, and for many years past has kept, as his concubine, one of his own slaves. Her name is SALLY. The name of her eldest son is TOM. His features are said to bear a striking although sable resemblance to those of the president himself. The boy is ten or twelve years of age...We hear that our young MULLATO PRESIDENT begins to give himself a great number of airs of importance in Charlottesville, and the neighbourhood...By this wench, Sally, our president has had several children. There is not an individual in the neighborhood of Charlottesville who does not believe the story, and not a few who know it...The AFRICAN VENUS is said to officiate, as housekeeper at Monticello." Callender's story was repeated in many Federalist newspapers and gained wide circulation.

In a letter to a fellow Federalist, Thomas Gibbons of Georgia wrote that Jefferson lived "in open defiance of all decent rule, with a Mulatto slave his property, named Sally." Gibbon referred to the children of Jefferson and Hemings as Tom, Beverly and Harriet.

Not all those who made reference to Jefferson's unsanctioned domestic situation were his enemies. In his diary, Jefferson's friend John Hartwell Cocke wrote that many Virginia plantation owners had slave families. "Nor is it to be wondered at," he continued, "when Mr. Jefferson's notorious example is considered."

Thomas Turner, a friend of the Randolphs, the family of Jefferson's son-in-law, wrote in 1805 that the president and "black, (or rather mulatto) Sally... have cohabited for many years, and the fruit of the connexion abundantly exists in proof of the fact.... The eldest son (called Beverly,) is well known to many."

The Unique Status of the Hemings Children
According to Madison Hemings, his mother's duties were, "up to the time of father's death, to take care of his chamber and wardrobe, look after us children and do such light work as sewing &c." As for Sally's offspring, Madison said, "we were permitted to stay about the 'great house', and only required to do such light work as going on errands." In their teenage years they were taught trades. Harriet Hemings learned to spin and weave, while her brothers were taught woodworking. Eston and Beverly were also musicians.

All of Sally Hemings's children were freed by the age of 21. Madison Hemings said that this was the result of a promise Thomas Jefferson had made to his mother. The children of Sally Hemings represent the only case at Monti-

Descendents of Thomas Jefferson and Sally Hemings pose for a group photograph at Monticello in 1999.
(Photograph by Leslie Close. AP/Wide World. Reproduced by permission.)

cello of an entire slave family being freed. In fact, the only female slave freed during the lifetime of Thomas Jefferson was Sally's daughter Harriet, who left Monticello in 1822 at the age of 21. According to the recollections of former overseer Edmund Bacon, Jefferson directed him to provide her with carriage fare plus $50. (Despite this unusual arrangement, Bacon still denied that Harriet Hemings was Jefferson's daughter).

Beverly Hemings left Monticello around the same time as his sister, and was not pursued. Upon Jefferson's death in 1826, Eston and Madison Hemings were freed in accordance with his will. Sally Hemings was never formally freed, but she too was allowed to leave Monticello after Jefferson died, and took up residence with her sons in Charlottesville. She remained there until her death in 1835.

The Weak Arguments Against a Jefferson-Hemings Relationship The Jefferson-Hemings story involves race, sex and slavery. These are highly sensitive issues even in our time, let alone over the 200 years that the controversy over the paternity of Sally Hemings's children has percolated.

Many historians of the South have denied the possibility of a relationship between Thomas Jefferson and one of his slaves, saying that such a relationship would have been out of character. There is little evidence for such an assertion. One might as well argue that owning slaves at all would be out of character for the man who wrote the Declaration of Independence, but Jefferson was a man of his time and place. He was a Southern plantation owner, and he kept slaves. Sexual relations and even long-term relationships between plantation owners and slaves were not at all unusual.

In fact, both Madison Hemings and other contemporary sources inform us that Sally Hemings and several of her siblings were the children of Jefferson's father-in-law, John Wayles. Their mother, Betty Hemings, whose family filled most of the coveted house servant positions at Monticello, was also the daughter of a white man. So Sally Hemings was probably three-quarters white and the half-sister of Jefferson's wife, Martha, who died in 1782. Isaac Jefferson described her as "mighty near white…very handsome, long straight hair down her back." Thus arguments by some historians that Jefferson's cultural conditioning would not allow him to find a black woman attractive are also not particularly relevant.

We of the modern world must wrestle to understand a brilliant man who could write so stirringly of freedom while owning his fellow human beings. Jefferson was by no means among the worst of slaveholders. There is every indication that his treatment of his slaves was as good or better than was common among his peers. We cannot know upon what mix of love and power his relationship with Sally Hemings may have been based. Certainly its apparent length and the treatment of her family suggests a certain amount of affection.

Thomas Jefferson never acknowledged a relationship with Sally Hemings or her children. He never specifically denied the relationship either, at least in writing. A letter referring to unspecified "allegations against me" has been interpreted by some as a denial of the Hemings affair by some, while others view it as a reference to other matters. The only account of Jefferson's responding directly to being linked with one of his slaves came from his granddaughters, who told a story of their mother Martha Jefferson angrily showing the president a bawdy poem on the subject. He is said to have laughed. —SHERRI CHASIN CALVO

Viewpoint:
No, the DNA testing is inconclusive and, at best, proves only that any of Thomas Jefferson's male relatives could have fathered a child with Sally Hemings.

A long-standing legend suggests that Thomas Jefferson, author of the Declaration of Independence and third president of the United States, fathered children by one of his slaves, Sally Hemings. Hemings, a light-skinned African American likely herself of mixed race, went to Paris with Jefferson when he served as ambassador to France. As the story has been passed down, she returned pregnant to Jefferson's Monticello plantation. Speculation about Hemings and Jefferson began soon after he returned to the United States, and later included accusations made by James Thomson Callender that Jefferson fathered Hemings's son, Tom, born in 1790, and other children of hers. Historians have written that Jefferson may have alienated Callender when he refused to appoint him postmaster in Richmond, Virginia, and that Callender's accusations were aimed at revenge.

Sally's last child, born in 1808, was named Eston. That Thomas Jefferson was also Eston's father has been rumored for almost two centuries. According to some historians, Eston bore a close resemblance to President Jefferson and easily entered white society in Madison, Wisconsin, as Eston Hemings Jefferson. These accounts have served over centuries as anecdotal "proof" that Thomas Jefferson was Eston's father.

There has also been a counter-tradition that either Peter or Samuel Carr, the sons of Jefferson's sister, could have been Eston's father.

The Search for an Answer In the 200 years since Jefferson's time, science has developed ways of looking at deoxyribonucleic acid (DNA) a molecule often called the messenger of life, to discover the genetic "fingerprints" of inheritance. Recently, some scientists felt that advances in DNA testing could either establish Jefferson's paternity or exonerate him once and for all in the case of Sally Hemings's son, Eston.

In 1998, retired pathologist Eugene A. Foster and colleagues attempted to prove or disprove paternity in the Jefferson-Hemings debate by examining the XY chromosomes in the known male descendants of Thomas Jefferson, the Carrs, and others. To help with the effort, Foster enlisted Jefferson genealogist Herbert Barger, who helped identify Jefferson male descendants who were appropriate for testing.

Testing required identifying that part of the study participants' DNA that would reveal the configuration of their Y chromosome. Every male has two inherited sex chromosomes, the X chromosome from his mother and the Y chromosome from his father. The Y chromosome is passed unchanged from father to son. In their genetic study, Foster and his colleagues found that the Y chromosome in the Jefferson line was quite distinctive. Had it been common, conclusions would have been beyond reach and the Carrs, for example, could not have been excluded as candidates.

Their methodology contained a flaw, however. In testing living Jefferson descendants, they could not exclude from the data genetic traces of long-deceased male members of the

Jefferson family, of whom there were many. All, including President Jefferson, could have passed along their Y chromosome to men in the subsequent generations.

The Published Findings Study results were published in the British journal *Nature* on November 5, 1998, under the headline "Jefferson Fathered Slave's Last Child." As the story reappeared in other popular media, the study results became simplified so as to suggest that it was an unequivocal fact that Jefferson fathered a child with Hemings. The study data, however, do not point to this conclusion.

According to Foster: "We ... analysed DNA from the Y chromosomes of: five male-line descendants of two sons of the president's paternal uncle, Field Jefferson; five male-line descendants of two sons of Thomas Woodson; one male-line descendant of Eston Hemings Jefferson; and three male line descendants of three sons of John Carr, grandfather of Samuel and Peter Carr. No Y-chromosome data were available from male-line descendants of President Thomas Jefferson because he had no surviving sons."

Foster and his colleagues found that the descendants of Eston Hemings Jefferson did have the Jefferson haplotype. "This haplotype is rare in the population, where the average frequency of a microsatellite haplotype is about 1.5 percent," wrote Foster. "Indeed, it has never been observed outside of the Jefferson family."

Thus, the findings of Foster and his colleagues suggest that Thomas Jefferson could have been Eston's father. At the same time, Foster confirmed that the Carr family haplotypes "differed markedly" from those of the descendants of Jefferson. This finding makes it certain that neither Samuel nor Peter Carr fathered Eston.

However, Foster and his colleagues noted that they could not "completely rule out other explanations of our findings" based in "various lines of descent."

They concluded that "a male-line descendant of Field Jefferson could possibly have illegitimately fathered an ancestor of the presumed male-line descendant of Eston. But in the absence of historical evidence to support such possibilities, we consider them to be unlikely."

The Evidence Falls Short In response to the question of paternity, the answer is no, DNA testing has not proven that Thomas Jefferson had at least one child with Hemings.

The data merely suggest that a number of males related to Thomas Jefferson could have fathered Eston. In other words, Jefferson was not the sole guardian of his genetic makeup; the XY chromosome is a DNA "family fingerprint" shared by some of his male relatives, any one of whom could have been the father of Hemings's son, Eston, or later fathered male descendants of Eston.

Because only living persons were tested, the Jefferson XY chromosome could have entered the lineage from several of Thomas Jefferson's contemporary male relatives or at any point in the almost 200 years since the rumor started.

Soon after the Foster article was published, *Nature* received letters, from scientists as well as nonprofessionals, disagreeing with the study results and, especially, disagreeing with the way they were reported.

In a letter to *Nature* that appeared in the January 1999 issue, David M. Abbey, MD, chief of medicine at Poudre Valley Hospital, Fort Collins, Colorado, and associate clinical professor of medicine at the Health Sciences Center at the University of Colorado, responded to the Foster study. "The DNA analysis of Y chromosome haplotypes used by Foster, et al to evaluate Thomas Jefferson's alleged paternity of Eston Hemings Jefferson is impressive," wrote Abbey. "However, the authors did not consider all the data at hand in interpreting their results. No mention was made of Jefferson's brother Randolph (1757–1815) or of his five sons. Sons of Sally Hemings conceived by Randolph (or by one of his sons) would produce a Y chromosome analysis identical to that described by Foster, et al." Abbey recommended that more data are needed to confirm Thomas Jefferson's paternity.

Could Jefferson's younger brother, Randolph, be considered an equal (if not better) candidate for being Eston's father? According to historian Eyler Robert Coates, records show that Randolph Jefferson was invited to Monticello in August 1807, about nine months before Eston was born in May 1808. Coates adds that Randolph had become a widower in 1806 and did not remarry until 1809; Coates speculates that Randolph was more likely in this period to be "susceptible to a sexual liaison." Of course, speculation over whether Randolph Jefferson, rather than Thomas Jefferson, was Eston's father is not a fact verifiable by science. He, like some other Jefferson males, was simply in the right place at the right time bearing the family XY chromosome.

Gary Davis, another letter correspondent, added in a letter to *Nature* (January 7, 1999), that "any male ancestor in Thomas Jefferson's line, black or white, could have fathered Eston Hemings. Plantations were inbred communities," wrote Davis, "and mixing of racial types was probably common .. . it is possible that Thomas Jefferson's father, grandfather or paternal uncles fathered a male slave whose line later impregnated another slave, in this case, Sally Hemings. Sally herself was a light mulatto, known even at this time to be Thomas Jefferson's wife's half sister."

Willard Randall, author of *Thomas Jefferson: A Life* and member of the God and Country Foundation, a group that seeks to safeguard the reputations of the founding fathers, said that at the time in question there were "20 to 25 men within 25 miles of Monticello who were all Jeffersons and had the same Y chromosome. Of them, 23 were younger than 65 year old Jefferson."

Shortcomings Are Acknowledged Even the study's lead author, Foster, admits the evidence is not in any way conclusive about Thomas Jefferson's alleged relationship to Eston. After the controversy over his findings erupted, Foster said in a response letter to *Nature* (January 7, 1999): "It is true that men of Randolph Jefferson's family could have fathered Sally Hemings later children . . . . we know from the historical data and the DNA data that Thomas Jefferson can neither be definitely excluded nor solely implicated in the paternity of illegitimate children with his slave Sally Hemings."

As Abbey added, "a critical issue always facing science is confounding variables. It is the scientific standard to comment on such variables when presenting a study, and especially to note how such variables could impact results. It is surprising that the authors (in their original paper) did not even address other conclusions. Too, when the public is presented with authors disagreeing with the title of their own paper, and the press reports conflicting accounts as to the validity of the results, public confidence in the scientific process may be eroded and create unnecessary skepticism toward DNA research in general."

As reported in an article in the *Washington Post* (January 6, 1999), editors at *Nature* admitted that the headline was "unintentionally misleading" and confessed as well that more "alternative explanations" should have been included in their conclusions.

Foster was quick to point out the inconsistencies between the data, the conclusions, and the headline. In a follow-up letter in response to letters from Abbey and Davis, he reminded readers of their original objective: "When we embarked on this study, we knew the results could not be conclusive, but we hoped to obtain some objective data that would tilt the weight of evidence in one direction or another."

According to Jefferson historian and genealogist Barger, the evidence for Jefferson's paternity is not tilted in any direction by the data. In conclusion, the DNA XY chromosome testing shows only that Thomas Jefferson could have fathered Eston, but so could any of several of his male relatives. The science is inconclusive, putting the speculation about Jefferson and Hemings back into the category of gossip.

On April 8, 2000, the University of Richmond hosted an all-day symposium on the Jefferson-Hemings dispute. Although no publications came out of that symposium, the discussion was videotaped. The tape is at the University of Richmond. Eugene A. Foster, author of the *Nature* article, was among the participants. —RANDOLPH FILLMORE

Further Reading

Beck, W. S., Karel Liem, and George G. Simpson. *Life: An Introduction to Biology.* New York: Harper-Collins, 1991.

Brodie, Fawn. *Thomas Jefferson: An Intimate Biography.* New York: W.W. Norton, 1974.

Foster, Eugene A. et al. "Jefferson Fathered Slave's Last Child." *Nature* 396, no. 5 (Nov. 5, 1998): 27–28.

Gordon-Reed, Annette. *Thomas Jefferson and Sally Hemings: An American Controversy.* Charlottesville: The University Press of Virginia, 1997.

Lewis, Jan Ellen and Peter Onuf, eds. *Sally Hemings and Thomas Jefferson: History, Memory and Civic Culture.* Charlottesville: The University Press of Virginia, 1999.

Mayo, Bernard, and James A. Bear, Jr., eds. *Thomas Jefferson and His Unknown Brother.* Charlottesville: The University Press of Virginia, 1981.

Randall, Williard S. *Thomas Jefferson: A Life.* New York: Holt, 1993.

Thomas Jefferson Memorial Foundation. *Report of the Research Committee on Thomas Jefferson and Sally Hemings.* Monticello, VA: 2000.

MATHEMATICS AND COMPUTER SCIENCE

Should statistical sampling be used in the United States Census?

Viewpoint: Yes, statistical sampling offers a more accurate method—at much lower cost—for determining population than does physical enumeration.

Viewpoint: No, the Supreme Court ruled that statistical sampling to calculate the population for apportionment violates the Census Act of 1976.

Statistics—the mathematical science of analyzing numerical information—is vital to the practice of all the empirical sciences. No modern science tries to account for the complexity of nature without using statistical methods, which typically provide investigators with a numerical outcome along with an analysis of the margin of accuracy of that outcome. With the help of computers, statistical techniques for collecting and analyzing large, complicated data sets have become very sophisticated and have proved to be reliable and effective for scientific researchers, inventors, and engineers working on problems in such diverse fields as economics, physics, and pharmaceuticals.

The popular perception of statistics, however, starkly contrasts with its valued role in the sciences. Statistics has often been dismissed as an unreliable and sinister ("lies, damned lies, and statistics") strategy for manipulating data to support a pre–determined point of view. While statistical techniques are quietly and successfully being used in many areas of modern life, one that most people are familiar (and perhaps uncomfortable) with is polling. Because polls—which survey selected sample groups of people and then extrapolate the responses to a larger population—are often done on behalf of political causes or candidates, their interpretation can be controversial. Bitter arguments about the outcome of a poll can taint the understanding of the statistical methods that made the poll possible.

The census has become a particularly contentious area of debate over the use of statistics. The census project seems deceptively simple: The census aims to count the population of the United States. But the population is large, diverse, moving, partially hidden, and changing every moment. A physical "count" of the population could never be done, and even if it could it would only be accurate for a few seconds. Any effort to count the population will contain errors of identification and omission. The challenge that faces those who design and administer the census, then, is to proceed in a manner that will minimize those errors. But the census poses much more than a scientific problem. The census is a political, economic, and social project, and those who are most interested in its outcome often have little regard for technical issues surrounding errors and estimates.

As the population of the United States has grown and become more diverse, the census has become more difficult to administer. It is well known that the standard method for performing the census, which relies primarily upon citizens to report information about themselves and their households and secondarily upon visits by census–takers to the homes of those who fail to report, undercounts the population by a significant amount. The most obvious way to correct this problem seems to be to make use of statistical sampling methods, which could account for the variety within the population as the count is adjusted upward. The undercount seems to be distributed unevenly throughout the population, tending to come primarily from certain groups that are harder to contact and locate, such as renters, immigrants, the homeless, and children. These groups disproportionately tend to support Democrats rather than Republicans, thus leading to the primary political schism over the possibility of using statistical techniques to refine the census. Politicians view the undercounted groups either as potential supporters or potential opponents, and argue accordingly about how to count them.

Opponents of the use of statistical sampling to improve the census attack on several fronts and take advantage of public skepticism about the validity of statistical methods. They argue that the Constitution quite literally calls for a physical enumeration (a physical counting) of the population to be performed during each decade's census, and use this as a foundation to block any effort to incorporate statistical modifications. Various legal issues surrounding Constitutional interpretation have been argued all the way to the United States Supreme Court. Incorporated in these legal challenges are criticisms of the statistical methods that would be used to improve the accuracy of the census. While members of the National Academy of Sciences as well as other mathematical and scientific experts have generally endorsed the superior accuracy of statistical sampling over enumeration, laymen remain somewhat perplexed and skeptical. One reason for that concern may be a form of circularity in the argument of sampling's proponents; that is, in order to argue that sampling gives a more accurate count, they use evidence collected by sampling. What sampling's advocates call accuracy, its critics call bias. Courts are notoriously poor places to settle technical disputes, and the debate over census methods is no exception. Because the census affects the creation of political districts and the apportionment of financial resources, however, it is inevitable that any change in its methods will be assessed on political as well as scientific grounds. —LOREN BUTLER FEFFER

Viewpoint:

Yes, statistical sampling offers a more accurate method—at much lower cost—for determining population than does physical enumeration.

In general, there are essentially three key aspects in the debate over the extent of the use of statistical sampling in compiling census data. Although there are a number of interrelated concerns regarding the use of statistical sampling, all the issues can be reduced to either Constitutional arguments, political arguments, or scientific questions regarding the ability of statistical methods to render a more accurate count. Although the legal and political arguments can be tortuous and partisan, the scientific and mathematical considerations strongly favor the use of statistical sampling over physical enumeration.

Although there are exceptions, both proponents and opponents of the use of statistical sampling in the census generally agree that there are two independent measures that can be used to validate a particular statistical method known as integrated coverage measurement. Census officials, mathematicians, and statistical modeling experts contend that these two methodologies enhance the accuracy of census data by reducing differential undercount (a greater undercount in selected groups when compared to the count of the general population).

Reducing the Differential Undercount The first method involves determining the extent of a census undercount, which can be reasonably estimated by the use of existing demographic data reflecting birth, death, immigration, and emigration records already maintained by governmental agencies. A second method, conducted following the census in the form of post–enumeration surveys, also enables mathematicians and statisticians to draw conclusions regarding the consistency of data collected during the census itself. Both validation methodologies currently support the argument that there exists a chronic undercount of almost all groups, but a more significant, and therefore differential undercount, of minorities, children, renters, and other identifiable groups.

Another powerful argument favoring the use of statistical sampling for all census measurements is that the modifications to the present system of enumeration (e.g., an emphasis on advertising designed to reach target undercount groups such as Spanish–speaking residents) failed, as measured by demographic and post–enumeration surveys to increase census accuracy. In fact, for the first time in census history, a decennial census (the 1990 one) proved to be less accurate than prior decennial census measurements. The accuracy of the 2000 census awaits full assessment.

Legislative Actions In response to the disappointing performance of the 1990 census methodologies, Congress passed the Decennial Census Improvement Act in 1991, providing for a study of census methodologies by the National

KEY TERMS

DUAL SYSTEM ESTIMATION: An estimation methodology that uses two independent attempts to collect information from a household in order to estimate the number of people missed by both attempts.

ENUMERATION: In dictionaries contemporaneous with the signing of the Constitution, the term "enumeration" refers to an actual or physical counting, not an estimation of a quantity.

ERRONEOUS ENUMERATION: The inclusion of a person in the census because of incorrect information or error, such as being counted twice.

IMPUTATION: A method for filling in information about a person from a previously processed respondent with whom the person shares similar characteristics and who lives in close geographic proximity.

POSTSTRATUM: A collection of persons in the census context. These persons share some characteristics such as race, age, sex, religion, or owner/renter. The collection is treated separately in estimation.

SAMPLING ERROR: Errors in statistical analysis that result from an improper or unrepresentative sample draw.

SIMPLE RANDOM SAMPLING: A complex mathematical method for generalizing from a small sample without regard for any ordering embedded in the choice.

SYSTEMATIC ERROR: Errors in statistical analysis that result from bias, application of improper formulae, etc.

UNDERCOUNT: The number of persons who should be included in the census count but were not for some reason. Persons living in group settings such as dormitories, or without permanent addresses such as migrant workers, are likely to be missed by the census.

Academy of Sciences. The Academy was specifically asked to render an opinion on the scientific and mathematical appropriateness of using sampling methods to compile and analyze census data. After extensive hearings and investigations, two of the three expert panels convened by the Academy concluded that significant reductions on the undercount (i.e., improved and more accurate census data) could not take place without the use of statistical sampling. Moreover, all three Academy panels concluded that census data would be made more accurate by an integrated coverage measurement procedure that relied on statistical sampling. One panel of the National Academy of Sciences specifically concluded that it was "fruitless to continue trying to count every last person with traditional census methods of physical enumeration."

Data examined by the panels of the National Academy of Science and submitted to the United States Supreme Court also indicated that there was a probable undercount of 1.8% of the general population. The differential nature of the undercount is confirmed by the fact that 5.7% of self–described "Black" or "African–American" residents were undercounted.

The debate over the use of statistical sampling is ironic in that the arguments against the use of statistical sampling ignore the increasing trend of accuracy associated with statistical sampling. The current Census Act was enacted in 1954, and within three years, Congress amended the act to allow limited use of statistical sampling except for the "determination of population for apportionment." In 1964, Congress again revised the Census Act to allow data collection via questionnaire in place of a personal visit by a census–taker as long as the questionnaire was delivered and returned via the United States Postal service. The use of statistical sampling was further expanded by a 1976 revision to the Census Act that allowed the gathering of population and census data—but that did not specifically authorize its use for issues pertaining to apportionment. Upon this oversight to specifically allow statistical sampling to compile data used for apportionment, constitutional literalists and the majority of the Supreme Court rested their arguments against the broad use of statistical sampling.

For the year 2000 census, government officials estimated that 67% of households would voluntarily return census forms. Based upon the subjective data provided in the forms—correlated to past data, some itself derived from statistical sampling—census officials planned to divide the population into groups with homogenous (similar) characteristics involving economic status and residence. Instead of attempting a visit to all non–responding households by census–takers, however, census officials then planned to selectively visit households that would be statistically representative of the non–respondents, until a total of 90% of the

households in the target group had been surveyed by questionnaire or interview. Integrated coverage measurement, encompassing the two validation methodologies previously outlined, would then be used to make a final adjustment to the undercount in target groups, or "strata," based upon such demographics as location, race, and ethnicity. As an additional check, randomized physical counts would be used to measure the accuracy of the statistical corrections.

The Political Divide The political argument over this proposed procedure—the increasing reliance on statistical sampling—proved polarizing along traditional party lines. Democrats, representing a political party that historically benefits from greater minority representation and participation in government, argued that the undercount denies accurate and effective representation of undercounted groups including minority voters, children, and those who do not have a regular or stable residence. In addition to arguments based upon the potential of statistical sampling to at least partially correct these problems, the Democrats voiced a strong social justice claim that rested upon the essential requirement that in a democracy or a republic the representatives shall be fairly drawn from the population, so that the voice of the government is indeed the voice of the governed.

Republicans, a party that historically performs poorly with regard to garnering minority votes, have typically argued for a strict and literal Constitutional interpretation that would continue a reliance on a decennial census (the major United States census, Constitutionally mandated to take place every 10 years) via physical enumeration. Republican opponents of sampling particularly resist attempts to integrate enhanced sampling techniques specifically targeted at reducing the undercount among selected groups. Because mathematical arguments and performance records related to prior census collections favor statistical sampling, a Constitutional argument relying, as statistical sampling advocates will argue, on an outdated and narrow interpretation of the Constitution is clearly the strongest case physical enumeration advocates bring to the debate. On its face, and by historical precedent, the Constitution of the United States calls for an actual physical enumeration—a physical count of residents.

The Judicial Position In 1998, on appeal of the landmark case the *Department of Commerce et al. v. the United States House of Representatives et al.,* a majority of the United States Supreme Court—split along well-established political and ideological lines—affirmed a lower court's interpretation of both the Constitution and federal statues (e.g., the Census Act). According to the Supreme Court, the Constitution's census provisions authorize the appropriate Congressionally appointed agencies to "conduct an actual enumeration of the American public every 10 years (i.e. the decennial census) as a basis for apportioning Congressional representation among the States." The Court's ruling majority further cited narrowly existing provisions that specifically authorized the limited use of statistical sampling procedures "except for the determination of population for purposes of congressional apportionment." The ruling quashed the Department of Commerce's plans to use statistical sampling in the 2000 decennial census as part of an attempt to correct chronic and well-documented undercounts.

Significant, however, to the continuing argument regarding physical enumeration versus statistical sampling, the Court's ruling inherently recognized that there would be a difference in the outcome of the two counts—and, more importantly, that such a difference would be significant enough to alter the apportionment of Congressional representatives and state legislators. In its ruling the Court held that there was a recognized likelihood that voters in areas relying on physical enumeration would have their representation "diluted *vis-à-vis* residents of [areas] with larger undercount rates." Justice Sandra Day O'Conner, in writing the majority opinion, recognized that there was a traditional undercount of minorities, children, and other groups resulting from a census by enumeration. Although the Court's ruling majority recognized the power, utility, and validity of statistical analysis—they chose to rule on the law rather than on the acknowledged merits of statistical analysis.

In dissenting opinions (an opinion written by a Supreme Court justice who disagrees with the majority ruling), justices John Paul Stevens, David Souter, and Ruth Bader Ginsburg advanced the argument that the use of statistical sampling was authorized by a broader interpretation of prior authorizations allowing the use of statistical methods to compile and analyze census data. In particular, Justice Stevens argued that the planned safeguards and validation checks that are a part of the proposed integrated coverage measurement protocol were simply extensions of procedures previously authorized. Moreover, the use of integrated coverage measurement as a supplement to traditional enumeration-based procedures was "demanded" by the accuracy required by the intent and scope of the Constitution and existing census legislation.

Justice Stevens also pointed out that the census officials had long used various statistical estimation techniques—including the imputation of data—to adjust counts gathered by enumeration. Imputation allows census workers to guess at estimated data based upon prior data collected under similar circumstances (i.e., when

A census worker interviewing a woman in 1930, as her 10 children look on.
(U.S. Census Bureau, Public Information Office. Reproduced by permission.)

the demographics are similar, census officials are able to make reasonable estimates of certain responses, such as approximating the size of a household based upon the number of individuals in neighboring households). In 1970, imputation methodologies added more than one million people to the national population count, and in 1980, imputation methods added more than three–quarters of a million people to the national population count.

The Advantages of Statistical Sampling

Although opponents decry that statistical sampling favors some groups over others (e.g., increases the number of minorities and children in the overall count), it is exactly this aspect of statistical sampling that reduces the differential error of a differential undercount. Because minorities and children are undercounted by enumeration methods, statistically based corrections to enumeration data—if well designed to enhance census accuracy—should differentially correct these selected population counts. Non–differential analysis that would correct only for a general undercount (i.e., a 10% correction to all data) would, at a minimum, simply preserve under–representation based upon undercounts.

With regard to error and error analysis, statistical sampling prevails. Statistical sampling data can be cross–checked with existing demographic data to allow a more accurate estimate of both gross omissions (failure to count) and erroneous errors wherein individuals are assigned to improper addresses. For example, identification of fractions of a target group in follow–up samplings are compared to initial samples in an effort to estimate the existence and extent of an undercount. Mathematically, the dual system estimate (DSE) can be explained as follows: C_n represents the number in the census count; E_n the number of erroneous enumeration errors; P_n the number of selected group individuals as determined in a post–enumeration survey; and M_n the group number in the post-enumeration survey accounted for in the enumeration count. The DSE is related by the following equation: $DSE_n = (C_n - E_n) \times (P_n / M_n)$.

Although it can be fairly argued that statistical sampling is not perfect—and is also subject to error based upon bias contained in designing sampling collection protocols and formulae applications—the mathematical errors and bias in statistical analysis are more easily identifiable, quantifiable, and correctable, in clear contrast to errors in the enumeration process. Bias or prejudice in enumeration methodology, which is often based upon the difficulties of interviewing certain groups, or on the fears of census workers to carry out tasks in certain geographic areas, is not easily quantifiable because, in part, it is difficult to obtain data on such attitudes and fears. Moreover, any culturally pervasive bias or prejudice that exists among census workers must invariably increase the magnitude of errors in enumeration-based data.

Issues related to tabulation or potential computer programming errors affect both sides equally. It is just as likely that an error could exist in enumeration data-handling software as in programs designed to handle statistical sampling data. Moreover, there are reliable mathematical methods to estimate errors in statistical analysis. Recognizing the errors inherent in enumeration-based data, as a last resort opponents of statistical sampling often argue that neither methodology is perfect, and so, therefore, the increased accuracy of statistical sampling is not worth the effort to change. The assertion of this argument, however, ignores the fact that sampling is far less expensive than enumeration.

At a minimum, a strong argument for statistical sampling rests upon an enhanced ability to determine and control uncertainty and error. The use of sampling data does not eliminate error, but it does control the type of errors encountered. Statistical sampling offers enhanced sampling error and bias recognition. Although sampling errors usually average out—especially in large randomized census-sized samples—bias errors, which are more difficult to determine in enumeration-based methods, are usually directional errors that do not average out and that differentially degrade census accuracy.
—K. LEE LERNER

Viewpoint:
No, the Supreme Court ruled that statistical sampling to calculate the population for apportionment violates the Census Act of 1976.

Background The U.S. Constitution dictates that a census be taken every 10 years specifically for the purpose of apportioning representatives, and further specifies "actual Enumeration." Congress has dictated the particulars of exactly how the enumeration is to be carried out through the Census Act, which has been amended from time to time over the years (1976 being the latest version). The Secretary of Commerce oversees the "actual Enumeration" of the population. The Census Bureau was formed within the Commerce Department to conduct the decennial census. The law states that the Bureau "shall, in the year 1980 and every 10 years thereafter, take a decennial census of population as of the first day of April of such year." It further stipulates the timetable for completing the census—within nine months.

The subject of statistical sampling evolved from the problem of the undercount rate, which has plagued the census since 1940. A reasonable question is: How can an "undercount" be determined? If persons were not "counted" in the census, what evidence is there that they exist? Since 1940, the Census Bureau has been utilizing such methods as *demographic analysis* to produce an independent estimate of the population using birth, death, immigration and emigration records, and also a "Post-Enumeration Survey," which uses sampling to make an estimate of population. The numbers that result are compared to the actual census. Identifiable groups, which include some minorities, children, and renters, have had noticeably higher undercount rates in the census than the general population. Since the census is the basis for determining representation in Congress, some states may have been underrepresented as the result of undercount.

Concerned about the discrepancy in representation, Congress passed the Decennial Census Improvement Act of 1991. The Secretary of Commerce was instructed to contract the National Academy of Sciences to study the problem and arrive at a solution. Hence comes the statistical sampling controversy. The Academy was instructed in the Improvement Act to consider "the appropriateness of using sampling methods, in combination with basic data-collection techniques or otherwise, in the acquisition or refinement of population data."

Measuring a Changing Nation, Modern Methods for the 2000 Census was published by the National Academy Press in 1999, detailing the work of the Panel on Alternative Census Methodologies, Committee on National Statistics, Commission on Behavioral and Social Sciences and Education, and the National Research Council. On the issue of sampling for nonresponse follow-up, the panel "concluded that a properly designed and well-executed sampling plan for field follow-up of census mail nonrespondents will save $100 million (assuming an overall sampling rate of 75 percent)."

Sampling for nonresponse follow-up was predicted to reduce the Census Bureau's total workload, which would permit improvements in the control and management of field operations, and which would allow more complete follow-up of difficult cases, leading to an increase in the quality of census data collected by enumerators. The combined committees added, "of course, sampling for nonresponse follow-up will add sampling variability to census counts."

In spite of the potential for added errors from sampling, the Census Bureau responded to the findings and decided to use two sampling procedures (explained below) to supplement information collected by traditional procedures in the 2000 Census. Critics of the statistical sampling procedures brought their case to the courts, and in January 1999, the Supreme Court ruled that such procedures were in violation of the Census Act of 1976. As the constitutionality of statistical sampling was not brought up in the lower courts, that issue was not decided.

The Statistical Sampling Plans: NRFU and ICM The Census Bureau plan included a Nonresponse Followup (NRFU) program. NRFU is described in *Measuring a Changing Nation* as a field operation conducted by census enumerators in order to obtain interview data from people who failed to mail back their questionnaires. Data obtained from some, but not all, of these nonresponse follow-up cases are called "sampling of nonresponse follow-up" (SNRFU). It was the possible use of this data that was challenged in the courts.

Details of the Census Bureau's plans are included in the "Opinion of the Court" published by the Supreme Court. In one part of the program, the Census Bureau planned to divide the population into census tracts of approximately 4,000 people that have what they describe as "homogeneous population characteristics, economic status, and living conditions." From that list, the enumerators would visit a randomly selected sample of nonresponding households as *statistically representative* of all the group. This, in effect, would create what the *Chicago Tribune* described as "virtual people" (October 31, 2001, editorial titled "Census and Nonsense").

The second statistical sampling procedure that was challenged was one that would have been put in place after the first statistical sampling was completed. It was called Integrated Coverage Management (ICM). As described in the Supreme Court's Opinion, ICM uses the statistical technique called Dual System Estimation (DSE), a system that requires the Census Bureau to classify each of the country's seven million blocks (called strata) according to defined characteristics. It is a complex sampling process.

Using the 1990 census information, the characteristics used in the classification of strata include: state, racial and ethnic composition, and the proportion of homeowners to renters. The Census Bureau planned to select 25,000 blocks at random for an estimated 750,000 housing units. Enumerators would canvas each of the 750,000 units. Where discrepancies exist between information taken before the ICM sampling and the ICM information, repeat interviews were to be conducted to resolve the differences. The information from the ICM would be used to assign each person to a post-stratum, which is described in the National Academy of Sciences report as a collection of individuals in the census context that share some characteristics as to race, age, sex, region, owner/renter. The information was to be treated separately in estimation as part of the statistical sampling.

The Supreme Court's Opinion describes what happened next. A bill was passed in 1997 allowing the Census Bureau to move forward with their plans for the 2000 Census, but requiring the Bureau to explain any statistical methodologies that might be used. In response to the directive, the Commerce Department issued the Census 2000 Report. Congress followed with an Appropriations Act providing for the plan and also making it possible for anyone aggrieved by the Bureau's plan to bring legal action. This provision was the basis for the two suits and the ultimate denial of the use of statistical sampling by the Supreme Court. Since the denial was based on this Act, the issue of constitutionality was never decided.

After the 2000 Census The Census Bureau may have been prevented from using statistical sampling to determine population count for congressional apportionment, but it was able to use sampling to collect additional information, as it had done in previous censuses. About one in six households received a long-form questionnaire to obtain additional information. The form includes 53 questions about the person's life and lifestyle that were considered too intrusive by many users. In a May 2000 report sponsored by The Russell Sage Foundation and others, titled *America's Experience with Census 2000, A Preliminary Report,* the authors found that privacy concerns had a negative impact on cooperation among the households who received the long form. Questions concerning income and physical and mental disabilities ranked highest among those considered too personal for the census to ask.

In hearings in the House of Representatives prior to passing the Decennial Census Improvement Act in 1991, Congressmen expressed concerns that statistical sampling and subsequent adjustment would discourage voluntary public response through mail-back forms. The mail-back form is the most accurate, effective, and efficient source of census data, according to the General Accounting Office. The Congress also expressed concern that statistical sampling may discourage state and local participation in that it would remove their incentive for obtaining a full and complete count.

By law, within nine months of the census date, the apportionment population counts for each state are to be delivered to the President. That would be December 31, 2000, for the 2000 Census. The President then has the responsibility of delivering the information to the Clerk of the House, who, in turn, must inform each state governor of the number of representatives to which each state is entitled. Although the enumeration of the population for apportionment was fixed to the actual count data by the Supreme Court decision, what constitutes an actual count is still a fuzzy figure because it included a statistical procedure called *imputation* to "count" people about whom the Census Bureau had incomplete or no direct information.

It was the use of imputation-provided data that accounted for the reduction in net undercount in the groups of people missed in the 1990 census. 2.4 million people were reinstated using imputation data. To evaluate the success of the census, and to adjust the numbers for non-political purposes if needed, the Census Bureau conducts an Accuracy and Coverage Evaluation (A.C.E.). The A.C.E. is bound to a complex set of procedures and operations much like the census itself. As measured by A.C.E., the net undercount of the population was about 3.3 million or 1.2% of the population in 2000. This is an improvement over the 1990 census which had a 1.6% undercount. So, with imputation, but without using statistical sampling, the troublesome undercount of the 1990 census was reduced.

The Acting Census Bureau Director William Barron said in a news conference in October 2001 that the results of the A.C.E. survey did not measure a significant number of erroneous enumerations and so declared there would be no adjustment made in the 2000 Census figures for any purpose. He described the net national undercount as "virtually zero in statistical terms." Congressman Dan Miller (R-FL), chairman of the Subcommittee on the Census, applauded the ruling that the census accurately shows what he describes as "real people, living in real neighborhoods and communities in a very real nation."

However, the final chapter on the counting controversies in the 2000 Census may not be written for some time. Although imputation accounted for less than half a percent of the total U.S. population, Utah has been fighting an ongoing court battle on the issue of statistical "imputation." The state of Utah claims that the use of imputation has unfairly given the state of North Carolina a seat in Congress that rightfully belongs to them. This imputation issue and the question of the constitutionality of statistical sampling are likely to need clarification before they are used in the 2010 Census. Until all such issues are resolved, statistical sampling should not be used in the U.S. census.—M. C. NAGEL

Further Reading

Anderson, Margo. *The American Census: A Social History.* New Haven, CT: Yale University Press, 1988.

———, and Stephen Fienberg. *Who Counts? The Politics of Census-Taking in Contemporary America.* New York: Russell Sage Foundation, 1999.

Brown, L. D., M. L. Eaton, D. A. Freedman, et. al. "Statistical Controversies in Census 2000." Technical Report 537. Department of Statistics, University of California, Berkeley, October 1998.

Cohen, Michael, et al. *Measuring a Changing Nation: Modern Methods for the 2000 Census.* Washington, D.C.: National Academy Press, 1999.

Draga, Kenneth. *Fixing the Census Until It Breaks.* Lansing, MI: Michigan Information Center, 2000.

National Research Council. *Preparing for the 2000 Census: Interim Report.* Ed. A. White and K. Rust. Report of Panel to Evaluate Alternative Census Methodologies. 1997.

Panel on Census Requirements in the Year 2000 and Beyond. *Modernizing the U.S. Census.*

Supreme Court Opinion. *Department of Commerce et al. v. United States House of Representatives et al.* No. 98-404. <http://supct.law.cornell.edu/supct/html/98-404.ZS.html>.

United States Department of Commerce, Bureau of the Census. *Census 2000 Operational Plan.* 1997.

Wright, T. "Sampling and Census 2000: The Concepts." *American Scientist* 86, no. 3 (May–June 1998): 495–524.

Has the calculus reform project improved students' understanding of mathematics?

Viewpoint: Yes, reform-based calculus provides students with a better grasp of the real-world applications and context of mathematical principles, and it also increases the participation of student populations that have been underserved by traditional teaching methods.

Viewpoint: No, the calculus reform project purges calculus of its mathematical rigor, resulting in a watered-down version that poorly prepares students for advanced mathematical and scientific training.

Calculus is quite literally the language of science and engineering. While the concepts and formalisms of calculus are more than 300 years old, they have never been more centrally important than they are today. Educators face the challenge of preparing students for careers that increasingly depend on science and technology. Traditionally, at colleges and universities calculus has served as a prerequisite for the study of any kind of science or engineering. In practice, the high failure rate in introductory calculus courses served to filter out many would-be science and engineering majors. During the 1980s, as more and more scientific and technological fields offered promising career opportunities, college administrators became critical of mathematics departments for the high failure and attrition rates typical of introductory calculus courses. Some educators began to consider new ways to teach calculus, hoping to improve performance and to make calculus a "pump" for prospective science students, rather than a filter. But to others these reform efforts seem merely to weaken the calculus curriculum, substituting faddish pedagogy for rigor and hard work.

At the core of the calculus reform debate is a long-lived problem common to most mathematical education. Should mathematics be taught primarily as a toolbox for its many applications, or should it be taught for its own sake as a challenging and important intellectual achievement? Both positions have their risks as well as rewards, which have been quite evident in the struggles surrounding the calculus curriculum. Critics of calculus reform suggest that students fail to grasp the general concepts and rigorous proofs that are the essence of calculus, while those who defend an application-based approach argue that students who focus solely on mathematical theorems and proofs often fail to understand how, why, or when to apply their knowledge. A newer, related twist to debates about teaching calculus is determining the best way to use computers in the curriculum. Computers can be used to perform tedious calculations and algebraic manipulations, and also to illustrate complex concepts such as graphing functions. But the value of computers to teaching calculus is diminished by the risk that students will learn more about manipulating a particular computer program than about calculus in general. Furthermore reliance on computer-generated solutions to problems can obscure important ideas, such as the difference between approximations and valid proofs.

The calculus reform debate has taken place in a wider context of concern about disparities in educational performance and opportunities for minority students. In the case of mathematics, and the scientific and technical courses that rely upon it, the traditional student population in the United States tended to be white (or Asian) and male. Proponents of calculus reform suggest that conventional teaching methods contributed to the dearth of successful female, black, or Hispanic students in calculus courses. They argue that reform strategies that emphasize problem-solving, group cooperation, and verbal accounts of mathematical problems would be more inclusive and lead to a more representative population of calculus students. If this is indeed successful, this could have broad impact on the representation of women, blacks, and Hispanics in many scientific and technical fields.

There is a great deal at risk with calculus reform. Because calculus is so essential to the successful study of science and engineering, poor mathematical preparation can cripple even the most talented student in those fields. The traditional curriculum, with its emphasis on mathematical rigor, theorems, proofs, and calculations, served the best students well. Those with adequate mathematical preparation and a capacity for hard work could emerge from a calculus sequence with a satisfactory grade and a good knowledge of the subject's fundamental concepts and techniques. Those with poor preparation or a lack of discipline, on the other hand, would often give up or fail. One of the primary goals of calculus reform has been to make it easier for all students to do well in calculus. While this could be good news for those who might otherwise have failed the subject entirely, better students—the traditional pool of future scientists and engineers—may pass through the calculus sequence knowing far less than they would have otherwise. Critics charge that while reformers claim they are democratizing the curriculum, they are really just watering it down.

The impetus to reform the calculus curriculum in the 1980s and 1990s came from concern over the large number of students who failed to progress through calculus and therefore be eligible for further study and careers in fields such as physics, engineering, and computer science. The growing economic importance of these fields ensured that college administrators would support educational reform efforts aimed at increasing enrollment and success rates in that pesky prerequisite, calculus. Reform-minded educators transformed the calculus curriculum. Innovations included computer-based learning, group study, reliance on learning by concrete examples, and verbal analysis of mathematical problems. But careful study of the traditional methods of problem solving and the rigorous demonstration of theorems and proofs were largely put aside. Reformers cheered a larger, more diverse group of students achieving passing grades in calculus, while critics moaned that the new curriculum doomed all calculus students to a lowest common denominator of merely superficial knowledge. Judging the success of calculus reform requires consideration not only of the respective curriculums, but also of the goals of calculus teaching and its position in the university curriculum. Is it for the talented few or the mediocre many? The debate over calculus reform has inadvertently posed this dichotomy, but surely successful reform should achieve both goals—rigorous training in important concepts and techniques, accessible to any committed student. —LOREN BUTLER FEFFER

Viewpoint:

Yes, reform-based calculus provides students with a better grasp of the real-world applications and context of mathematical principles, and it increases the participation of student populations that have been underserved by traditional teaching methods.

Calculus is an elegant intellectual achievement that can reduce complicated mathematical problems to simple but precise rules and procedures. Unfortunately, calculus has failed to become an important part of most college students' education for the same reason.

Taught for decades as merely rules and procedures, calculus, as it is traditionally taught, has had little or no meaning for most students. In addition, while teachers in other scientific classes and disciplines were using new technologies like computers to enhance their students' skills and prepare them for a high-tech marketplace, the calculator's first appearance in the 1970s was the only major technological advance in calculus education. Not surprisingly, at about this same time-period, the number of math majors at universities declined by 60%. Over the following decades, nearly 50% of the students who did take traditional calculus would fail or receive low grades.

Calculus is the study of "continuously" or "smooth" changing quantities (think of planets changing position or the movement of a car). Its applications range from physical concepts, such as velocity and acceleration, to geometric ideas, such as graphing curves. The ability to interpret and calculate change (for example, population or economic growth) can be an important skill for college graduates as they pursue a variety of careers, including in math and physics, biology, the health sciences, and social and economic sciences. As a result, more and more students should be learning calculus. But traditional calculus, with its emphasis on memorizing theorems and formulas regardless of understanding the concepts behind them, has slammed the door shut for most students.

In 1986, leading math educators from around the country attended a conference at Tulane University to look at the way calculus was taught, with an eye toward making it a more valuable and practical skill that would appeal to students. More and more educators were coming to the conclusion that most students were not grasping important concepts and skills, and the math problems and techniques being taught were too limited.

Two years later, the National Science Foundation (NSF) initiated a program funding efforts to design a new approach to calculus education. Out of this a radical idea emerged. Why not teach students to understand calculus in relation to real-world problems? By emphasizing active learning, mathematics instruction could become what it really should be for most students: training ground for becoming qualified scientists and other professionals in a variety of endeavors. This emphasis on "active" learning is the basis of the calculus reform movement.

What's Wrong with the Old Way? The traditional approach to teaching calculus could be described as the "plug-and-chug" approach. The student is given a neatly packaged textbook problem and then trained to plug in the answers based on well-defined procedures. The ultimate goal for most students is not to understand the problem, but to get an answer that agrees with the one in the back of the book.

The results are two-fold. First of all, students who are required to enroll in a calculus course but have little prior preparation or aptitude for the field are failing calculus at an alarming rate. Secondly, even when students do chug along and learn how to do the calculations and succeed in the coursework, few can actually recognize a calculus-based problem outside of the classroom or even begin to think of applying calculus in real-world situations. This became more and more apparent as teachers of higher-level physics and engineering courses complained that the students entering their classes were not "prepared" to handle the calculus needed in their courses.

The essence of calculus is not to be able to do computations. Rather it is "proof," that is, a formal explanation of an observed pattern, like Sir Isaac Newton's proof explaining that the planets' orbits are subject to gravity. Newton certainly used calculations as the means of developing the proof; but they are not an end in and of themselves. Newton is not remembered as a great calculator but as a developer of the theory of gravity.

> **KEY TERMS**
>
> **FUNCTION:** A mathematical relation in which each element of a set is assigned to exactly one element of (the same or) another set.
> **RULE OF THREE:** A central part of the calculus reform approach that states that every topic should be approached graphically, numerically, and algebraically, rather than the algebraically dominated traditional approach.
> **THEOREM:** A mathematical proposition that is provable on the basis of explicit assumptions or deductive logic. Some famous mathematical theorems include the Pythagorean Theorem and Fermat's Last Theorem.

In the traditional calculus approach, the emphasis is placed on memorization of theoretical definitions. There is usually one right answer and a "preferred approach" to getting a solution, which is not reflective of real-world situations, whether it is in physics, economics, or medicine. As a result, students learn to do computations but not to understand. Repetition does instill learning, but with its emphasis on formulas and rigid rules, traditional calculus teaching falls dismally short of educating students in the realm of applying what they learn. In the end, many of those students who do well in the traditional calculus course achieved their success by memorizing rules and procedures to plug into homework and answers to exam questions.

Algebraic manipulation is important, but the thinking process is paramount. Asking students to memorize theoretical definitions with no real understanding leads to passivity. Nevertheless, most traditional calculus teachers expect their students to rise to a higher level without addressing students on a level that most of them can understand. In other words, students are expected to take giant steps in intellectual understanding without first having established a solid foundation.

Traditional calculus textbooks also have few problems in them that relate to real-world situations. This leads to disinterest in students who are not pursuing a career in mathematics or physics. In the end, traditional calculus education has undergone reform because it has committed a cardinal sin of education—failure to make calculus interesting or exciting to most of its students.

How Is the Reform Approach Better?
Reform-based calculus uses a radical new

approach to teaching calculus based on real-world problems, computer technology, and more active participation by students (both individually and as a group). The goal is to make calculus more accessible to a wider group of students and to improve students' learning and comprehension of the field, moving away from the "chalk-and-talk" approach in which teachers lecture and students listen. Instead of the traditional approach in which students often work in isolation, more emphasis is also placed on interactions, both among students and between students and teachers.

For many students taking a traditional undergraduate college calculus course, a nagging question usually filters through their minds: "What on earth does this have to do with anything?" Calculus reform's emphasis on real-world problems helps to pique students' interest, motivate them, and increase their conceptual understanding. For example, real-world scenarios can be generated from numerous fields, including biology, economics, and even military science. They are often based on important issues that could impact everyone, like the growth of populations or the spread of epidemics. It is the investigation of practical problems that leads to formal definitions and procedures and not vice versa.

Integral to the new approach to teaching calculus is the "Rule of Three," that is, making sure students understand the interaction among numerical, symbolic, and graphic representations as opposed to the emphasis on algebraic solutions alone, as has been the custom in traditional calculus courses. (Some calculus reform proponents refer to the "Rule of Five," adding written and oral communication to the three basic rules.) The Rule of Three fosters the concept of looking at things from more than one perspective, which is valuable in more than just mathematics. It encourages understanding but gives students with weak manipulative skills the opportunity to grasp the basic concepts of calculus while strengthening their backgrounds.

One example of a problem that allows students to use multiple strategies to solve real-world problems would be population growth comparisons, or exponential growth. For example students may be asked to make a table showing the changing population of two cities and to determine in which year one city's population first exceeded the other city's population based on each city's populations in a specific year (say 1980) and growth percentages for all the following years (for example, 1.5% for city A as opposed to 4.5% for city B). The problem can be solved several ways, including using arithmetic to create one table one line at a time, numerically using tables together, graphically, and traditionally using logarithms. By using these separate approaches, students would know "automatically" that something in their solution of the problem was wrong if the answers used in different approaches did not match.

The additional emphasis on students being able to talk and write coherently about the "problem" and its "solution" also increases their interpretive and communications skills. It enables students to explain in a language he or she understands rather than using a foreign language, which is what mathematics is for many students. For example, students may be presented with the problem of determining how to choose the best telephone savings package and how much money would be saved by choosing a specific plan. In addition to the use of step functions to solve the problem, they then may be asked to write a letter to convince their telephone company to give them better rates and to include a written explanation of the mathematical justification for their request.

Technology Traditional calculus education has lagged behind most other disciplines in integrating technology into the classroom. Most of the algebra needed to perform calculations can now be done by a computer, eliminating the need for the laborious and, for the most part, boring rote, paper-and-pencil calculus of traditional calculus learning.

Computer technology also improves learning through its use of graphics. Educators recognize that students comprehend things more readily when they can visualize it. With computers, students access well-conceived interactive computer graphics and can create their own graphics of almost any function as well. With software such as Mathematica and Matlab, students can quickly produce graphs of most functions, fostering a geometric approach to calculus, better understanding of concepts, and stronger connections between graphical and formulaic representations.

Several evaluations of students using computer algebra systems to do computations compared with students using traditional approaches showed that both sets of students had about the same manipulative skills for completing problems. However, students using the computer performed slightly and sometimes significantly better on conceptual problem solving.

Admittedly, technology in and of itself is not necessarily better. But the emphasis of reform is on understanding and not computer technology. For example, the old square root algorithm is no longer taught in grade school because it can be done with a calculator. But that does not mean that students can't be taught or grasp what the square root means. It is clear that computers help students to break free of the meaningless plug-and-chug approach to solving problems by fostering students' understanding of the interaction among numerical, symbolic, and graphic representations. The addition of the computer "laboratory" in many calculus reform programs also fosters communication as students work more in groups to solve problems.

Overlooked Benefits An often overlooked but significant aspect of calculus reform is that it fosters an ongoing dialogue about the best ways to teach calculus. It is important to have the right mix of computer technology and hands-on approach, and many questions have been raised regarding this balance. How much technology should students use? How much of the mechanical skills should be discarded? How does a teacher develop a proper balance between theory and application? By probing and discussing these issues, calculus education can't help but improve.

Although the "demographics" of who studies and does well in science and math are changing, they traditionally have been white males. The reason why females and other minority students have not done as well may include social influences and access to quality education. Calculus reform can help these students. With less emphasis on formula memorization and traditional testing, even students with weak algebra skills can succeed in calculus and build confidence. The symbolic computational skills that are necessary are then more easily learned. Although more formal data is needed on outcomes due to reform calculus, many reform calculus teachers are reporting improved retention and passing rates.

The Future: A Synthesis Throwing out the baby with the bath wash is not the intent of calculus reform. It is also not about technology. Calculus reform is about making calculus more accessible to more students and, as a result, improving the learning process. Although the content in many reform calculus courses has been greatly revised, these courses still focus on the traditional nature of calculus: applying it to problems of change and motion and developing precise definitions combined with rigorous statements of results. Reform calculus is also not a dilution of calculus education. Students discover that the courses are more challenging and demanding, but in an interesting way.

According to the authors of *Assessing Calculus Reform Efforts*, over 95% of the colleges and universities that incorporated calculus reform into their curriculums continued to use reform texts the next year, and very few faculty members who have tried the reform approach have gone on to abandon it. By emphasizing interactive, numerical, and geometric reasoning combined with communication skills, reform calculus has incorporated new approaches and materials to improve students' conceptual understanding and overall ability to learn calculus. Calculus reform represents a shift from an approach that seemed intent on "weeding out" students, focusing only on those who appear suitable for the rigors of calculus and scientific careers. In contrast, reform calculus "equalizes" the learning environment, enabling more and more students to appreciate the fundament nature of calculus, while opening up new vistas of understanding and career opportunities. —DAVID PETECHUK

Viewpoint:
No, the calculus reform project purges calculus of its mathematical rigor, resulting in a watered-down version that poorly prepares students for advanced mathematical and scientific training.

The essential difference between the reform and traditional approaches to teaching calculus is the emphasis placed on solving essential mathematical problems using a rigorous and sophisticated approach. Calculus reform requires only a superficial use of mathematical skills. It emphasizes real-life problems but does not believe that students must first be able to understand concepts and actually perform the algebraic equations necessary to do calculus. In contrast, traditional calculus focuses on teaching students mastery of basic calculus skills, including understanding, developing, and using theory and proofs.

In the late 1980s, with 40% of undergraduates failing introductory calculus, educators began to question their approach to teaching calculus. Not only was the poor student performance raising college administrators' eyebrows, but calculus teachers' own colleagues in science and engineering academic departments were complaining that students, after taking basic college calculus courses, were still unprepared for the calculus needed in their courses. Their solution, in the form of calculus reform, was to make the courses "easier" and to condemn computational skills and mathematical proofs as "boring" and "unnecessary." Increased emphasis was placed on computer learning and more conceptualization, including students writing about the problems and how they came up with solutions.

Trying to usurp basic skills with conceptual understanding is putting the horse before the cart. For example, while an emphasis on students writing about their calculus problems seems to be a good idea, the very essence of calculus is that it's a symbolic representation that presents a different, more scientific approach than everyday human language to describing and learning about the world.

Traditional calculus does not ignore conceptual understanding. However, it does emphasize problem-solving skills as the way to conceptual knowledge. Basic computational skills and understanding are inextricably linked in the world of mathematics. For example, on the basic arithmetic level, adding, subtracting, multiplying, and dividing reveal patterns and relationships that are absolutely vital to understanding algebra. Although modern, streamlined, computerized society may eschew the one-step-at-a-time approach, it is the ability to perform basic tasks that leads to progress, and this is especially true in mathematics. Once the "basic" computational skills are learned, students are then "free" to pursue conceptual understanding in a meaningful way.

High-Tech Math Most adherents of calculus reform claim that the increasing use of computers in the classroom allows students to experiment and be creative in dealing with more open-ended questions, most of which have more than one single right answer. While the computer may give students more to talk about, it does little to enhance the computational skills needed for real understanding. For example, many reform calculus computer programs help students to avoid using l'Hôpital's rule for calculating limits or even to do calculations by integration by parts or partial fractions, all of which are basic to a true understanding of calculus.

Overall, the use of computers has not provided students with enough practice in standard algebraic manipulations. In addition, computers, coupled with poorly thought out calculus reform textbooks, often underexpose students or gives them no exposure at all to such basics of algebra and calculus as:

- The definition of limits—a theoretical foundation of calculus that enables students to know what "e" to the "x" really means
- The definition of continuity—a fundamental idea in mathematics that is often undertreated by calculus reform texts and computer programs
- The intermediate value theorem—a theorem that is important in both theory and the practical use of finding roots of functions
- Partial functions—a basic calculus topic that is necessary for science and engineering majors in later courses

Both computers and standard calculus reform texts also emphasize open-ended problems, which only furthers students' misunderstanding concerning the idea of proofs. In general, calculus reform proposes that if a mathematical proposition is true for a finite number of cases, then it is true in general. This leads to confusion between when something has been shown to be true and when it is *likely* to be true. By the overuse of general conclusions, calculus reform software programs and texts are giving students a false picture of mathematics. Telling students that something is always true because of three or four examples is a "lazy" arbitrary approach to mathematics that actually hinders students' critical thinking and understanding.

In addition to relying on programs to perform basic tasks, computers rob students of time that could be used to learn calculus instead of learning how to use the computer software. For most students, their short time learning calculus is the *only* time they would ever use such a program. Computers also take away from the time spent on human dynamics of teaching, such as adjusting the material to the audience, setting a pace for students to learn, and instilling critical thinking and reasoning. Not less but more teacher-student interaction is needed to help students become engaged in the process of learning calculus and to set standards for what understanding calculus really means. Finally, establishing expensive computer "labs" for calculus courses also drains much needed monetary resources that could be better used for faculty training and seed money for overall department improvements.

Race, Ethnicity, and Gender The calculus reform movement also draws legitimacy from its proposal that the reform movement helps "level the playing" field for people of both sexes, regardless of race or culture. By de-emphasizing mathematical rigor, say the reformists, students from lower socioeconomic groups who did not have access to quality education or women who did not learn calculus because of social biases will now be able to "compete" with white and Asian males who dominate the math and engineering fields in the United States.

Culture undoubtedly plays an important role in academic achievement and in the opportunities and personal inclinations to pursue college and higher education. Nevertheless, racial and gender theories of learning must be viewed with caution. Is it "logical" to make courses "easier" so that more students do well rather than maintaining a high standard of integrity in a discipline that is admittedly difficult for many students? The reform movement seems to have gone astray by focusing on how to make calculus education easier. What is needed is a long, hard look at the discipline of mathematics education and how to improve teaching to obtain the same end results in terms of knowledge, mathematical ability, and understanding of calculus.

A famous example of how traditional educational approaches can be effective in helping minorities, women, and other groups thrive in mathematics is Jaime Escalante, whose success story in high school mathematics was made into the movie *Stand and Deliver*. Escalante did not seek to "dumb down" his teaching approach. The high school where he taught had a large minority population of Hispanics and working-class students and an extremely poor performance rate by its students in testing. In fact, the school was in peril of losing it accreditation in mathematics. Escalante turned the tables by insisting on the basics but making it exciting for his students. He went on to create a calculus class, with a significant number of his students going on to pass an advanced placement calculus exam. Even more important, more of his students went on to universities equipped with a strong understanding of basic algebra and calculus. The fact is that calculus reform movement came about, in part, because high schools failed to teach minorities, women, and others the basic algebra needed for calculus. Part of the answer to teaching calculus in college is to improve high school mathematics education and to set higher standards for teachers. No one is served by being required to learn less, which, in the case of minorities and women, only perpetuates stereotypes.

Calculus reform's failure in addressing specific student populations is further compounded by the educational reform notion of all for one. Universities using calculus reform are tending to use reform courses for all students, including those looking to pursue careers in math, engineering, and the physical sciences. Requiring science majors to take a three-semester calculus course that provides comprehensive coverage of theories and proofs is absolutely fundamental to their future performances, both in academia and in their chosen fields.

Real-World Problems A good idea associated with calculus reform is the emphasis on real-world problems. Few would argue that students show more interest in learning something if they can apply it in a practical way. This is true not only for non-science majors in college, but also for students pursuing engineering and other science-based fields. However, using real-world problems still requires mathematical rigor. Just like the real-life example of Jaime Escalante in *Stand and Deliver*, the search should focus on how to better motivate students so that they will learn algebra and calculus as opposed to decreasing expectations by not requiring mathematical skills. Calculus is an analytical tool that represents the world in symbolic forms, and analysis is rooted in the practical.

Many of the newer calculus textbooks are beginning to recognize this fact and, as a result, are shying away from the most "radical" ideas of calculus reform. Instead, they are committed to the "roots" of calculus education in understanding and using mathematical models to describe and predict physical phenomena through the ability to use convergence, limits, derivatives, integrals, and all the other fundamentals of algebra and calculus. Real-world problems should be used to focus students' minds on practical problems, but in a way that motivates them to learn and use algebra and mathematics in calculus, not discard them.

Where's the Proof? Calculus reform is not reform at all but a fad, that is, something followed with exaggerated zeal. Too many universities and calculus faculty have accepted calculus reform as being universally superior to traditional calculus. But there is little evidence to support this belief. Why are colleges and universities in such a rush to discard traditional methods of teaching calculus when the efficacy of calculus reform classes has not been established? For example, no one is sure if calculus reform is really increasing students' critical thinking and their understanding and mastery of mathematics. And there is little evidence that calculus reform programs are really making students excited about calculus and mathematical science. Many universities, such as the University of California, Los Angeles (UCLA), and the University of Iowa, have either flat out rejected calculus reform after developing and using reform courses, are in the process of moving away from reform courses, or are instilling a stronger traditional approach to them.

While the reform movement takes away from mathematical content in calculus courses, its emphases on group learning, computers, activities and projects, and writing has led to the evaluation of students and grading systems that are based more and more on the "processes" of learning as opposed to understanding content. In addition, grading approaches vary greatly from one class to another and from one university to another so that "standards" are practically non-existent. The result is that calculus reform results in large variations in the quality of teaching calculus and tends to confuse teachers on how to differentiate between success and failure in the course.

The trend to water-down calculus by purging it of its mathematical rigor in order to reduce failure rates and increase overall grade point averages is a dire mistake. It is not possible to provide students with an understanding of calculus by minimizing the component skills.

The way to improve students' understanding and achievement in college calculus is to maintain standards based on the fact of life that hard work leads to success and results. Calculus should not be expected to apologize for being difficult for many students. Instead, calculus education should adhere to the principals of traditional calculus to provide students with skills in computation, conceptualization, and theory—skills that will help students succeed.
—DAVID PETECHUK

Further Reading

Heid, M. K. "Resequencing Skills and Concepts in Applied Calculus Using the Computer as a Tool." *Journal for Research in Mathematics Education* 19, no.1 (1988): 3-25.

Klein, David. "Big Business, Race, and Gender in Mathematics Reform." In *How To Teach Mathematics.* American Mathematical Society, 1999.

———, and Jerry Rosen. "Calculus Reform—For the $Millions." *Notices of the AMS.* 44, no. 10: 1323-1325.

Leitzel, James R. C., and Alan C. Tucker, eds. *Assessing Calculus Reform Efforts.* The Mathematical Association of America, 1994.

McCallum, William. "The Goals of the Calculus Course." <www.math.arizona.edu/~wmc?Research?Goals/Goals.html>.

Owen, Thomas. "Some Contradictions of Education Reform: Constructivism, High-Tech, and Multi-Culti." <http://people.delphi.com/vlorbik/talk/amstalk.html>.

Rosen, Jerry, and David Klein. "What Is Wrong with Harvard Calculus?" <http://mathematicallycorrect.com/hc.htm>.

Smith, David A. "Trends in Calculus Reform." *Preparing for a New Calculus. MAA Notes* no. 36 (1994): 3-13.

Uhl, J., et al. *Calculus & Mathematica.* Addison Wesley, 1994.

Wilson, Robin. "'Reform Calculus' Has Been a Disaster, Critics Charge." *The Chronicle of Higher Education.* (February 7, 1997): A12.

Wu, H. "Basic Skills Versus Conceptual Understanding." *American Educator* (Fall 1999): 1-6.

Do humans have an innate capacity for mathematics?

Viewpoint: Yes, recent scientific studies suggest that we are born with at least some mathematical ability already "hardwired" into our brains.

Viewpoint: No, mathematics involves not just counting or simple arithmetic but also abstraction, which can only exist in the presence of language skills and symbolic representation.

The debate over whether humans have an innate capacity for mathematics often hinges on two semantic questions: 1) What do we mean by "innate?" and 2) What cognitive skills are to be classified as mathematical?

In common usage we often speak of people having innate abilities in a particular area, or being a "natural" at some skill. This generally means that they have an aptitude for it; they seem to learn it quickly and easily. However, philosophers define *innate knowledge* as that which is not learned at all, but which is present at birth. An influential school of philosophy known as empiricism holds that innate abilities do not exist. Empiricism, championed by the English philosopher John Locke (1632–1704), regards all knowledge as learned, and the newborn as a "blank slate."

Yet experiments have shown that many behaviors and instincts are apparently inborn in humans as in other species. And a few of these seem directly related to mathematics. For example, researchers have presented babies with cards bearing two black dots. After looking at these for a while, the babies lose interest. But they begin to stare again when the two dots are replaced by three. Changes in color, size, or brightness of the dots do not elicit the same response.

In general, people can perceive how many objects are in front of them without counting them, if the number is small; say, up to three or four. This ability is called *subitization.* Subitization is not an ability unique to humans; other primates, rodents, and even birds have demonstrated it in experiments. Scientific evidence indicates that subitization and other basic abilities related to mathematics are innate. They seem to be controlled by the left parietal lobe of the brain. Stroke victims with damage to this region of the brain may be unable to distinguish two objects from three without counting them, even if they are otherwise unimpaired.

Yet whether these basic abilities constitute "mathematics" is another question. Some scholars maintain that true mathematics must involve abstract concepts, such as the relationships between numerical and spatial entities. This definition excludes not only the understanding of small quantities and the ability to compare two quantities, abilities that are apparently innate, but also basic learned skills like counting and simple arithmetic.

In order to manipulate abstractions, language and symbolic representation (that is, a number system) must exist. These are cultural phenomena that evolved differently in the various regions of the world and spread along human migration and trade routes.

Obviously, babies do not have the knowledge or skills to perform mathematical calculations. No one is born with the ability to do calculus. Environment, encouragement or discouragement by parents or teachers, and the presence or absence of "math anxiety" all affect the individual's likelihood of acquiring mathematical skills and abstract reasoning abilities. —SHERRI CHASIN CALVO

Viewpoint:

Yes, recent scientific studies suggest that we are born with at least some mathematical ability already "hardwired" into our brains.

John Locke (1632–1704), an English philosopher, famously argued in *An Essay Concerning Human Understanding* that "No man's knowledge here can go beyond his experience." Locke's position, known as empiricism, holds in essence that all human knowledge is acquired through experience and experience alone. Thus, Locke believed that humans possess no "innate ideas" at birth. As such, Locke likened the minds of humans at birth to a *tabula rasa*, or soft tablet (sometimes translated as a "blank slate"), upon which knowledge is written. Thus, for Locke and other empiricists, knowledge is a matter of nurture, not nature.

The complexities of mathematics seem to present an obvious and intuitive example of Locke's thesis against "innate ideas." No one, after all, is born knowing how to solve quadratic equations and other complex mathematical functions. Indeed, acquiring the ability to solve equations—let alone understand what a mathematical equation is—takes years of education and practice, and even then many continue to have enormous difficulty. Math, according to this view, is a matter of nurture, not nature: one has to *learn* how to "do math."

Some people, however, seem to have a knack for mathematics, even at a very young age. Every so often we hear reports in the media about children who are able to perform calculus and solve sophisticated mathematical equations. What accounts for the different mathematical abilities in these children? Do such prodigies indicate that math is more a matter of nature rather than nurture?

As with many issues, the answer appears to be somewhere in between. Indeed, despite the intuitive appeal of Locke's position, there now appears to be strong evidence that humans—and even higher-order mammals—possess an innate mathematical ability and can perform some rudimentary mathematical operations such as simple addition and subtraction. So, although we have to *learn* how to solve a quadratic equation, recent scientific studies suggest that we are born with at least some mathematical ability already hardwired into our brains. The purpose of this essay is to look at this "innate" mathematical ability that all humans share.

What Does It Mean for Math to be Innate?

At this juncture, it is worth commenting on the definitions of "innate" and "mathematics." According to Webster's Dictionary, something is innate if it exists "from birth," or is "inherent in the essential character of something." Likewise, mathematics is defined as "the systematic treatment of magnitude, relationships between figures and forms, and relationships between quantities *expressed symbolically*." Certainly, no one contends that an infant has, at birth, the ability to *express symbolically* anything, let alone magnitudes, figures, forms, and quantities. Nurture, not nature, is required in order to understand the system of mathematics.

But does this mean that mathematics is not innate? The answer, of course, depends on how narrowly one interprets and defines mathematics. In a rather technical, but uninteresting, sense, mathematical ability is not innate; we cannot recognize the meaning of the symbols 2 + 2 = 4 at birth. We must, of course, learn the symbols of mathematics before we can express mathematical relationships, i.e., before we can "do math" as is commonly understood. This essay, however, adopts a broader view of mathematical ability. Although infants cannot express anything symbolically in terms of writing down mathematical or linguistic expressions, they nevertheless can, and do, reveal rudimentary understanding of such mathematical concepts as number, addition, and subtraction.

At this point it is important to recognize that there is a difference between understanding the meaning of—and being able to manipulate—the symbols 2 + 2 = 4, and the concepts underlying those symbols. In other words, although infants may not recognize that "2" means "two things" and + symbolizes the mathematical operation of addition, infants nevertheless can express behaviorally that they understand such concepts as quantity, addition, and subtraction. How they express this understanding is discussed below.

Infants and Mathematics

For many, just a glance at this WORD will reveal that it has four letters. One need not count the individual letters to arrive at a number—it seemingly is an automatic function of the brain. However, in the case of a word with NUMEROUS letters, it is not so easy to ascertain quickly the number of letters present—one may in fact be forced to count the letters in order to determine that "numerous" has eight letters in it.

This ability to discriminate among a collection of things, at least when the number of things in question is of a relatively small amount, is known as "subitizing." Although it is not clear how the brain operates when it subitizes, it is clear that this ability requires the brain to distinguish between quantities—in other words, to recognize that there is a difference between • • and • • • and between • • • and • • • •.

Research conducted throughout the 1980s and 1990s indicates that human infants as young as three or four days old have the capacity to subitize. In other words, they can recognize a difference in a collection of things, when the difference concerns the number of things present and not the things themselves. Thus, the concept of "number" appears to be innate. Within a few months, infants begin to exhibit behavior indicating that they understand that one plus one is equal to two, and that two minus one is equal to one. And as they become older, research indicates that they understand even more complex mathematical operations.

Researchers working on the arithmetic capabilities of infants are able to come to these conclusions by exploiting what is known in developmental psychology as the "violation-of-expectation paradigm." An example of an experiment that uses this paradigm to study the mathematical ability of infants is described below:

A researcher places an infant on her mother's lap facing a small stage. The researcher then sets a puppet on the stage immediately in front of the infant, making sure that she sees it. The researcher then places an opaque screen in front of the stage so that the infant can no longer see the puppet. While the screen is up, the researcher, in a manner completely visible to the infant, then places a second puppet behind the screen. After a moment, the researcher drops the screen to reveal one of two things: either two puppets or just one puppet (in the latter case, the researcher surreptitiously removes one of the puppets before dropping the screen).

What researchers have noticed is that when two puppets remain after the screen is dropped, infants tend not to take much notice. However, when the screen is dropped and there is only one puppet when there should be two, the infants tended to "startle" or stare longer at the stage than otherwise. Repeated experiments with various different objects and numbers of objects have tended to produce the same outcomes, providing support to the idea that not only do infants understand the concept of "number," they also have the capacity to under-

KEY TERMS

ABSTRACTION: The transformation of a concrete object into an intangible idea.
EMERGENT PROPERTIES: Properties of a complex system that are greater than the sum of the properties of the parts of that system.
EMPIRICISM: View that all knowledge is acquired through the senses; generally opposed to the view that any knowledge is innate.
INFERIOR PARIETAL CORTEX: Region of the brain thought to be responsible for mathematical ability.
NEUROSCIENCE: Science of the study of the brain and its functions.
PLACE VALUE SYSTEM: A system of numerical notation, where the value of a symbol varies depending upon its position in the number.
SUBITIZATION: The innate ability to comprehend the size of small numbers (1–3) without having to count. The "knowledge" of numbers.

stand that one thing added to another thing makes two things, and that one thing taken away from two things leaves one. This capacity seems to indicate that some mathematical abilities are hardwired in the human brain. The tablet of the brain, apparently, is not as "soft" as Locke presumed.

Mathematics and the Human Brain Some neuroscientists believe they have identified the particular area of the brain responsible for our ability to understand mathematical concepts and perform arithmetic calculations. Neuroscientists have long been familiar with how lesions on the brain affect certain mental abilities of patients. In particular, neuroscientists have discovered that when legions occur on the *angular gyrus* within the inferior parietal cortex—a rather small area on the side of the brain toward its rear—mathematical ability is severely affected. Thus, mathematical ability appears to be localized in a different area of the brain than speech or language.

Another piece of evidence for a special "mathematics" area of the brain comes from studying perhaps the greatest mind of all time: that of Albert Einstein. When Einstein died in 1955 at the age of 76, his brain was removed and preserved in order to study it. In June of 1999, neuroscientists concluded studies of the first-ever detailed examinations of Einstein's brain, comparing the structure of Einstein's brain to the brains of 35 men and 56 women of normal intelligence. What they discovered was amazing. Although Einstein's brain was, for the most part, no different from "normal" brains, the inferior parietal lobe regions of his brain—the supposed seats of mathematical ability—were significantly larger, about 15% wider than normal. As a result of this finding, researchers have proposed studying the brains of other exceptionally intelligent people in order to better understand the relationship between "neuroanatomy" and intelligence.

The human brain is not a "soft palette" or a blank slate upon which all knowledge, including mathematics, is written. Just as the processor chips in computers have basic arithmetical operations hard-wired into them, there appears to be software already preprogrammed into the brain that allows it to acquire more knowledge. As discussed in this essay, humans do posses innate mathematical abilities, which, in turn, prepare us to understand more abstract mathematical concepts and problems. —MARK H. ALLENBAUGH

Viewpoint:
No, mathematics involves not just counting or simple arithmetic but also abstraction, which can only exist in the presence of language skills and symbolic representation.

What Is Mathematics? Before we can address the question of whether or not our mathematical abilities are innate or acquired, we must define exactly what we mean by mathematics. This is not an easy task, but we can start by looking at what mathematics is not. It is not comprehension of quantity, which, as we will see, is a skill innate not only to humans but also to many animals. Nor is mathematics counting or simple arithmetic. These abilities are found in most human cultures, whereas higher mathematics is relatively rare. *The Oxford English Dictionary* defines mathematics as "the abstract science which investigates deductively the conclusions implicit in the elementary conceptions of spatial and numerical relations, and which includes as its main divisions geometry, arithmetic, and algebra." This raises two important points: the abstract nature of math (see below) and the idea of math as relationships between numbers. We might also say that math focuses on changes in numbers, and on the operations that allow us to perform these changes.

Mathematics involves abstraction. The number 5, for example, can be used as an adjective to describe a group of objects, but it does not become a mathematical concept until we are able to abstract it—that is, separate it from its connection to the physical world. This is not easy to do. Although we have developed a sophisticated system for representing and manipulating large numbers, we have no real feeling for their magnitude, unless we visualize it as something concrete, say the population of Los Angeles, or the number of people in a sports stadium. We have to learn abstraction. Small children learning arithmetic begin by using real objects, often their fingers. The next step is to imagine the number applied to reality—a child may struggle to, say, subtract 1 from 4, until prompted by being asked to imagine a situation with real, familiar objects, such as pieces of candy. Children do not naturally use numbers in an abstract sense.

Does mathematics exist independently of the human mind? The Platonic view holds that it does, that it is a body of concepts existing in an external reality. This implies that math is gradually discovered and mapped out, much as an unknown country might be. The opposing view is that math is internal, created by humans as the product of our minds and the way they function, so that we would describe mathematical innovations as inventions. This view is advocated by George Lackoff and Rafael Nunez in their *Where Mathematics Comes From: How the Embodied Mind Brings Mathematics into Being* (2001).

This article draws on what might be termed an intermediate view, which comes from the mathematician Reuben Hersh. He holds that math is neither entirely physical nor entirely mental but, like law or religion, a social or cultural phenomenon. It is external to the individual but internal to society. This is known as the humanist philosophy of mathematics and has much in common with the views of another famous mathematician, Ian Stewart, who describes math as a "virtual collective." Each individual mathematician has his or her own system of mathematics inside his or her head. These systems are all very similar because all mathematicians have similar training, and they communicate their ideas with one another.

The Evolution of Mathematics In order to see how mathematics has developed as a cultural phenomenon, rather than as an innate capability, we must look at its development in human history.

Number Sense. Humans have the ability to subitize—this means we can see two or three objects and "know" how many there are, without having to count them. We can also add and subtract these small numbers. Experimenters have shown that pre-verbal human babies have these same skills, which suggests they are hardwired in the brain. Many other species, including chimps, raccoons, rats, and pigeons, also appear to be able to subitize, suggesting that the ability has ancient evolutionary origins.

In all species, subitization works only up to 3. Above this, a counting system becomes necessary. However, there are societies that have managed to survive without counting above 3 or 4. Many aboriginal societies lack words for numbers over 3. The Warlpiris in Australia, for example, have words for only 1, 2, "some," and "a lot." The people in such tribes manage without the need for counting. For instance, they may identify each sheep or cow in their flocks individually, so they would know if one is missing, or they may use the correspondence principle, whereby the herdsman uses a bag of pebbles, with each pebble representing a sheep.

Unlike language, counting systems are not universal. They possibly arose from religious or spiritual ritual encompassing concepts such as duality, pattern, and sequence. The historian Abraham Seidenberg, who made extensive studies in the 1960s, speculates that counting was invented in a few, or even only one, civilized centers, for ritual purposes and from there spread to other cultures. There is some evidence to support this idea. There are many superstitions and taboos regarding numbers. Many peoples believe counting brings misfortune or angers the gods. For example, some African mothers will not count their children, for fear this would bring them to the notice of evil spirits. There appears to be no practical origin for such taboos, so it seems unlikely that they arose spontaneously in so many different places, supporting the view that counting arose in relatively few places and then dispersed to other tribes and cultures.

Counting begins with body parts, especially fingers. There are different methods of finger counting. Some cultures start counting one on the thumb, some on the little finger. Some start with the right hand and others the left. Certain societies extend the fingers from a closed hand while others fold them into the palm from an open hand. The distribution of these different methods adds weight to the argument that counting developed in a few places and then spread to other cultures.

Counting existed for thousands of years before it progressed into mathematics in just a few civilizations. Several developments had to have occurred before the field of mathematics could be established. First, a symbolic representation of numbers, or notation, was required. This began with a one-to-one correspondence, such as one pebble for each sheep, or one knot in a piece of string for each sack of corn. Bones as old as 30,000 years old, marked with notches, which appear to have served as a tally, have been found in the ruins of ancient civilizations. Individual symbols for particular numbers developed much later. More sophisticated number representation also required a place value system, a system of numerical notation in which the same symbol can have different meanings, depending on its position in the number—that is, 1 can mean 1 or 10 or 100, depending on its position. This also necessitated the use of a base number (in our number notation, the base is 10), with successive places representing successive powers of the base. Zero became necessary as a "place holder." Leaving a space in a column can be ambiguous, but inserting a zero makes the representation clear. While base numbers and forms of notation developed widely, very few societies developed a place value system and the concept of zero. These innovations, along with sophisticated language skills, gave us true mathematics.

Mathematics and Language Archaeological evidence indicates that language predates numeracy. Our complex language skills have allowed us to develop symbolic number systems. This, in turn, has made possible abstraction in mathematics.

Examples of the close relationship between language and math are not hard to find. Consider learning the multiplication tables. For most people, these are learned, not as abstract patterns of number but through language, almost like a poem, and it is not unusual for people to recite the table in order to find the answer to a mathematical problem.

Asian people, especially the Chinese, are often particularly good at math because of their language. The system of words used for numbers in Chinese is far more clear and logical than in Indo-European languages. There are Chinese words for the numbers 1–9 (*yi, er, san, si, wu, liu, qi, ba, jiu*), plus multipliers 10 (*shi*), 100 (*bai*), and 10,000 (*wan*). There are no special words for the numbers 11–19 or multiples of ten (20, 30, and so on). Thus 10 is *shi yi* (ten one). The number 35 is *san shi wu* (three ten five). Note that the words relate more closely to the symbols, making them easier to understand. Kevin Miller and his colleagues found that four-year-old Chinese children, on average, could count to 40, while American children of the same age could count only to 15. This delay is due to the difficulties with the "teens" numbers, which do not follow the same pattern as other numbers. While American children do eventually catch up, by that time, Chinese children would have had several years more experience handling larger numbers than their American counterparts.

The conciseness of the Chinese number words is also an advantage. Our memory span for a list of numbers relates directly to the length of the words used for those numbers. Native Chinese speakers can routinely remember strings of nine or 10 numbers, whereas native English speakers can only manage six or seven.

Fluent bilingual people can often think in their second language—they can understand

and answer a question without translating it into their first language. However, anecdotal evidence suggests that this ability breaks down when the question is an arithmetic problem. In such cases, the person solves the problem in his or her native language, translates it, and then gives the answer in the second language.

Mathematics and the Brain Stanislas Dehaene and his colleagues have recently shown that there is a significant difference in brain activity in subjects performing exact and approximate arithmetic and that the use of language appears to be important. Exact arithmetic involves parts of the brain normally used in word association, and multilingual subjects have been found to perform better on exact arithmetic problems if the problems are posed in their native language.

Approximate arithmetic, where the subject is presented with a problem and a set of possible answers and is asked for the solution closest to the correct answer, is language independent, and creates activity in the parietal lobes, an area usually associated with visuo-spatial processing. This non-language-dependent ability may be the same as the number sense observed in babies and animals. Evidence in support of this idea comes from the British cognitive scientist Brian Butterworth, who has identified a small area of the left parietal lobe, which he calls the "number module," believed to be the location of our number sense.

It seems clear that we have an intuitive sense of quantity for small numbers and an ability to manipulate them to a limited extent. The fact that we share this with many other species has been used to support the hypothesis that mathematical ability is innate, but these experiments suggest that "number sense," while necessary for higher mathematical reasoning, is not sufficient proof of innate mathematical ability.

Using positron-emission tomography (PET) and magnetic resonance imaging (MRI) studies, Dehaene, in his book entitled *The Number Sense*, states that when we perform complex tasks that are not part of our evolutionary behavioral repertoire, many different, specialized areas of the brain become activated. It would be interesting to analyze the brain activity of a skilled mathematician performing complex calculations to see how the brain operates when faced with a highly abstract problem.

Conclusion Humans have certain innate abilities such as number sense, pattern identification, and spatial awareness that contribute to mathematical reasoning. We can see that mathematical ability is built on these fundamental units, but we must learn how to organize and use them. The development of symbolic notation and the ability for abstraction were essential for the emergence of mathematics. Thus, mathematics itself is not innate but rather a cultural acquisition. In fact, we might say that mathematics is an emergent property of the human mind and culture combined, in as much as the immense intellectual system that is mathematics today is far greater than the sum of the properties that form it. —ANNE K. JAMIESON

Further Reading

Barrow, John D. *Pi in the Sky: Counting, Thinking and Being.* New York: Little Brown, 1993.

Butterworth, Brian. *What Counts: How Every Brain Is Hardwired for Math.* New York: Free Press, 1999.

Cardoso, Silvia Helena. "Why Was Einstein a Genius?" *Brain and Mind* (Oct.-Dec. 2000). <http://www.epub.org.br/cm/n11/mente/eisntein/einstein.html>.

Dehaene, Stanilaus. *The Number Sense: How The Mind Creates Mathematics.* New York: Oxford University Press, 1997.

———, et al. "Sources of Mathematical Thinking: Behavioural and Brain-Imaging Evidence." *Science* 284 (May 7, 1999): 970–74.

Dennett, Daniel. *Consciousness Explained.* New York: Little Brown, 1993.

Devlin, Keith J. *The Math Gene: How Mathematical Thinking Evolved & Why Numbers Are Like Gossip.* New York: Basic Books, 2000.

Hersch, Reuben. *What Is Mathematics, Really?* London: Random House, 1997.

Lackoff, George, and Rafael E. Nunez. *Where Mathematics Comes From: How the Embodied Mind Brings Mathematics into Being.* New York: Basic Books, 2001.

Motluk, Alison. "True Grit." *New Scientist* no. 2193 (July 3, 1999): 46–48.

Pinker, Steve. *The Language Instinct.* New York: William Morrow, 1994.

Stewart, Ian. "Think Maths." *New Scientist* no. 2058 (November 30, 1996): 38–42.

Does private strong encryption pose a threat to society?

Viewpoint: Yes, while it's true that data encryption protects and furthers important civil liberties, it also poses a threat to public safety.

Viewpoint: No, private strong encryption contributes to American society by safeguarding basic rights, personal information, and intellectual exchanges. It protects individuals from over-zealous law enforcement agencies.

The increasing sophistication, interconnectedness, and ubiquity of communication and computer technologies have made information society's most valuable commodity. Protecting information is of vital importance to commerce, government, and private citizens. Technologies such as the Internet, which has greatly facilitated the transaction and exchange of all kinds of information, require a high level of protection to keep these exchanges secure and private. Coding and encrypting information has traditionally been the province of spies and their governments. Today, it is a part of everyday life for nearly all U. S. citizens. How do we balance the need for privacy and security with government's role in protecting its citizens from illegal and criminal activities?

The need for data encryption is undeniable. Private information about individuals and businesses is stored on computers and exchanged over computer networks, whether by health care providers, banks, insurance companies, or purveyors of commerce over the Internet. Without effective encryption of this information, any one of these transactions would be vulnerable to criminal interference. Identity theft, credit card fraud, and other kinds of "cyber-crime" would make Internet commerce impossible.

While data encryption protects the rights of citizens and their ability to conduct business on computer networks, it also provides security for criminals, petty crooks, and international terrorists as well. Law enforcement agencies have sought ways to gain access to encrypted data in order to prevent or prosecute criminal activity. Since the 1990s, the U. S. government has tried to make strong encryption techniques illegal, to force users of encryption to give "keys" to government agencies, and to restrict the export of encryption products. Each of these efforts has run into difficulties.

Critics of the government's efforts to control encryption have repeatedly raised two issues. Foremost, the critics argue, it violates various freedoms guaranteed in the Bill of Rights—for example, they suggest that encryption programs are protected by the First Amendment's provision on free speech, and that decrypting encoded transmissions would be unreasonable search and seizure in violation of the Fourth Amendment. The second charge is that government efforts to control encryption are technologically outdated and hopelessly flawed. By allowing government control of encryption, the entire system might crumble—criminals could exploit its many flaws, while innocent citizens would lose the protection that private strong encryption has afforded.

While it is important to protect privacy and commerce from government interference, it is equally important for government to protect its citizens from illegal and criminal activities. Clearly the government must be able to extend its traditional crime-fighting roles into cyberspace, and to do so will require some kind of access to encrypted information. Negotiating the competing interests of strong encryption is a complicated problem, but one whose solution is essential. —LOREN BUTLER FEFFER

Viewpoint:

Yes, while it's true that data encryption protects and furthers important civil liberties, it also poses a threat to public safety.

Introduction On the bright, clear, and otherwise unassuming morning of Tuesday, September 11, 2001, an unprecedented and nearly unimaginable coordinated series of terrorist hijackings occurred aboard four commercial airliners within the United States. Within an hour of 9:00 A.M., teams of hijackers overpowered the flight crews of the airliners, herded them into the back with the frightened passengers, and then flew two of the airliners at nearly 400 miles-per-hour (644 kph) into each of the two towers of the World Trade Center, causing their eventual collapse and the deaths of thousands of innocent civilians. A third slammed into the Pentagon, killing a few hundred, and a fourth crashed into a field south of Pittsburgh, Pennsylvania, after some passengers apparently fought back. That last airliner is widely believed to have been targeting either the White House or the Capitol building. President George Bush, along with many others in the federal government, called these despicable hijackings "acts of war" and vowed to retaliate by declaring a war on terrorism.

A commercial airliner had not been hijacked within the United States in over 10 years, and never before had multiple, near-simultaneous hijackings occurred anywhere on earth for the purpose of using the airliners essentially as guided missiles. Historically, hijackings occurred for the purposes of extorting money from authorities, or for coercing authorities into complying with other demands such as releasing certain prisoners from jails. No one had imagined that hijackers would hijack airliners for the sole purpose of committing murder-suicide on such a massive scale. No one had imagined that numerous, perhaps dozens, of international terrorists would be able to coordinate such a sophisticated series of terrorists strikes without being detected, which many experts believe would have taken many years to plan. As the hijackers are suspected of being part of a large international ring of terrorists, they had to communicate with each other over long distances. So, how did they do it without revealing themselves?

Almost immediately after the attacks, the FBI, CIA, the National Security Agency, and other law enforcement and investigatory agencies from around the world began the arduous task of identifying the hijackers and their co-conspirators in an effort to bring all those responsible to justice, as well as to head-off any possible further attacks by other co-conspirator terrorists who may be laying in wait. As with virtually all criminal investigations of today, the focus of the investigation immediately turned to computers and the Internet. Indeed, it now is believed that the terrorists exploited the secrecy afforded by electronic communication via encrypted messages and files sent over the Internet in order to plan, organize, and finance their attacks.

The purpose of this article is to examine how data encryption acts both as the backbone of the Internet as well as its Achilles Heel in the sense that data encryption both protects and furthers important civil liberties, but may also present a threat to public safety. Part II provides a brief overview of data encryption, and Part III discusses in more detail the dual nature of data encryption and the respective benefits and burdens it presents. Part IV then briefly discusses a current proposal by the United States government for balancing the sometimes conflicting interests between civil libertarians on one hand and law enforcement officials on the other.

What Is Data Encryption? The term "data" simply refers to information. It is the plural form of the singular Greek word *datum*, which means a "gift" or a "present." A *datum* thus is a piece of information that is given and from which a conclusion may be inferred; *data* are a collection of *datum* (although we often speak as if *data* are a singular thing, as in "Where is the data?", correct usage is to speak of *data* in the plural, as in "Where are the data?"). The text on this page, for example, can be considered data from which you, the reader, now are inferring information.

Encryption simply is the process of taking information, or data, and translating it into a code. Here is an example of an encryption. Let's say you have the following message: "JACKIE." You also have chosen the following convention as your code: the letter 'A' is equivalent to the number '1'; 'B' to '2'; 'C' to '3'; and so on. Thus, when you translate or encode JACKIE into the code you have chosen, JACKIE becomes 10-1-3-11-9-5. Now, if you were to ask the average person on the street what this seemingly random group of numbers mean, few would be able to guess. However, given a few minutes of reflection, probably many would be able to decipher or decode the meaning of the numbers. Each number corresponds to a letter in an alphabetical arrangement.

Encryption at this level obviously is not very secure, as the "key" to the encryption can be easily discovered. A "key" simply is the clue that unlocks, or deciphers, the encrypted code. The key in the above example simply was the discovery that each number corresponds to a letter of the alphabet. Whoever has this key—or has discovered the key—can then decrypt the encrypted message. The point of encryption thus is to make the key extremely difficult for any one but the intended end-user to know. Cryptography, therefore, essentially is the art and science of hiding the meaning of messages from unintended users. So, when one purchases an airline ticket on-line using a credit card, the information is encrypted, and if it is intercepted by a hacker, the encrypted credit card information simply would appear as gibberish.

The type of encryption exemplified above is known as symmetric encryption. Symmetric encryption, however, has within it a fundamental flaw that makes encrypted messages susceptible to decryption no matter how sophisticated the encryption. That flaw is the key. Although the example above is rather simplistic, it does require, as noted, a key in order to decipher the scrambled message, namely, that 1=A, 2=B, and so on. The problem is that even though one may have an encrypted message that is absolutely impossible to decipher, if the intended recipient is to decipher the encrypted message, he or she will need the key to the encrypted message in order to decipher it. This requires the sender not only to deliver the encrypted message to the intended recipient, but also the key. Because someone always could intercept the key, one therefore has to figure out a way of getting the message to the recipient in a foolproof manner, i.e., without allowing the key to be intercepted. This dilemma raises an interesting question: if one could get a message to the recipient without it being intercepted, then why encrypt the message in the first place?

An alternative to symmetric encryption was announced in August 1977. Asymmetric encryption, or RSA cryptography, does not require the sender to transport a key to the recipient. (RSA represents the last names of Ronald Rivest, Adi Shamir, and Leonard Adleman, the three mathematicians and computer scientists who devised this form of cryptography). RSA is one of the more popular forms of so-called "strong" encryption freely available today, and is virtually impossible to crack.

In a nutshell, RSA works because there are two keys available for encrypting and deciphering messages. One key is known as the private key and is kept in secret by the recipient. The second key is known as the public key and can be known by anyone who wants to send the recipient an encrypted message. RSA cryptography works by exploiting a certain attribute of very large numbers, namely, that they are incredibly difficult to factor. (A factor of a number simply is a smaller number that divides evenly into it. For example, 2 is a factor of 6, but 4 is not). Before any message is sent to the recipient, the recipient chooses two very large prime numbers and keeps them secret. These two numbers constitute the private key. The recipient then multiplies these two numbers together to get an even larger number, which represents the public key. Anyone wishing to send the recipient an encrypted message then uses the public key to encrypt the message.

KEY TERMS

CIVIL LIBERTIES: Includes those rights and freedoms contained in, or derived from, the United States Constitution, such as the right to privacy and the freedom of expression.

CRYPTOGRAPHY: The science of encrypting messages or concealing meaning of messages.

CYBERCRIME: Generally, computer crimes involving hacking or authoring viruses, worms, and logic bombs. Cybercrimes are generally engaged in for the purposes of financial gain or for intellectual excitement.

CYBERSPACE: The aggregate of the thousands of cable and telephone lines and computers that are networked together to form the Internet.

CYBERTERRORISM: Generally, computer crimes involving the disruption of a state's computer infrastructure for the purposes of intimidation. Cyberterrorists engage in cyberterrorism for political purposes.

DECRYPTION: The process of unscrambling a message back into its readable format.

ENCRYPTION: The process of scrambling a message into an unreadable format.

KEY: A collection of bits, usually stored in a file, which is used to encrypt or decrypt a message.

KEY ESCROW: A process that requires a copy of all decryption keys be given to a third party, such as a government entity, so that encrypted messages may be decrypted as required by a law enforcement agency.

PRIVATE ENCRYPTION: Single-key encryption system known only to the two parties communicating with each other.

STRONG ENCRYPTION: Process using a 128-bit algorithm to convert plain text into a disguised file or message.

Without going into the details of how the process actually works, suffice it to say that asymmetric encryption is a one-way process (hence, its name) such that once the sender uses the public key to encrypt the message, the sender cannot then decrypt it (as he or she would be able to do under a symmetric encryption scheme). Indeed, the encrypted message is now nearly impossible to decrypt by anyone. Why? Because once the message is encrypted according to RSA methods, one must then have access to the private key in order to decrypt it, and presumably no one would if the sender has been diligent about guarding the private key. Someone, of course, could try to break the encryption by guessing the two primes numbers that make up the private key. But this would require factorizing the public key. For very large numbers, the factorization process quite literally could take billions of years. So, although RSA cryptography is not in principle unbreakable, for all practical purposes, it is. And this is what has governments, especially the United States government, worried. Because RSA cryptography is freely and publicly available, criminal conduct in cyberspace can potentially go unnoticed and undetected forever. Although strong encryption may give confidence to the public that their privacy is being protected in cyberspace, strong encryption also renders traditional methods of criminal investigation and monitoring of criminal conduct essentially worthless.

The Problem of Encryption Without encryption, *e*-commerce, i.e., the millions of commercial transactions that take place over the Internet daily, likely could not exist, for no one would engage in transactions over the Internet if doing so presented a significant risk that their information could be intercepted and decrypted by others. Because strong encryption largely prevents decryption by third parties from occurring, more and more members of the public are engaging in commercial transactions online, and using the online world to send private communications to people all around the globe. It is not surprising that in light of the many benefits provided by computers and the Internet, a significant and growing proportion of all commerce today is *e*-commerce. Indeed, virtually every aspect of our society is inextricably linked to computers—from writing reports and memoranda and performing research, to checking the weather and stock quotes, communicating with loved ones, and storing the most personal information about ourselves on the computer.

So complete is our reliance on these electronic devices that without computers and the Internet, our society simply could not function in the manner it does now. Thus, inasmuch as encryption makes travel on the "information superhighway" safe and convenient, encryption protects one of our most fundamental social values—privacy—by making it difficult, if not impossible, for others to gain access to any information we deem private. With our privacy protected, we are free to engage in many intellectual, financial, spiritual, and commercial enterprises without fear of oppressive governmental oversight and regulation, or the prying eyes of nosy neighbors, or worse, the theft of our personal identities by criminals. Encryption protects liberty.

This same encryption that protects liberty, however, also can be—and is being—exploited by criminals and terrorists not only to facilitate crimes such as fraud and money laundering, but in alarmingly more cases, to engage in political acts of violence. Crime and terrorism, of course, are threats to liberty. Thus, encryption also harms the very same liberty it otherwise is intended to protect. As former Vice-President Al Gore has stated, "[u]nlawful criminal activity is not unique to the Internet—but the Internet has a way of magnifying both the good and the bad in our society."

In addition to the many criminals and terrorists who exploit encryption for nefarious ends, many foreign states also are suspected of launching cyber-attacks against the computer infrastructure of the United States, if for no other reason than to test our weaknesses and responses. Indeed, many analysts now believe that the "first-strike" in any major modern war likely will take place not on a battlefield, but in cyberspace. Under such a scenario, an enemy state would first "launch" viruses, worms, and logic bombs contained in file attachments over the Internet. These programs would be downloaded by the unsuspecting masses, stay hidden for a time, and then simultaneously begin erasing data, slowing networks, and crashing critical systems. Our ability to communicate and function would be greatly hindered, and while we were distracted and disoriented, the enemy state could then launch a series of terrorist attacks with various biological agents or even nuclear weapons. No massive air and land assaults would be needed to bring our country to its knees, just some well-written code encrypted within various emails and files would open the door to financial and social calamity. Although just few years ago such a scenario would be relegated to the realm of Hollywood science fiction and fantasy, the real possibility of this scenario is now part of our everyday reality.

Thus, and as expected, many clues behind the recent airliner hijackings are already emerging from the various Internet transactions that the terrorists engaged in prior to their suicide missions. Indeed, the FBI already has enlisted the assistance of various Internet Service Providers (ISPs) for purposes of ascertaining where and when the suspected hijackers accessed the Internet and for what purposes. The FBI, of course, is interested in who the terrorists emailed, and the contents of those emails. Further, the FBI wants to know what sites the terrorists visited and for how long, and whether they purchased anything or made deposits into or transfers out of their bank accounts.

The FBI and other agencies will also likely pore over the hard-drives of any computers and floppy disks they may come across during their investigations. They will look to see what programs and files, if any, are on the hard-drive and if any files have been deleted. If they find the remnants of deleted files on the hard drives they will attempt to recover those files to see what they contain. However, if any of these transactions or files have been encrypted, it will be very difficult to recover the information.

Thus, as mentioned earlier, the problem with encryption essentially is that it provides both a benefit and a burden. As a result, debates regarding encryption center on resolving the seemingly irresolvable tension between encryption as an essential element in the protection and advancement of our civil liberties, and encryption as a tool for criminals, terrorists, and rogue states for wreaking disharmony, chaos, and disaster on the public. The key, of course, is to find a balance between personal liberty and public safety.

Encryption Recovery and Law Enforcement
In light of the horrific terrorist attacks discussed in Part I above, Congress recently passed the Combating Terrorism Act of 2001, which makes it easier for the FBI, CIA, and the Department of Justice to monitor and investigate persons suspected of engaging in serious offenses, not the least of which is terrorism. Similarly, several Congressmen as well as the Department of Justice have voiced concerns over strong encryption for the reasons discussed in the previous section. Many wish to curtail or even outlaw such encryption. Given their fears about how it could be exploited by criminals and terrorists, one cannot blame them for such proposals.

Nevertheless, our nation fundamentally is built upon notions of civil liberties such as privacy, freedom of speech, freedom of association, and freedom of expression. These individual freedoms, by their very nature, provide a check on government action. The government simply cannot decide whether it will or will not honor these values—it must. Some civil libertarians argue, however, that many of the current proposals in response to criminal exploitation of strong encryption technology unnecessarily infringes on civil liberty. Of course, civil libertarians readily will admit that another cherished value of our society is the freedom from unnecessary harm and fear. To that end, we believe that it is the government's responsibility to ensure as much as possible that this value, too, is respected and protected. Thus, the government is left with the problem of balancing the various liberties against each other.

One current proposal that appears to be gaining momentum involves requiring manufacturers of encryption software to build in a "back-door" into the encryption. Under this proposal, the government would be given a "government key" that could be used to decrypt alleged criminal communications. As the government first would have to obtain permission from a court in order to use the key, it would not be unlike our present search warrant procedures. Of course, some argue that this still leaves room for abuse by the government with respect to the indiscriminate and potentially illegal use of its key. Further, what would happen if the government's key ever became public knowledge, especially by criminals? Massive problems could potentially ensue.

Another similar proposal would be to require users of encryption software to give copies of their private keys to so-called "trusted third parties" or TTPs. Rather than the government holding onto back-door keys, which could potentially be used to access information unbeknownst to anyone, these nongovernmental third-parties would be in possession of the private keys and would only give them out to the government upon court order. Although this proposal provides an added layer of protection from potential government abuse, that potential still remains. The government, some have said, may try to strong-arm TTPs into giving out private keys. (Although presumably such TTPs would have recourse to the courts to block such government tactics).

Conclusion It is clear that we must make a decision on strong encryption. One option simply is to do nothing and pray that criminals and terrorists will cease exploiting encryption for their evil ends. That, of course, is dangerously naive. Conversely, another option is to outlaw strong encryption. As encryption is already firmly rooted in cyberspace, this option does not seem to have much promise. Indeed, it is ludicrous since it may essentially kill *e*-commerce.

A sort of middle-ground alternative option is to live with some weakened form of encryption that would allow the government to decrypt citizens' e-mail and computer files, but only in certain specified situations and only for legitimate reasons. To paraphrase Simon Singh in *The Code Book*, how weak a version of encryption we can live with will depend on whom we fear more—criminals and terrorists,

or the government. As of this writing, it is clear that we fear criminals, especially international terrorists, far more than we do our own government. —MARK H. ALLENBAUGH

Viewpoint:
No, private strong encryption contributes to American society by safeguarding basic rights, personal information, and intellectual exchanges. It protects individuals from over-zealous law enforcement agencies.

Nearly as long as engineers have connected computers into networks, an essential tension has existed between those who write computer code and those charged with protecting the interests of national security. Philosophically, members of the computing community have believed in the free exchange of their ideas and programs, while the American companies employing many computer professionals have demanded the right to compete freely in the global marketplace by using encryption to safeguard the content of their wares. At the same time, however, certain federal agencies have desired the ability to infiltrate networks at will in order to catch criminals. Yet, the reality has been that encryption programs have been disseminated much more quickly than the government's capability to control them. As a result, contemporary networks depend upon private strong encryption. This system benefits private citizens in multiple ways: as they choose between the products of American and international businesses, as they are protected from consumer theft, and as their Bill of Rights freedoms are upheld.

The Problems with Governmental Controls
There are legitimate security concerns raised when strong encryption is held in private hands. For example, a criminal organization or individual could encrypt records of illegal activities and stymie law enforcement agencies attempting to examine those records for criminal evidence. Or, a terrorist group could shroud its arms trafficking and plots against American citizens. However, the solution is not turning control of the keys to strong encryption over to the federal government. Throughout the 1990s, the executive branch attempted to establish an escrow system which would require all makers of encryption systems to deposit their decryption keys with the government. That way, law enforcement and national security officials could ask the courts to allow them to decrypt any information they deemed necessary for preventing or prosecuting crime.

Not surprisingly, these federal attempts to control strong encryption have failed repeatedly. One of the more notorious examples is the Clipper Chip initiative endorsed by the Clinton administration in 1993. Designed by the National Security Agency (NSA), the chip was to be installed in devices requiring the encryption of conversations, such as secure telephones. A related chip called "Capstone" would be used for data encryption in computers. These chips contained a key deposited with the Commerce and Treasury Departments which allowed the government to retrieve the message being transmitted. Essentially, the Skipjack algorithm used in Clipper and Capstone would ultimately allow the United States government to eavesdrop on any exchange in which it believed a crime had occurred.

When the federal government holds copies of encryption keys, as it would have if any version of the Clipper Chip initiative were implemented, a number of freedoms guaranteed by the Bill of Rights are endangered. The most obvious intrusion caused by the eavesdropping in cyberspace possible with the Clipper Chip is upon Fourth Amendment rights preventing unreasonable searches and seizures, including wiretapping, and the Fifth Amendment protection against self-incrimination. Both of these Amendments provide for Americans to be secure in their persons, papers, and, by extension, their computer code. Additionally, on April 4, 2000, the United States Court of Appeals for the Sixth Circuit ruled in *Junger v. Daley* that encryption hardware and software are protected by the free speech provisions of the First Amendment "because computer source code is an expressive means for the exchange of information and ideas about computer programming."

Members of the computing community argue as well that some dissenting groups could feel stripped of their anonymity when they were required to deposit decryption keys in an escrow system. They would not be able to voice their opinions without fear of authoritarian reprisal and thus would lose their First Amendment rights to free speech and assembly. Some critics of the government's policies have also claimed that, since encryption tools are treated as arms by the International Trafficking in Arms regulations, they have a right to own and use strong encryption under the Second Amendment's guarantee that private citizens may keep and bear arms.

In addition to impinging on the Bill of Rights, Clipper illustrated a number of troubling characteristics of federal security agencies with respect to encryption. Complete control has been beyond the government's capability from the beginning. In fact, strong encryption had already been disseminated over the Internet and was available for purchase in other nations before the federal government even proposed its first escrow system. In addition, the Federal Bureau of Investigation (FBI) and NSA tried to push through their form of key escrow even though it was technically unsound. Its limitations prevent Clipper from demonstrating conclusively that one specific person in fact made the telephone call or sent the data in question. Criminal defendants could argue that the communications intercepted by the government were forged, and there would be no way to disprove their claim. Further, in the 1990s, there were already ways for criminals to evade the proposed escrow system while innocent citizens were not shielded from the prying eyes of Clipper.

NSA and the FBI have also demonstrated an intransigent attitude of self-interest toward strong encryption. NSA has continually attempted to circumvent the Computer Security Act of 1987, which gave the civilian agency, the National Institute of Standards and Technology (NIST), responsibility for the security of unclassified, non-military governmental computer systems. Not satisfied with undercutting NIST's authority, NSA then tried to control encryption in private industry by developing Clipper. NSA and the FBI have historically disregarded the valid concerns of industry leaders and shown an unwillingness to compromise. As Jim Barksdale, President and CEO of Netscape Communication Corporation, wrote in the *Wall Street Journal* on September 26, 1997, "The criminals will still be able to buy advanced encryption technology outside the United States . . . [but if] we and the network operators couldn't guarantee the government immediate access to data on everyone's computer, the federal government could put us in jail." People like Louis Freeh, former director of the FBI, have further damaged the image of the executive branch by making irresponsible public statements. For instance, Freeh used the crash of TWA Flight 800 in July 1996 to justify his call for federal control of encryption even though that tragedy was caused by mechanical failure and was not a terrorist act.

Governmental key escrows also feed the temptation for agencies to overstep their bounds. The federal government hoped that phones equipped with the Clipper Chip would be sold abroad, which would have raised questions about the United States monitoring activity outside its borders and the legality of these devices under the encryption laws of other nations. The FBI has also been prone to use its powers to intimidate rather than police. These tendencies have admittedly been exacerbated by

Louis Freeh
(Photograph by Stephen Jaffe. © AFP/Corbis. Reproduced by permission.)

the contradictory American encryption laws, under which an encryption program written in a book could be exported but a computer file containing the same program could not. Still, other nations have devised acceptable solutions for balancing the interests of law enforcement and private enterprise. For instance, Germany rejected restrictions on the availability of strong encryption in June 1999, while the United Kingdom abandoned its efforts in key recovery as detrimental to its desire to be a world leader in electronic commerce in May of that year. Unrestricted export of strong encryption was legalized in Finland in 1998. Even France has eased its long-standing ban on the import and export of encryption products.

Private Encryption Protects Corporations and Consumers In the United States, legislators such as Rep. Bob Goodlatte of Virginia helped lead the way toward a more reasonable governmental approach to strong encryption. Goodlatte introduced in 1998 and re-introduced in 1999 the Security and Freedom through Encryption (SAFE) Act. SAFE prohibited any mandatory key escrow systems while it criminalized the use of encryption in the commission of other crimes. SAFE also allowed the export of encryption products after a one-time, 15-day technical review. Although the law was weakened during its two tenures in various committees of the House of Representatives and was

Jim Barksdale, president of Netscape, testifying on Capitol Hill before the Senate Judiciary Committee.
(AP/Wide World Photos. Reproduced by permission.)

not without other drawbacks, it was designed to prevent hackers and thieves from accessing digital information and communications, to allow American companies to retain their international technological and economic lead, and to advance national security by promoting democracy and the free exchange of ideas about encryption. A group of major computer companies, led by Cisco Systems, also volunteered to compromise with the NSA and FBI by submitting a "private doorbell" plan in 1998. Under this proposal, law enforcement and national security officials with a warrant could ask a network operator to intercept and record data at the point either just before encryption began or immediately after it ended.

The Clinton administration finally accepted the inevitable and issued a directive about private strong encryption in January 2000. These regulations eased restrictions on the export of strong encryption products by stating that open source code could be posted on the World Wide Web as long as the URL address was given to the Department of Commerce. The directive also instituted the one-time technical review of the SAFE law for encryption programs sold in retail or to other governments. The only prohibited customers were in Cuba, Iran, Iraq, Libya, Sudan, Syria, and North Korea, the nations on the State Department's terrorist list. This move made the SAFE law irrelevant. As a consequence, no further legislative action was undertaken on SAFE or on the "private doorbell" proposal. These plans, though, did demonstrate that American companies are willing to cooperate with law enforcement and national security interests by providing their technological expertise as long as governmental policies allow the companies to continue to innovate.

Indeed, the January 2000 directive enabled American companies to compete freely in global markets. They no longer had to make two versions of software products—one for domestic sales and a weaker version to distribute internationally. They could feel free to develop new, even stronger encryption products to make their data even more secure from outside attacks. After the European Union decided to create a "license free zone" for most encryption technologies in July 2000, the Department of Commerce further liberalized the export and re-export of encryption products to the 15 Economic Union members and Australia, the Czech Republic, Hungary, Japan, New Zealand, Norway, Poland, and Switzerland. Still, the licensing rules that any potential American exporter of encryption software must follow remain complex and especially onerous for small companies. These guidelines also fail to exempt academic researchers from the paperwork.

Nevertheless, consumers are better protected with this access to private strong encryption than without it. Previous encryption schemes, such as the Digital Encryption Standard (DES) are no longer sufficient for safeguarding data. By 2000, the 56-bit DES could be broken in 22 hours and 15 minutes, while it required 10,000 years or more to crack 128-bit strong encryption. Credit card, banking, and telecommunication companies must be able to assure their customers that private information will not be exposed to identity thieves. These companies need powerful encryption tools and the capability to keep their decryption techniques out of the hands of third parties. In other words, they need free access to strong encryption to inspire confidence in their customers that their products are safe. As larger and larger segments of the economy come to depend on Internet transactions, it is imperative that these exchanges remain secure.

Private strong encryption is not a magic solution for every information security issue. The security of information on the Internet is a complex issue which will require constant and diligent attention. Allowing companies to hold their own encryption keys will not prevent unauthorized disclosures of sensitive financial or medical data. However, allowing companies to retain control of their keys and academics to freely exchange their ideas is the best of the available alternatives. The costs law enforcement and national security agencies would incur by demanding key escrow far outweigh any benefits they would reap in nabbing a few criminals. The difficulties raised by private strong encryption are not insurmountable, while cherished American Bill of Rights freedoms are upheld when private strong encryption is encouraged by the federal government. Far from a threat to society, private strong encryption is a necessity of the Internet Age. —AMY ACKERBERG-HASTINGS

Further Reading

Brin, David. *The Transparent Society: Will Technology Force Us to Choose Between Privacy and Freedom?* Reading, MA: Addison-Wesley, 1998.

Center for Democracy and Technology. Washington, D.C. <www.cdt.org/crypto/>.

Electronic Frontier Foundation. <http://www.eff.org>.

The Electronic Frontier: The Challenge of Unlawful Conduct Involving the Internet. Washington, D.C.: President's Working Group on Unlawful Internet Conduct, 2000.

Electronic Privacy Information Center. Washington, D.C. <www.epic.org>

Landau, Susan, et al. "Crypto Policy Perspectives." *Communications of the ACM* 37 (August 1994): 115–21.

Levy, Steven. *Crypto: When the Code Rebels Beat the Government—Saving Privacy in the Digital Age.* New York: Viking Press, 2001.

Miles, Wyman E. "Encryption." *Networker* 6 (summer 1996).

Newton, David E. *Encyclopedia of Cryptology.* Santa Barbara, CA: ABC-Clio, 1997.

Sanger, David E., and Jeri Clausing. "U.S. Removes More Limits on Encryption Technology." *New York Times* January 13, 2000.

Schneier, Bruce. *Applied Cryptography.* 2d ed. New York: John Wiley & Sons, Inc., 1996.

Singh, Simon. *The Code Book: The Evolution of Secrecy from Mary Queen of Scots to Quantum Cryptography.* New York: Doubleday, 1999.

United States Department of Justice, Computer Crime and Intellectual Property Section of the Criminal Division. <http://www.cybercrime.gov>.

MEDICINE

Does the addition of fluoride to drinking water cause significant harm to humans?

Viewpoint: Yes, the evidence is clear that significant concentrations of fluoride in drinking water cause harm to humans.

Viewpoint: No, the addition of small amounts of fluoride to drinking water does not cause significant harm to humans and provides several benefits.

Throughout history, dental caries—that is, decay of the teeth—has been the principal problem of dentistry. During the 1880s, Willoughby D. Miller, a student of the great German bacteriologist Robert Koch, showed that microorganisms are involved in the development of dental caries. Miller's findings were published in his book *Microorganisms of the Human Mouth* (1890). Following the path established by Miller, J. Leon Williams then demonstrated that bacteria in dental plaque produce the acid that attacks tooth enamel. Williams believed that good dental hygiene would prevent decay. Although until the twentieth century dentists could do little but remove affected teeth, the concept of preventive dentistry had been promoted as early as the eighteenth century. Since the 1930s dentists have attempted to prevent caries by reducing the acid–forming microorganisms in the mouth, or by using chemicals to inhibit the formation of acid and make the teeth less susceptible to acid. Researchers have investigated various chemicals, including fluorides, ammonium compounds, and penicillin. The best known and most controversial is sodium fluoride, which has been added to the drinking water of many communities.

What Is Fluoride? Fluorine, a member of the halogen family in the periodic table of elements, is a very reactive chemical element. In nature, fluorine is primarily found in the widely distributed mineral called fluorspar (calcium fluoride) and in combination with aluminum in cryolite. Fluoride, a binary compound of fluorine, is found in varying amounts in drinking water and in foods, but scientists estimated that the intake of fluoride by the average adult was between 0.7 and 3.4 mg per day. In areas where the concentration of fluoride was relatively high, characteristic signs usually appeared on the teeth. These include dull white patches, pitting, or brown stains and a mottled appearance. The addition of high levels of fluoride to the diet of experimental animals affects calcium and phosphorus metabolism and produces fragility of the teeth and bones. However, the addition of small amounts of fluoride seemed to strengthen tooth enamel and prevent the development of dental caries. No other element seemed to have such an effect on tooth enamel. Thus, animal studies suggested that appropriate levels of fluoride might prevent tooth decay either by making the structure of the tooth more resistant to decay, or by inhibiting the bacterial action on food particles that promotes decay.

The relationship between fluoride and mottling of the teeth was noted by Frederick Sumter McKay when he moved to Colorado Springs, Colorado, in 1901 to establish a dental practice. About 90% of the people who had grown up in the Colorado Springs region had strange brown stains on their teeth. Although area dentists had little interest in the condition, McKay found evidence that people with mottled brown teeth had fewer cavities. In the 1920s, McKay and collaborators found a correlation between the brown stains and fluorides in the water supply. Further studies of the relationship between fluoride concentration and staining were carried out by Henry Trendley Dean, chief of the Dental Hygiene Unit at the National Institute of Health. Dean and his associates found that when the level of fluorides exceeded 1 part per million (ppm), fluoride began to accumulate in tooth enamel.

In 1938 the United States Public Health Service (USPHS) conducted a study of fluorides and dental conditions in two communities in Illinois. The children in Galesburg, where tooth mottling was common and the water contained fluorides, had fewer cavities than those in Quincy. After extensive testing in animals to determine that the addition of fluorides to water was safe, the USPHS began testing the effect of adding sodium fluoride to the water systems in Newburgh, New York, and Grand Rapids, Michigan. Researchers concluded that the use of fluorides reduced dental caries by about 60%. Based on several comparative studies, the USPHS recommended that all communities add fluorides to their public water supply so that the level would be 1 ppm. In 1951 the American Dental Association (ADA) officially endorsed fluoridation.

The Opposition Speaks Out Although the dental, public health, and medical communities strongly supported fluoridation, many groups were actively opposed. Some surveys indicated that about 30% of Americans objected to fluoridation of their water supply. Some people opposed the addition of "unnatural" chemicals to their water, arguing that fluorides were toxic and caused many different diseases. Among the reasons cited was the fact that there is little margin of safety between the level of fluoride that is supposed to prevent dental decay and the level that causes deleterious health effects. Now that fluoride has been added to other products, such as toothpaste and mouthwash, and is found in air and food contaminants, critics contend that total intake is unpredictable and excessive. Moreover, critics believe that dangerous amounts of fluoride can accumulate in the body because of long–term exposure to numerous sources of fluorides, including insecticides, Scotchguard, and Teflon.

Opponents of fluoridation claim that fluorine is a toxic agent that causes serious health problems. People with diabetes and kidney disorders could be at special risk. High levels are said to cause mottling of the teeth, deformities of the spine, joint problems, muscle wasting, neurological defects, arthritis, osteoporosis (a decrease in bone mass), hip fractures, infertility, genetic mutations, Down's syndrome (a form of congenital mental retardation), and all forms of cancer. Lower levels are said to cause eczema, dermatitis, headache, chronic fatigue, muscular weakness, mouth ulcers, lower urinary tract infections, and the aggravation of existing allergies. In addition to associating fluoride with various disease conditions, critics also charge that fluoridation does not really reduce tooth decay. Certain studies are cited as proof that children who drank fluoridated water actually had more decay than did children residing in areas without fluoridated water.

Much of the debate about fluoridation has been political rather than medical. Indeed, during the 1950s, at the height of the Cold War (a period characterized by diplomatic tensions between the United States and the former Soviet Union), critics argued that the fluoridation of American water supplies was part of a Communist plot. Whether or not fluoridation causes significant harm or not, many people object to compulsory medication as an infringement on personal liberty.

Supporters Emphasize the Merits of Fluoridation Public health workers, in contrast, consider fluoridation to be one of the greatest public health achievements of the twentieth century. Some authorities believe that fluoridation of public water supplies deserves to be ranked with pasteurization of milk, purification of water, and immunization against infectious diseases. All of these measures met with fierce opposition and triggered controversies that involved both scientific and political issues. Public health measures, by their very nature, must reach almost the whole population, and many such actions must be compulsory to achieve a public good. For example, immunizations are required before a child can enter public school, and people suffering from contagious diseases may be quarantined. People who object to adding chemicals to water ignore the fact that drinking water must be treated in order to make it safe. In part, such objections reflect the success of public health measures. By the beginning of the twenty–first century, few people remembered that contaminated drinking water caused deadly epidemics of typhoid fever and cholera, and unpasteurized milk was a source of tuberculosis.

Those who support the fluoridation of drinking water argue that many years of study have shown that the process is safe and effective in reducing tooth decay. They charge opponents with using safety issues as a screen for other kinds of objectives that are more political than medical. In answer to claims that fluoridation causes many serious diseases, public health officials argue that carefully controlled comparative studies have not revealed any significant difference in the health, growth, and development of those who drank fluoridated water and those who did not. The only statistically significant difference appeared to be a reduction in tooth decay. Statistical studies of large populations, however, cannot rule out the possibility that some individuals might be sensitive to fluoridation, just as some individuals are sensitive to strawberries, or peanuts, or penicillin.

According to the ADA, fluoridation of community water supplies and the use of fluoride–containing products constitute safe and effective approaches to preventing tooth decay. The ADA has continuously endorsed these measures for over 50 years. Nevertheless, the addition of fluoride to drinking water continues to provoke significant opposition. Despite reassurances by the medical, dental, and public health communities, opponents of fluoridation argue that the practice is neither safe nor effective. —LOIS N. MAGNER

Viewpoint:

Yes, the evidence is clear that significant concentrations of fluoride in drinking water cause harm to humans.

Fluoride has never received United States Food and Drug Administration (FDA) approval and is listed as a contaminant by the Environmental Protection Agency (EPA). Fluoride is the agent used by cities all over the world to fluoridate their municipal water supplies. According to the Centers for Disease Control and Prevention (CDC), today over 62% of United States cities fluoridate their water supplies. In 1944, the federal government instituted fluoridation of municipal water supplies as a public health measure to help prevent tooth decay. The "fluoride intake" standard for optimal benefit for teeth was set between 0.7 and 1.2 ppm (parts per million) or mg/L (milligrams per liter). At that time, fluoride was not available from other sources, such as toothpaste or mouth rinses, as it is today. In addition to fluoridated water and fluoride-enhanced dental products, fluoride can be found in food, beverages, fluoride–based pharmaceuticals (e.g., Prozac [fluoxetine]), air emissions, and the work place. The use of fluoride over the past 50 years has been linked in government and scientific reports to some very serious national health problems—dental fluorosis (irreversible mottling, staining, and pitting of teeth); crippling skeletal fluorosis (deformities of the spine and major joints, muscle wasting, and neurological defects); arthritis; osteoporosis (a decrease in bone mass); hip fractures; infertility; and all types of cancers.

Maximum Contaminant Levels (MCLs) In 1986, the EPA set new maximum contaminant levels (MCLs) for fluoride in water. The agency specified that with fluoride levels greater than 2.0 mg/L children are likely to develop dental fluorosis, that with fluoride levels greater than 4.0 mg/L individuals are at risk of developing crippling skeletal fluorosis, and that it is against federal law to fluoridate water above 4.0 mg/L. Shortly thereafter, the EPA's union of professional employees who are responsible for setting standards attempted to file suit in federal court to overturn the new standard, charging that the EPA had ignored scientific evidence that revealed adverse health effects.

Fluoride is an acute toxin with a rating slightly higher than that of lead. This fact was presented in the fifth edition of *Clinical Toxicology of Commercial Products,* published in 1984, in which lead was given a toxicity rating of 3 to 4, while fluoride was rated at 4 (3 = moderately toxic, 4 = very toxic). In 1992, the new EPA maximum contaminant level (MCL) for lead was set at 0.015 ppm, with an ultimate goal of 0.0 ppm. The MCL for fluoride is currently set at 4.0 ppm—an MCL that is over 250 times greater than the permissible level for lead.

A 1998 survey showed that 30% of Americans were opposed to community water fluoridation. Opponents base their arguments on six major problems with fluoridation: uncontrolled random dosages, no margin for safety, excessive intake, carcinogenicity (cancer–producing), ineffective reduction of tooth decay, and dental and skeletal fluorosis.

Six Major Problems with Fluoridation of Drinking Water *One: Fluoridation Leads to Uncontrolled Random Dosages.* Because individuals are unique in their sensitivity to medication and in their body size and weight, consistently adding a substance such as fluoride to drinking water can lead to adverse health effects due to uncontrolled random dosages. Such compulsory mass fluoridation is poor medical practice because it does not allow tailoring of dosages to individuals' unique needs, and it does not consider the varying levels of water consumed. Factors that have a bearing on individuals' level of water consumption include general health, lifestyle, age, dietary habits, and climate. Athletes and physical laborers who drink large quantities of water are dosed with more fluoride than the sedentary elderly who drink considerably less water. Patients with diabetes insipidus, a rare disorder of the pituitary gland that causes the release of large amounts of urine, consume enormous amounts of water due to an excessive thirst precipitated by the disease. Infants fed with formula prepared with fluoridated water receive proportionately higher doses of fluoride than adults because of the smaller body weight of the former group and because of its total dependence on fluid nourishment. Individuals who live in warmer climates tend to consume more water than individuals who live in cooler climates.

KEY TERMS

FLUORIDATION: The process of adding fluoride to the water supply of a community to preserve the teeth of its inhabitants. Two fluoridating agents are sodium fluoride (NaF), a colorless crystalline salt, and hydrofluorosilic acid (H_2F_6Si), a direct byproduct of phosphate fertilizer production. Sodium fluoride is used largely as a basis for research on the risks of fluoridation, while hydrofluorosilic acid is used in the majority of municipal fluoridation procedures.

FLUOROSIS: In simple terms, fluorosis means fluoride poisoning. Fluorosis is the first visible sign that the body is being poisoned by excessive fluoride. Dental fluorosis is caused by drinking water with a high fluoride content during the time of tooth formation (from birth to six years of age). Dental fluorosis is characterized by defective calcification that gives a white chalky appearance to the enamel, which gradually undergoes brown discoloration. Skeletal fluorosis, like dental fluorosis, is caused by drinking water with a high fluoride content. It is the action of the fluoride on the bone that causes the bone to decrease in density and to become brittle. Because fluoride's effects are cumulative, the changes that take place during skeletal fluorosis range from pain and stiffness of joints in the early stages to reduced mobility, skeletal deformities, and increased risk of bone fracture in the later stages.

MARGIN OF SAFETY: In industry, science, and medicine, the margin of safety is the range of dosage between the therapeutic and toxic levels of a substance. The accepted margin of safety in medicine between therapeutic and toxic doses is often a factor of 100 or more.

MAXIMUM CONTAMINANT LEVEL (MCL): MCL is the highest level of a contaminate that is allowed in drinking water. MCLs are set as close to the maximum contaminant level goal (MCLG) as feasible using the best available treatment technology. The MCLG is the level of a contaminate in drinking water below which there is no known or expected risk to health. MCLGs allow for a margin of safety.

PARTS PER MILLION (PPM) OR MILLIGRAMS PER LITER (MG/L): With regard to fluoridation, parts per million (ppm) means one part fluoride per million parts water—a ratio that corresponds to one minute in two years or a single penny in ten thousand dollars. One ppm equals one mg/L, or one milligram per liter.

THERAPEUTIC LEVEL: Therapeutic level is the level at which a substance is supposed to provide benefit. With regard to fluoridation, the therapeutic level is the level at which fluoride is supposed to benefit teeth by minimizing tooth decay.

TOXIC LEVEL: The level at which a substance begins to do harm to an organism. With regard to fluoridation, the toxic level is the level at which fluoride causes dental and skeletal fluorosis and is associated with other maladies, such as cancer and hip fractures.

Two: There Is No Margin for Safety for Fluoride Exposure. Even supporters of fluoridation do not dispute the relatively narrow gap between the therapeutic dose—the level at which fluoride is supposed to benefit teeth—and the toxic dose—the level at which fluoride begins to do harm. Today, the optimum level set by the government and endorsed by dental authorities is 1.0 mg/L; the maximum contaminant level (MCL) as prescribed by the United States EPA is 4.0 mg/L. In a number of countries, severe skeletal fluorosis has been associated with a fluoride level of 0.7 mg/L—a level that is below both the recommended optimum and MCL levels. Several studies have noted a relationship between consumption of water with a 1.6 mg/L level of fluoride and inhibition of bone healing and with osteoporosis and osteosclerosis (abnormal and weak bone foundation). In 1992, a study in the *Journal of the American Medical Association* (*JAMA*) found that even low levels of fluoride may increase the risk of hip fracture in the elderly.

A margin of safety of zero—when toxicity due to fluoride occurs at or below claimed therapeutic levels of fluoride—is difficult to justify when the accepted margin of safety in medicine between therapeutic and toxic doses is often a factor of 100 or more.

Three: Total Fluoride Intake Is Excessive. Besides fluoridated water and fluoride–enhanced dental products, intake of fluoride includes environmental sources such as petroleum refin-

Citizens protesting the fluoridation of New York City's water supply in 1965.
(© Bettmann/Corbis. Reproduced by permission.)

ing, vehicle emissions, and household exposures, such as the use of Teflon or Tefal pans and fluorine–based products (e.g., Scotchguard products and insecticides). Also, some foods, such as grapes (due to production with fluoridated water or to pesticide application) and some fish contain fluoride. Grape juice has been found to contain 6.8 mg/L of fluoride, wine, 3.0 mg/L, and canned soup, 6.0 mg/L. Soy infant formula and tea also have been found to surpass the optimal dose of 1.0 mg/L. In 1997, the Academy of General Dentistry (AGD), representing 35 thousand dentists, admonished parents to limit their children's intake of juices because of the juices' high fluoride content. In 1991, the *Physicians' Desk Reference, Forty–Fifth Edition* warned of toxic reactions to prescription supplements with fluoride for infants and children.

In 1991, the U.S. Public Health Service reported that the range in total daily fluoride intake exceeded 6.5 mg/day. Recently, EPA data revealed that some individuals drink as much as 5.5 liters of 4.0 mg/L fluoridated water per day, which computes to a daily dose of fluoride of 22 mg/day. Such a dosage exceeds the crippling dosage of 10 to 20 mg/day published in 1993 by the National Research Council's Board on Environmental Studies and Toxicology in *Health Effects of Ingested Fluoride*. The 10 to 20 mg/day dosage was associated with crippling fluorosis when ingested over a period of 10 years.

Fluoride amounts as minimal as 0.04 mg/day have been shown to cause adverse health effects.

It is important to point out that the effects of fluoride depend not only on the total intake—ingestion (food, water, beverages), inhalation, absorption (via the skin)—but also on the duration of exposure (since the effects of fluoride are cumulative) and other factors, such as nutritional status and general health. Poorly nourished individuals and those suffering from kidney malfunction and diabetes tend to become affected after shorter exposures. Recently, a number of studies have suggested that the kidneys of adult males excrete more fluoride than the kidneys of adult females.

When fluoride intake and/or exposure periods become excessive, adverse health effects, such as dental and skeletal fluorosis, bone disorders and fractures, infertility, and cancer, tend to occur. With less excessive intake or exposure, individuals may experience eczema (a disease of the skin); dermatitis (another disease, or abnormality, of the skin); epigastric (or abdominal) distress; headaches; excessive thirst; chronic fatigue; muscular weakness; mouth ulcers; lower urinary tract infections; and the flare–up of old allergies—symptoms that tend to disappear relatively quickly when fluoride intake and exposure are decreased or eliminated.

Four: Fluoride Is a Carcinogen (a Cancer–Causing Substance). Findings from the

1970s studies of Dr. Dean Burk, former head of the U.S. National Cancer Institute's cell chemistry section, and Dr. John Yiamouyiannis, biochemist and president of the Safe Water Foundation (SWF), suggested that water fluoridation is linked to about 10 thousand cancer deaths yearly. As a result of these findings, the United States Congress ordered studies to determine whether fluoride could be carcinogenic. The studies, which were conducted by the National Toxicology Program (NTP) under the auspices of the U.S. Public Health Service (PHS), confirmed what the Burk–Yiamouyiannis studies had suggested 13 years earlier—that fluoride is carcinogenic. The results of these studies were not made public for some time.

Studies at St. Louis University, the Nippon Dental College (Japan), and the University of Texas have consistently shown that fluoride has the ability to induce tumors and cancers and to stimulate tumor growth rate. In 1991, the SWF obtained studies in which Proctor and Gamble had examined the relationship between carcinogenicity and sodium fluoride. A close look at Proctor and Gamble's studies revealed that dose–dependent increases in carcinogenicity had been observed in every parameter tested. The results of these studies were not submitted to the PHS for four years. Further studies by the SWF found a five–percent increase in all types of cancers in fluoridated communities, while studies by the New Jersey Department of Health confirmed a 6.9–fold increase in bone cancer in young males.

Five: Fluoride Does Not Reduce Tooth Decay. More and more research is showing that fluoridation is not effective in reducing tooth decay, but rather that fluoridation is associated with increased tooth decay. The most comprehensive United States review was carried out in 1994 by the National Institute of Dental Research on 39 thousand school children, ages five through 17. While the review showed no significant differences in terms of decayed, missing, and filled teeth, it showed that cities with high rates of tooth decay had 9.34% more decay in the children who drank fluoridated water. The review also showed that nine fluoridated cities with high rates of decay had 10% more decay than nine equivalent non–fluoridated cites. Observations also revealed a 5.4% increase in students with tooth decay when 1.0 mg/L fluoride was added to the water supply.

Other United States studies showing that increases in fluoride are accompanied by increases in tooth decay rates include one 1994 study of four hundred thousand students whose rate of tooth decay increased 27% with a 1.0–mg/L–fluoride increase in drinking water, and a second 1994 study of 29 thousand students whose rate of tooth decay increased 43% with a 1.0–mg/L–fluoride increase in drinking water. In a 1972 study in Japan, fluoridation was associated with decay increases of 7% in 22 thousand students. As has been the case for some time, the real route to reducing tooth decay remains to be regular dental hygiene and a nutritious diet.

Six: Excessive Intake of Fluoride Causes Dental and Skeletal Fluorosis. Fluorosis, which in simple terms means fluoride poisoning, is the first visible sign that the body is being poisoned by excessive fluoride. Dental fluorosis is caused by the ingestion of toxic amounts of fluoride during the period of calcification of the teeth in infancy and early childhood (from birth to age six). Most often, permanent teeth are affected, while baby teeth usually are not. Mild dental fluorosis is characterized by erosion of the tooth enamel that leaves the teeth mottled and discolored with brown and opaque white spots that are insensitive to whitening treatments that use bleach. Severe dental fluorosis is characterized by more significant enamel erosion, tooth pain, and impairment of chewing ability. According to the CDC, at least 22% of all American children have dental fluorosis as a result of ingesting too much fluoride. That rate sometimes rises to as high as 69% in children from high socioeconomic–status families. Similar to the CDC statistic of 22%, the U.S. Public Health Service estimates that one in five children has dental fluorosis. That dental fluorosis is rarely seen in California has been related to the fact that California is the least fluoridated state, with less than 16% of the population drinking fluoridated water.

Skeletal fluorosis is caused by long–term exposure to fluoride at levels that are much higher than those to which average individuals are exposed. Skeletal fluorosis, most often found in the spine, pelvis, and forearm, is a progressive but not life–threatening condition in which the bones decrease in density and gradually become brittle. Because fluoride's effects are cumulative, the changes that take place during skeletal fluorosis range from pain and stiffness of the joints in the early stages to reduced mobility, skeletal deformities, and increased risk of bone fractures in the later stages. As noted earlier, fluoride's effects depend not only on total dosage and duration exposure, but also on other factors such as nutritional status and kidney function. In 1993, the National Research Council's Board on Environmental Studies and Toxicology issued a statement specifying the amount of fluoride and the duration of exposure that would be conducive to skeletal fluorosis to be 10 to 20 mg/day for 10 to 20 years. Applying these figures to a lifetime of 55 to 96 years reveals that just 1.0 mg/day (the amount of fluoride in one liter of water) for each 55 pounds of body weight could conceivably be a crippling dosage.

Conclusion: Fluoride Is Damaging It is not debatable that some dosages of fluoride used to fluoridate water can cause significant harm to humans. The real question is whether the fluoride concentrations added to the drinking water consumed by the majority of Americans contribute to that significant harm. The answer is undoubtedly "yes" because factors such as variation in the amount of fluoridated water consumed; the amount of fluoride ingested, inhaled, and absorbed from other sources; duration of exposure period; and individual differences make it a statistical certainty that some of the millions of individuals consuming fluoridated water will be harmed. —ELAINE H. WACHOLTZ

Viewpoint:
No, the addition of small amounts of fluoride to drinking water does not cause significant harm to humans and provides several benefits.

The addition of fluoride to drinking water is a highly beneficial process with no known harmful effects, as shown by numerous studies and years of public and scientific scrutiny. Opposition to fluoridation, which has existed since it was first introduced, is generally based on mistaken assumptions, incomplete data, and unwarranted fears. Those who oppose fluoridation on health grounds have little, if any, evidence and are often pursuing other agendas. Hundreds of trials and statistical surveys have shown time and again that the benefits of fluoridation to individuals and society are real, and that there are no health risks to the general population.

Fluoridation and Its History Fluoridation is the process of adding small amounts of fluoride to public drinking water. The additive is usually in the form of either sodium fluoride (NaF) or hydrofluorosilic acid (H_2F_6Si), a direct byproduct of phosphate fertilizer production. Sodium fluoride is used largely as a basis for research on the risks of fluoridation, while hydrofluorosilic acid is used in the majority of municipal fluoridation procedures. Fluoridation occurs naturally in some areas; indeed, it was from observations of a community drinking naturally fluoridated water that the benefits of fluoride in dental health were first discovered.

The promotion of fluoride to prevent tooth decay began in Europe in the 1800s, but it was observations in the United States that led to the widespread acceptance and understanding of the utility of fluoridation. In the early 1900s a dentist, Frederick Sumter McKay, noticed a strange staining of teeth in many of his Colorado Springs patients. Eventually the cause was traced to the high levels of fluoride in the local water, but McKay was surprised to find that his patients' mottled teeth were very resistant to decay. Tests and trials were done to determine a level of fluoridation that would provide protection against tooth decay while avoiding the undesirable mottling effect. A level of 1 ppm was found to be ideal, and early trials showed a dramatic, and extremely rapid, reduction in dental decay in the trial areas. Fluoride is also used in toothpaste and in mouth rinses, and is often taken as a pill. Some foods are also a significant source of fluoride. Fluoride is added to public drinking water all over the world, and is supported and promoted by the vast majority of international and governmental health bodies, including the World Health Organization.

Opponents Cite Health Risks However, opposition to fluoridation has occurred almost everywhere it has been introduced. Possible health risks are often cited as reasons to oppose fluoridation, but other more emotive issues are often involved as well. Public concern over the process of adding fluoride to drinking water has led to hundreds of trials and studies being carried out by government and independent groups in dozens of countries. In all of the laboratory and fieldwork carried out, however, no evidence linking the recommended levels of fluoridation to health risks has been shown.

Public concern with fluoridation has often focused on the toxic nature of the additives used. Indeed, large doses of sodium fluoride, the most commonly used additive for fluoridation, are fatal. However, fatal levels require consuming several grams in a short period of time, whereas the levels used in fluoridation are usually only 1 ppm or less. In order to ingest fatal levels of sodium fluoride in drinking water one would have to drink so much water he would actually drown long before the levels of sodium fluoride became a problem.

The most common medical charge made by opponents of fluoridation is that it can lead to skeletal fluorosis, causing problems in bone development and strength. It is well known that high levels of fluoride ingestion can cause skeletal problems. For example in India, where much of the drinking water is naturally fluoridated to high levels (often more than 10 ppm), the incidence of skeletal fluorosis is very high. However, the minimum levels of fluoride associated with skeletal fluorosis are an order of magnitude 10 times greater than the levels used in fluoridation. Studies have shown that there is no statistical increase in the incidence of skeletal fluorosis in populations consuming water fluoridated at 1

POWER LINES, CANCER, AND THE MEDIA

The media scare that power lines cause cancer illustrates how science can have problems getting across information in the popular press. In 1989 the media widely reported fears that high-voltage power lines are a major source of cancers, in particular childhood leukemia. The media reports were based on research dating back a decade when a study in Denver, Colorado, suggested that children who live close to power lines have a greater chance of contracting leukemia from the high magnetic fields.

Serious concerns were raised regarding the methods used in the study, however, because it did not follow standard scientific methods, such as double-blind trials. But in 1988 another study in Denver, which did use double-blind studies, seemed to support the earlier findings.

What puzzled many scientists, however, was how power lines could possibly cause cancer. Cancer-inducing agents all damage DNA, such as the chemical carcinogens in tobacco smoke or ionizing radiation in x rays. The magnetic field from a power line is related to the amount of current flowing through it. High-voltage lines, which became the main target of media and public campaigns, use very little current. Every day our bodies are subjected to a much more powerful magnetic field than any power line emits: that of the earth itself.

In 1996 the American National Academy of Sciences released the results of a three-year review looking into the possible health risks from exposure to residential electromagnetic fields. They found none. The report found no evidence that power lines are a health risk. Seventeen years' worth of studies had been evaluated, over 500 individual studies, and none showed any link between power lines and health risks. In 1997 the American National Cancer Institute completed a large epidemiological study, and it also found no link between leukemia and power lines. In 1999 a Canadian epidemiological study was released, confirming the work of the 1997 report. These large studies, and many smaller ones, have all come to the same conclusions. From a scientific viewpoint, power lines do not cause cancer. There is no known health risk from high-voltage lines, except for climbing up and touching them.

Yet the popular belief still exists that there is a link between power lines and cancer. The media were quick to report the initial possibility of the health risks, and many special programs and feature articles were written about the issue. Yet the scientific reports denying the link were not given much attention at all. It is the nature of the media to focus on sensational and interesting events. Scientific reports denying the power lines-cancer link are not riveting news, whereas the original scare was just that.

—David Tulloch

---

ppm compared with those drinking non–fluoridated water. It has also been shown that there has been no rise overall in the incidence of skeletal fluorosis in countries that have introduced fluoridation. Indeed, in the Western world, the causes of skeletal fluorosis are generally recognized as excessive thirst and kidney failure.

Another medical concern raised by opponents of fluoridation is the occasional discoloration or mottling of teeth. Opponents argue that mottling of teeth is proof that fluoridation can cause bone problems. While no one denies that mottling of teeth occurs, and is a direct effect of fluoride, it is at worst a cosmetic problem and not a sign of bone damage. Again, a number of studies have shown no link between mottling of teeth and bone disorders or weakness, or any link between fluoridation and bone development or the incidence of bone fractures.

Some critics have claimed that fluoride can cause allergic or intolerance reactions. However, there is no conclusive evidence of such reactions to the levels associated with public water treatment. The major work on this topic was done by George L. Waldbott and is often used by critics of fluoridation. However, Waldbott did not consistently use double blind trials (experiments in which neither the tester nor the subjects are made aware of the factors involved—to ensure objectivity), and refused to make his unpublished data public, casting doubt on his overall findings. Major studies done since have not corroborated Waldbott's work, including trials done by allergy research organizations such as the American Academy of Allergy. Intolerance and allergenic reactions to fluoride have been noted in higher concentrations, but there is no evidence of such reaction from fluoridated water of 1 ppm.

Some of the most radical and potential worrying health claims made by anti–fluoridationists are those concerning genetic mutations and cancer. Work done with mice in the 1950s suggested that fluoridated water might speed up the development of cancer in those predisposed to it, and some opponents of fluoridation attempted to link fluoride with Down's syndrome. In the 1970s more claims were made concerning cancer and fluoride, with a statistical study seemingly showing a link between them. However, the statistical analysis of the 1970s has been severely criticized on mathematical grounds. Since then, many other studies have failed to find any link between cancer rates and fluoridation. While some more recent studies with mice and fluoride have shown higher rates of cancer, these were with very high levels of fluoride, well above the levels associated with public drinking water. To date, no evidence has been found for an association between public water fluoridation and cancer incidence or mortality in humans, including bone and joint cancers or thyroid cancers. Many comparative studies also have been done between populations drinking fluoridated water and others with unfluoridated water. The only statistically significant difference in health (or growth and development) that has been found is a reduction in the number of tooth cavities (caries).

Despite the overwhelming medical and statistical evidence for the safety of fluoridation, some doubt remains in the cases of isolated individuals who may have developed health problems due to fluoride. No medical treatment can be considered safe for all patients, and statistical analysis does not account for extremely small numbers of potential sufferers. So it is possible that a handful of people worldwide have suffered adverse effects from fluoridation. However, such cases have generally been minor in nature, and the presence of other factors cannot be ruled out. Scientific and medical standards are concerned with large populations, not every single individual, and so from a scientific and medical standpoint fluoridation has been shown time and time again to be not only completely safe, but positively beneficial to public health.

Opposition to fluoride is also justified on grounds other than potential health risks. Some critics have claimed that fluoridation is not an effective treatment. Philip R. N. Sutton, for example, attempted to show that the early trials of fluoridation were flawed, and that fluoridation did not necessarily reduce dental decay. However, while some of Sutton's claims regarding the trials done in the 1940s were substantiated, his work in no way proved that the addition of fluoride to drinking water was ineffective. Before and after Sutton's work, experiments have shown the mechanism by which fluoride strengthens teeth, and comparative studies have shown that tooth decay is reduced in fluoridated areas.

Social and Political Opposition to Fluoride
Political and ethical issues are also invoked as reasons for opposing fluoridation. Many politicians and political groups have used the fluoridation debate for their own ends. In the 1950s there were attempts to link fluoridation with communism. Pamphlets with headlines such as "Fluoridation is Communist Warfare," "Communism via the Water Tap" and "Fluoridation is MASS MURDER" appeared in many Western countries. The link was claimed so often that the Stanley Kubrick film *Dr. Strangelove* could parody the issue, with the insane Colonel Jack Ripper lecturing his second–in–command on the communist evils of fluoridation. Even today fluoridation is a popular political target for those hoping to capitalize on people's fears of an issue they do not fully understand. Invoking the possibility of cancer and genetic mutations is a very effective method of influencing the voting public, and no matter what the scientific evidence may be such fears are often lingering and all pervasive.

The major ethical issue that is raised by fluoridation is that of individual freedom and rights. This debate is particularly prevalent in the United States. Indeed, fluoridation is the only public health measure to be decided by public vote in the United States. Many health measures are compulsory in many countries, such as the isolation of individuals with highly contagious diseases, or immunization measures. There are even individuals who may experience negative outcomes from such measures, but such measures are justified on the grounds of the greater public good. The compulsory wearing of seatbelts, for example, causes a number of injuries every year; however, many more serious injuries are avoided, and many lives are saved, by wearing seat belts, and the small incidence of negative outcomes is far outweighed by the overwhelming positive effects. Indeed, there are no statistically significant negative impacts of fluoridation, unlike immunization or seat belts, yet it is opposed with more passion and vocal strength than many other compulsory health measures. The issue of fluoridation often suffers in the public arena, as the positive benefits it provides are not so obvious, or as graphic, as with other compulsory health and safety measures.

Another common criticism of fluoridation is that it is an unnatural additive to something that should be pure. However, water is treated in many other ways, such as chlorinating to kill harmful bacteria. Both processes are designed to help the end user. Also, some drinking water is naturally fluoridated; indeed, that was how the effect was first noticed. Some natural water is

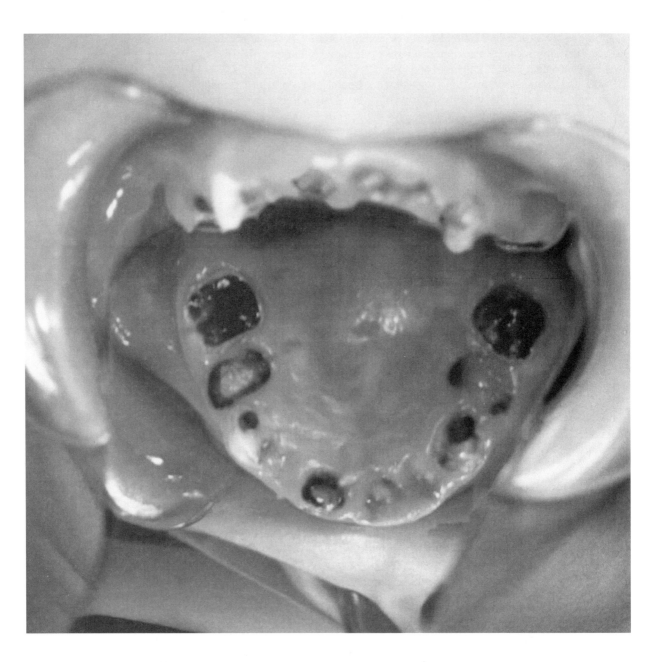

The rotted teeth of a boy from Utah. The boy's community does not fluoridate its water supply. (Photograph by Val Cheever. AP/Wide World Photos. Reproduced by permission.)

overly fluoridated, and must have the levels reduced to avoid widespread mottling of teeth, or more significant problems.

Fluoride Reduces Tooth Decay While tooth decay is not life threatening, it is economically costly. Many recent studies have shown that fluoridation is still a cheap and effective method of lowering the incidence of tooth cavities. Therefore, the result of not fluoridating public drinking water is an increase in costs to the state and the individual. While there may be some strictly philosophical violation of individual rights, there are wider benefits to be gained for society as a whole, and for individuals within a society. People should have the right to good teeth because of fluoridation, and other approaches do not provide this as widely and as cost effectively.

One potentially valid criticism of fluoridation that has arisen in recent years is that the overall dental health in many western countries has been increasing to the point where fluoridation may no longer be necessary. General dental hygiene practices are more widely practiced, and fluoride is available from many other sources, such as toothpaste. However, cost–benefit analyses done since 1974 in a dozen countries have shown that there is still a significant difference in dental cavities between populations drinking fluoridated water and those not doing so. Also a number of recent studies have shown that the incidence of dental cavities increases in areas where fluoridation has been discontinued.

Fluoridation was recently named "One of the Ten Greatest Public Health Achievements in the Twentieth Century" by the U.S. Centers for Disease Control and Prevention. Its utility and cost–effectiveness, as well as its ability to offer dental protection across social and economic divides, has gained it support from dental associations across the world, by many governments, and from many international health bodies such as the World Health Organization. While potential health risks in any treatment should always be considered seriously, decades of research, and hundreds of studies and population analyses have shown no link between health problems and fluoridation. Indeed the only statistical difference noted between populations drinking fluoridated water and those not doing so is a reduction in dental decay. Fluoridation is not only a safe process, it is a cost–effective and beneficial method of improving dental health. —DAVID TULLOCH

Further Reading

"Adverse Health Effects Are Linked to Fluoride." Preventive Dental Health Association. October 1998. <http://www.emporium.turnpike.net/P/PDHA/fluoride/adverse.htm>.

Beeber, Paul S. "New York Coalition Opposed to Fluoridation." 2001. <http://www.orgsites.com/ny/nyscof/>.

Carmichael, William. "Truth Can't Be Hidden: Fluoride Causes Cancer." 1992. <http://www.hghoralspray.com/harmful_ingredients/fluoride_toxic.htm>.

"Fluoridation Status: Percentage of U.S. Population on Public Water Supply Systems Receiving Fluoridated Water." January 2001. <http://www.cdc.gov/nohss/FSWaterSupply.htm>.

"Flouridation: The Overdosing of America." *Health and Happiness Newsletter* 4.1C. International Center for Nutritional Research (INCR), Inc. <http://www.icnr.com.FluoridePres/FluoridationOverdose.html>.

"Fluoride: Nature's Cavity Fighter." The American Dental Association (ADA). July 2001. <http://www.ada.org/public/topics/fluoride/artcl-01.html>.

"Fluorides and the Environment: The EPS's Fluoride Fraud." *Earth Island Journal*, No. 3 (Summer 1998).

Kennedy, David C. "Fluoridation: A 50–Year Old Blunder and Cover–Up." 1998. <http://emporium.turnpike.net/P/PDHA/fluoride/blunder.htm>.

Landes, Lynn, and Bechis, Maria. "America: Overdosed on Fluoride." June 2000. <http://www.zerowasteamerica.org/fluoride.htm>.

"Leading Edge Master Analysis of Fluorides and Fluoridation." <http://www.futuredynamicadvantage.com/trufax/fluoride/f3.html>.

Martel, Lillian Zier. "Fluoridaton Debate Won't Fade Away." February 2000. <http://fluoride.orahealth.org/papers/wichitaeagle020200.htm>.

Martin, Brian. *Scientific Knowledge in Controversy: The Social Dynamics of the Fluoridation Debate*. State University of New York Press, 1991.

"National Center for Fluoridation Policy and Research (NCFPR)." <http://fluoride.oralhealth.org/>.

"National Health Surveillance System Glossary." <http://www.cdc.gov/nohss/GLMain.htm>.

Rothbard, Murray N. *Fluoridation Revisited*. 8.25 (1992). <http://www.thenewamerican.com/departments/feature/2001/121492.htm>.

Schuld, Andreas. "Fluoride—What's Wrong with This Picture?" Parents of Fluoride Poisoned Children (PFPC). Fall 2000. <http://www.bruha.com/fluoride/>.

"Skeletal Fluorosis" National Pure Water Association. <http://www.npwa.freeserve.co.uk/skeletal.htm>.

"Some Scientists and Professionals Opposed to Fluoridation." (Letters presented to Calgary's Operations and Environmental Committee in October 1997). <http://www.fluoridation.com/calgary.htm>.

"The Story of Fluoridation." <http://www.nidr.nih.gov/flouride.htm>.

Susheela, A. K. "Control of Fluorosis." February 1996. <http://www.fluoridation.com/calgaryl.htm#Fluorosis%20Control%20Cell>.

Sutton, Philip R. N. *Fluoridation: Errors and Omissions in Experimental Trials*. 2nd edition. Melbourne University Press, 1960.

Whaley, Monte. "Fluoridation Foes Bare Teeth." June 2001. <http://denverpost.com/cda/article/detail/0,1040,53%7E48344%7E36%7E%7E,00.html>.

Woffinden, Bob. *Earth Island Journal*. June 1998. <http://www.Britannica.com/magazine/print?content_id=71516>.

Should xenotransplants from pigs raised at so-called organ farms be prohibited because such organs could transmit pig viruses to patients—and perhaps into the general population?

Viewpoint: Yes, xenotransplants from pigs should be prohibited because the risks of transmitting pig viruses to patients and the general population are too great.

Viewpoint: No, xenotransplants from pigs raised at so-called organ farms should not be prohibited because the risks of disease transmission are relatively small and are vastly outweighed by the need for donor organs.

The ancient idea of using the parts or products of exotic animals as tonics and medicines has been replaced by modern attempts to transplant living cells, tissues, and organs from animals to humans. A transplantation of cells or organs from one species to another is known as a xenotransplant. Scientists have been experimenting with xenotransplantation as a means of coping with the chronic shortage of donor organs. Every year tens of thousands of Americans suffer from conditions that could be alleviated by organ transplants, but the supply of human organs cannot meet the demand. Xenotransplants are also seen as a way to improve the lives of people with spinal cord injuries, Alzheimer's disease, Parkinson's disease, and diabetes. Although the field of xenotransplantation has been evolving very rapidly since the 1990s, researchers still face many practical, theoretical, and ethical obstacles.

The experimental physiologists who first attempted to transfuse blood into animals and humans in the 1660s could be considered the precursors of modern xenotransplant scientists. In both cases, the major obstacle was finding ways to deal with the immunological mechanisms that help the body distinguish between "self" and "nonself." Unlike twentieth-century surgeons, however, seventeenth-century scientists could not appreciate the immunological barriers between species and individuals.

In the 1660s Jean-Baptiste Denis transfused blood from a dog to a dog and from a calf to a dog, before using animal blood to treat human diseases. Denis suggested that animal blood might actually be safer than human blood. Because humans could readily assimilate the flesh of animals, it was reasonable to assume that animal blood could also benefit humans. After two apparently successful experiments, a third patient died. When human heart transplantation began about 300 years later, the shortage of donor human hearts inspired attempts to use the organs of chimpanzees, baboons, pigs, and sheep.

The same problem that defeated seventeenth-century scientists, the body's rejection of foreign materials, ensured the failure of the early heart transplants. Unlike Denis, however, doctors in the 1960s were well aware of the powerful immunological barriers between individuals and species. Nevertheless, attempts have been made to transplant kidneys, hearts, livers, and other animal tissues into humans. In one famous case, a child known only as "Baby Fae," who was born with a malformed heart, survived briefly after receiving the heart of a baboon. A few patients have survived briefly after receiving baboon livers, and in 1995 bone marrow from a baboon was transplanted into a man with AIDS in an attempt to restore his immune system. Optimistic transplant surgeons pointed out that blood transfusion had once faced seemingly impossible obstacles, and they predicted that organ transplants would one day be as commonplace as blood transfusions. Advocates of xenotransplants from pigs specifically raised as organ donors have similar hopes that this procedure will eventually become safe and routine. Indeed, heart valves from pigs and blood vessels from cows were used successfully to repair damaged human heart valves and blood vessels before the introduction of synthetic materials such as Dacron and GORE-TEX.

Scientists believe that advances in biotechnology will make it possible to avoid organ rejection by transforming animal organs and suppressing the immune response. By establishing strains of transgenic animals that produce appropriate human antigens, animal organs might be sufficiently "humanized" to avoid or diminish the immune responses that lead to rejection. Although immunological obstacles to xenotransplantation may be overcome, there may be other problems with xenotransplants.

Critics, especially virologists and epidemiologists, have argued that xenotransplants could pose a danger to patients and to society as a whole because animal organs might harbor viruses that would allow new diseases to cross over into the human population. The potential danger of transmitting new diseases is highly contested among scientists. Some scientists think that pig organs pose less of a threat for transmitting viruses than primate organs, but unknown retroviruses, which might be similar to HIV (the virus that causes AIDS), are a source of concern.

Critics in the animal rights movement object that it is unethical to use animals as a source of organs for humans. Members of People for the Ethical Treatment of Animals (PETA) are opposed to the use of animals for "food, clothing, entertainment, or experimentation." In 1999 PETA petitioned the U.S. Food and Drug Administration (FDA) to ban all xenotransplantation experiments. Most Americans, however, believe that animal research is essential to scientific progress against disease. Indeed, the Humane Society of the United States acknowledges that "biomedical research has advanced the health of both people and animals." Scientists and ethicists, however, generally agree that nonhuman primates should not be used as xenotransplant donors because of their endangered status, their closeness to humans, and their tendency to harbor potentially dangerous pathogens. Pigs that have been specially bred and raised at carefully maintained organ farms could provide a good source of donor organs. Pig cells contain the so-called porcine endogenous retrovirus (PERV) incorporated into their DNA. Researchers have not yet determined whether these viruses could spread to transplant patients or to people who come in contact with such patients.

Some critics suggest that recruiting more human organ donors would be a better approach to the organ shortage than pursuing the uncertain prospect of xenotransplants. Another important ethical concern raised during the debate about xenotransplants is the broader issue of social justice. The United States remains the only industrialized country in the world that does not have some form of national health care. Some people consider it unethical to spend huge amounts of money for xenotransplants to help a few patients when millions cannot afford basic medical care. —LOIS N. MAGNER

Viewpoint:

Yes, xenotransplants from pigs should be prohibited because the risks of transmitting pig viruses to patients and the general population are too great.

In medicine today, transplants of organs such as hearts and kidneys are taken for granted. Following such transplants, many people who would otherwise be dead without these procedures are leading full and productive lives. However, there is a chronic shortage of donor organs. Most deaths do not occur in appropriate circumstances for organ harvesting; of those that do, the individual may not have volunteered to donate his or her organs, or the next of kin may not be willing to give consent. Given this shortage, the need for new organ sources is extreme.

Many people—both scientists and members of the public—see xenotransplantation as the solution to the problem of donor organ shortage. Since early experiments on primates proved problematical for both ethical and practical reasons, pigs have become the animal of choice in xenotransplantation. This is mainly because pig organs are an appropriate size for the human body, and pigs are easily available. Unlimited supplies of suitable organs are expected to save the many thousands of people across the world

KEY TERMS

ALLOTRANSPLANT: Transplant between genetically different individuals. Used to describe human–to–human transplants.

ENDOGENOUS RETROVIRUS (ERV): A retrovirus that has permanently integrated into the host genome and is transmitted through generations as a genetic trait.

EXTRACORPOREAL PERFUSION: A means of treating a person's blood outside his or her body, whether through a machine or an organ.

HYPERACUTE REJECTION: A fast and widespread immune antibody response to "foreign" cells in the body.

RETROVIRUS: An RNA–based virus that replicates itself as DNA and integrates into the host genome, where it highjacks the host's cellular machinery for viral proliferation.

XENOZOONOSIS/XENOSIS: An animal disease that infects humans as a result of xenotransplant. Currently a theoretical occurrence.

ZOONOSIS: A disease normally affecting animals (with or without symptoms) that can be passed on to humans.

---

that await life–saving transplant surgery. However, there are considerable obstacles to the fulfillment of this vision. Not the least of these is the risk of infection arising from the foreign transplant, not only to the recipient, but also to the population at large.

Risks of Xenotransplantation All transplantation poses risks, for several reasons. First, the transplanted tissue is equivalent to a direct injection of donor cells, overriding all the body's gross defenses, such as the skin and digestive juices. Second, cells from the graft (transplanted living tissue) may migrate from their original site to other areas of the body, which may be more susceptible to infection. Third, transplant patients require drugs to suppress their immune system, to prevent rejection of the grafted tissue. This drastically reduces the hosts' ability to fight any infection that might be inadvertently introduced with the transplant.

Xenotransplantation carries other risks, in addition to those mentioned above. Like all other mammals, pigs carry a wide range of bacteria and viruses. Many of these are harmless to the pigs themselves, who have developed defenses against them over millions of years; however, they can cause disease in other species, who lack this protection. Such diseases are known as zoonoses. For example, the virus responsible for the outbreak of "Spanish flu," which killed over 20 million people worldwide in 1918–19, originated in pigs, where it did not cause any harmful symptoms. While pigs used for xenotransplantation would be born and raised in sterile conditions to avoid such outside infections, there are more difficult problems, such as retroviruses, to be resolved.

Retroviruses Retroviruses are a group of viruses that spread via RNA (ribonucleic acid). The best–known example of a retrovirus is HIV, or human immunodeficiency virus, which causes AIDS (acquired immunodeficiency syndrome). Retroviruses enter the host cell through receptors on the cell surface and use an enzyme called reverse transcriptase to make a DNA (deoxyribonucleic acid) copy of their RNA. This viral DNA is then integrated into the host cell genome and is copied into new RNA by the cellular machinery of the host. These new viral transcripts are packaged into pockets of the host cell membrane and enter the system to infect additional cells. The results of retrovirus infection are variable. Some retroviruses, such as HIV, are fatal. Others may be completely harmless. Still more can cause mild symptoms at first and then become dormant. Symptoms may be triggered long after infection, at times of stress, or may arise only if the virus infects a certain cell type or infects individuals who have a genetic susceptibility.

PERVs The retroviruses that are of most concern in xenotransplantation are called endogenous retroviruses, or ERVs (in the case of the pig, porcine endogenous retroviruses, or PERVs). These viruses have succeeded in integrating themselves permanently into the host genome, so that they are passed onto the offspring in each succeeding generation as an inherited genetic trait. Examples of such viruses have been identified in all mammals that have been tested, including humans. These viruses may be a legacy of virulent plagues that decimated populations in the distant past until the virus became integrated into certain species' egg and sperm cells, or germline. Once an ERV has become integrated, it is in its evolutionary interests that the host species survives. This leads to an evolutionary armed truce, so to speak, where the host transmits the virus without causing any detrimental effects. However, if the ERV suddenly finds itself in a new host with no evolved defenses, the truce may be called off.

Most ERVs have become inactive over many generations, losing the ability to be passed on and cause infection, but some remain potentially infective. One ERV, for example, does not affect chickens but can cause disease in turkeys and quail.

ERVs might be responsible for autoimmune diseases (a collapse of the body's immune system causing the immune system to attack the body's own organisms). Like all genes, the ERVs can move around the genome during recombination. Depending on the new site, the retroviral proteins may be activated and expressed in certain cells, causing the immune system to attack them. They could also be inserted into functional genes, making them inactive, or they could activate proto–oncogenes (genes that produce tumors), leading to cancers. Introducing PERVs into the genome via xenotransplantation could activate human ERVs, causing similar symptoms. Recombination with human ERVs could create previously unknown infectious viruses, for which we have no diagnostic tests.

Virologist Robin Weiss and his colleagues have discovered up to 50 PERVs in the pig genome, and their experiments have shown that three of these have the ability to infect human cells in culture. The likelihood of infection varies depending on the human cell type involved and the length of time of exposure. This obviously has implications for xenotransplantation. The risk of infection depends on the type of transplant, and migration of cells from the original transplant site to other areas may increase the risk. In a whole–organ transplant, the organ would be expected to function for months or years, so long–term exposure to PERVs could occur.

Another researcher, Daniel Saloman, has shown that pig pancreas cells transplanted into mice with lowered immunity can cause active infection in the mouse cells *in vivo*. Virus production in the pig cells actually increased—probably as a result of the stress of the transplant procedure—and virus production continued for the full duration of the three–month experiment. The infection produced no symptoms in the mice and appears to have become dormant after a few cycles of replication. Researchers in France have identified 11 types of PERVs in major organs of the pig, including the heart, liver, pancreas, and kidneys: all prime candidates for transplantation.

Viruses closely related to PERVs, found in dogs, cats, and mice, can cause lymphoma and other diseases, so it is likely that similar illnesses could result from activation of the viruses in humans.

Engineering Disaster? In xenotransplantation, the risk of viral infection is higher than in other transplantations. This is because not only are the T cells (the cells that congregate at the site of infection and mount a direct attack on the foreign matter) artificially suppressed, as in human transplants, but the antibody–producing B cells must also be suppressed, to avoid hyperacute rejection.

A pig farm in Spain.
(© Owen Franken/Corbis. Reproduced by permission.)

Hyperacute rejection involves a specific sugar molecule that is expressed on the membrane of pig cells and is recognized by the human immune system as foreign, invoking a massive and rapid immune attack. PPL Therapeutics (the company that created Dolly the sheep) has engineered a "knockout" pig that lacks the enzymes to make this sugar, so the human immune system will not see the cells as foreign. The problem lies in the fact that if the PERVs become active and bud from the cell, they will be surrounded by this "disguised" membrane, which will further hamper the body's attempts to attack the virus. In effect, the new virus particles will have a "free ride" to infect the host's system. The next planned step in this technology is to add three other human genes to the pig cells that will help to combat a longer–term immune response. This will even further protect the virus from immune attack.

Unknown Enemies While the problems of PERVs may possibly be overcome, given time and research, the problem of the unforeseen will persist. We have sophisticated tests for known viruses but no way to detect anything new and unexpected. We cannot predict what mutations and recombinations might occur in these viruses once they are established in the human body. ERV–type viruses can be very elusive. The HTLV virus, for instance, is infectious in humans but has an incubation period of decades. It causes lymphomas and other diseases but only in about 4% of infected people, and the symptoms do not show for 20 to 40 years after initial infection. It is possible that a new virus arising from xenotransplantation could have similar characteristics. We would have no tests to detect it, and by the time the symptoms became apparent it could be widespread in the population.

Worst–Case Scenario The present state of knowledge suggests that changes in infectious agents such as PERVs will not be particularly infectious or virulent, but they evolve very quickly, given their short life cycles. Experiments with successive generations of PERVs exposed to human cells have shown that the PERVs do become more prolific over time. Virologist Robin Weiss has speculated that, given the nature of ERVs, they may well become more active in a new host that has had no time to develop specific defenses.

If these viruses were to become infectious and virulent, transmission would probably be via body fluids and sexual contact. Based on our knowledge of related viruses, the incubation period might be very long, and the symptoms could include lymphoma, neurodegenerative disease, or autoimmune disease. The result could be a plague to equal AIDS. This may not happen, but the risk is not calculable, and thus it is not meaningful to talk of making informed decisions at this time.

Informed Consent A patient who will die within weeks without a transplant may, for example, find an undefined risk of developing lymphoma in 20 or 30 years worth taking. However, if there is a risk of passing the infection on to family members and close contacts, and further into the general population, then the issue becomes much more complex.

Very close monitoring of such patients would be required. They would have to undergo regular tests of blood, saliva, and, possibly, semen to test for viral infection. The conditions under which they would need to live would be similar to those for HIV–positive individuals, who need to practice safe sex and record their sexual partners and who cannot give blood donations. Decisions about having children would be affected for the patients. This monitoring would continue for many years, possibly for life. In the event of an infectious, virulent virus occurring, quarantine measures would need to be considered. The problem lies in enforcing such stringent conditions. At present there is no legal structure to oblige a patient to continue with a program, and he or she could drop out at any time.

Alternatives Although there is a great need for transplants at the current time, there are alternatives to xenotransplantation, with all its risks and ethical dilemmas. Improved health education and preventative care could dramatically reduce the incidence of a great deal of heart disease and other disorders, and so decrease the need for transplantation. At present, only a very small percentage of people carry donor cards, and their wishes can be overridden at the time of death by their next of kin. There is great scope for increasing awareness of the need for donor organs and improving this situation.

In the long term, where there is no alternative to transplantation, stem cell technology holds out hope for the creation of human cells and/or organs especially for transplants. Cloned cells from the patient could be used to create the required organ, without risk of rejection or infection. This technology lies far in the future at present, and many workers in the field believe that xenotransplantation will be necessary in the meantime. However, a great deal of research into the nature and behavior of PERVs, to allow more informed debate regarding the risks, is required. Medical progress is not, and can never be, entirely risk–free, but public understanding and involvement in decision–making is essential. Only then should tightly controlled clinical trials even be considered. —ANNE K. JAMIESON

Viewpoint:
No, xenotransplants from pigs raised at so-called organ farms should not be prohibited because the risks of disease transmission are relatively small and are vastly outweighed by the need for donor organs.

Every day in the United States, at least 10 people die while waiting for an organ transplant that could save their life. Countless more battle diabetes or Alzheimer's disease, or lose their mobility to spinal cord injuries. All these people can benefit from the various forms of xenotransplants. The procedures are not risk–free. Like most advances in medicine (and many other areas of science), we must weigh the potential risks of xenotransplants against the benefits the procedures offer.

An Abundance of Transplant Candidates Obviously, any person on a transplant waiting list would benefit from an unlimited supply of organs. But many other people might benefit from xenotransplants as well. There are people with failing organs who never make it to the

Dr. David Sachs, a surgeon and immunologist at Massachusetts General Hospital, poses June 28, 2001, at his offices in Boston. Sachs is developing a strain of miniature pigs as a potential source of organs for human transplant.
(AP/Wide World Photos. Reproduced by permission.)

transplant list. The current organ shortage necessitates strict rationing of available organs to the patients who can benefit the most from them. Some patients have conditions that make them unsuitable candidates for a transplant procedure. In their book *Xeno: The Promise of Transplanting Animal Organs into Humans* (2000), Drs. David K. C. Cooper and Robert P. Lanza list several examples of what they call "borderline candidates." One such example is the patient with mild heart disease who requires a liver transplant. Such a patient is at risk of suffering a heart attack during the transplant surgery. In the worst–case scenario, the patient will die as a result of a heart attack, and the liver will have been wasted. If the patient survives, he will not get the full benefit of his healthy liver as a result of his heart condition. The transplant surgeon, faced with a limited supply of livers, will likely pass this patient over in favor of a recipient whose only problem is a failing liver. An unlimited organ supply from pigs will allow even the borderline patients a chance at life.

In countries where allotransplants (transplants between human donors and recipients) are taboo, the situation is even more dire and has given rise to various morally objectionable practices, including organ brokerages. Many more people die for need of an organ transplant in those countries than in places like the United States and Europe. Thus, the benefits of xenotransplants to people in need of organ transplants can be measured in many thousands of lives saved each year.

People with diseases or conditions that cannot be successfully treated today might benefit from xenotransplants. Many conditions that involve cell destruction may be improved or cured with cell implants from pigs. For example, fetal neural cells from pigs have been implanted in a small number of stroke victims, and notable improvements followed in some of these patients. Researchers are eyeing Huntington's chorea patients (persons with a hereditary disease of the brain), among others, as candidates for trials with pig neural cells. A study done on rats with severed spinal cords showed that it is possible to restore nerve fibers using nerve cells from pigs. Scientists may be on the brink of being able to repair spinal cord injuries, and xenotransplant may be one way to achieve this goal.

Extracorporeal Liver Perfusion Robert Pennington is a living example of the benefits xenotransplantation can bring. In 1997, Pennington lay in a deep coma at Baylor University Medical Center in Texas. He was in acute liver failure. His condition is known as fulminate liver failure, an irreversible destruction of the liver. When the liver ceases to function, toxins that the body creates during normal biological processes accumulate in the blood. These toxins make their way to the brain, eventually causing coma and death in a few days. Fulminate liver failure can be the result of a disease, such as hepatitis, or an exposure to certain chemicals or toxins, such as the toxin found in *Amanita* mushrooms. Fulminate liver failure often requires a speedy liver transplant. It has a mortality rate as high as 95%.

Pennington's name was placed at the top of the liver transplant list, but his condition was deteriorating rapidly. His transplant surgeon, Dr. Marlon Levy, suggested a new experimental procedure. Pennington's blood was routed through a pig liver that was taken from a specially bred transgenic pig. Transgenic pigs are pigs that have human genes in them. These genes code for human cell–surface proteins that then coat the pig organ, to minimize the risk of the organ's rejection by the human immune system. The pig liver was able to provide a sufficient amount of cleaning to keep Pennington alive until a suitable human liver donor was found. Pennington is alive and healthy today. Neither he nor his family show signs of any viral infection acquired through this procedure, known as extracorporeal liver perfusion.

The FDA (Food and Drug Administration) halted the perfusion trials shortly after Pennington's successful treatment. A specific pig virus known as porcine endogenous retrovirus (PERV) was found capable of infecting human cells in laboratory conditions. The discovery of PERV's prevalence in pigs, and the fact that there were at least two different strains of the virus, alarmed many scientists and regulators. The clinical trials resumed when detection methods for PERV in humans and pigs were refined. Nevertheless, the PERV controversy focused attention on the issue of viral infections resulting from xenotransplants.

When it comes to individual patients, it is easy to see why many will grasp at the chance to live, despite the risk of an unknown, potentially deadly infection from the donor animal. These patients have nothing to lose—they are dying. Pennington's family took a chance on what was described as "uncharted territory" so that he might live. The problem with xenotransplants is that we do not know if an unfamiliar animal infection contracted by the human patient can mutate and become another acquired immunodeficiency syndrome (AIDS)–like pandemic. This possibility puts not only the patient but also society at large at risk. The risk is real and warrants further investigation and careful control measures, according to everyone involved in xenotransplants. It is also, according to many virologists, a rather small risk. When weighed against the potential benefits of xenotransplants, in these scientists' opinion, the remote likelihood of a novel infection does not warrant banning the research and eventual implementation of xenotransplant.

The Dangers of Retroviruses Animals who routinely carry an infectious agent are called hosts for the agent (virus, bacteria, or parasite). Endogenous retroviruses are viruses whose DNA (deoxyribonucleic acid) sequence is integrated into the host's DNA in each cell of the host. We as humans carry our own endogenous viral sequences in our DNA. Because the viral sequence is integrated into the host's DNA, it is extremely difficult, and often impossible, to eliminate the virus from the host. There is concern that by transmitting PERV to humans, especially immunosuppressed individuals, the virus can become "hot" and cause infection. Or perhaps a PERV particle, or even an unknown virus that has not yet been detected in pigs, might combine with some of the human endogenous viral DNA to form a new, possibly infectious, virus. The concern is a valid one and should be investigated. This concern also makes a good case for strict follow–up of xenotransplant patients and their families. But the PERV situation is not unique. The current MMR (measles, mumps, rubella) vaccine, made with chicken cells, contains particles of an endogenous avian (bird) retrovirus. Because the vaccine is a live one, there is a possibility of combination between the avian virus and the MMR infectious particles. To date, no infections of any kind have been reported as a result of the MMR vaccine. The chance of a recombination event between retrovirus particles is far less likely to occur between non–homologous sequences (sequences that share little similarity to one another) such as pig and human retroviruses.

A study of 160 patients who were exposed to living pig tissues or organs for lengthy periods showed no evidence of PERV infection, or infection with any other known pig viruses. Patients in this study, many of whom were immunosuppressed during their treatment periods, were followed for more than eight years post–treatment. The study is not a guarantee that such infections have not or will not occur, especially in individuals who receive heavy doses of immunosuppression drugs. Nonetheless, this study is an encouraging sign. Dr. Robin Weiss, a virologist specializing in retroviruses, estimated in an interview for Frontline's *Organ Farm* that the chances of a human PERV epidemic infection are remote. Other scientists support his view. In addition, Dr. Weiss noted that a currently available anti–HIV (human immunodeficiency virus) drug has proven very effective against PERV. In a worst–case scenario, scientists already have at least one drug that can fight PERV infection, should one occur. Drs. Walter H. Günzburg and Brian Salmons, in a 2000 paper assessing the risk of viral infection in xenotransplants, pointed out that safety techniques used in gene therapy today can be successfully adapted to control a "hot" PERV in humans.

A BRIEF HISTORY OF XENOTRANSPLANTS

The idea of xenotransplants is not new. In fact, quite a few of our myths involve creatures such as centaurs, which are a combination of different species' anatomy. Actual xenotransplants from animals to humans date back as far as the 1600s. At that time, a nobleman in Russia with a skull disfigurement had the defect repaired with fragments from a dog's skull. The Russian church, however, threatened to excommunicate the nobleman, and he had the fragments removed. Sheep blood was transfused into humans as early as 1628 and was used throughout the following centuries.

The breakthrough that allowed true organ transplants to become a reality came from a French surgeon named Alexis Carrel. In 1912 Carrel won the Nobel Prize in physiology and medicine for developing the techniques that allowed blood vessels to be joined together. Interestingly, Carrel had also predicted, in 1907, the use of transgenic pigs in xenotransplant work.

Interest in xenotransplants declined following many failures in the early twentieth century. But as medicine advanced, interest increased again. In 1963 Dr. Keith Reemtsma, then at Tulane University in Louisiana, transplanted chimpanzee kidneys into 13 patients. Remarkably, one of the patients survived for nine months. It is thought she died of an electrolyte imbalance brought on by a difference in the functioning of human and chimpanzee kidneys. An autopsy showed no signs of kidney rejection. The other 12 patients survived for periods of between 9 and 60 days. A year later, three years prior to the first cardiac allotransplant, Dr. James Hardy, working in Mississippi, performed the first cardiac xenotransplant, using a chimpanzee heart. The heart proved too small, and the patient died two hours following surgery. In London in 1968 Dr. Donald Ross tried using a pig's heart as a "bridge" to keep a patient alive while the patient's own heart recovered. The pig's heart was immediately rejected. But the bridging idea stuck, and it was tried several times later. In 1984 Dr. Leonard Bailey and his team in Loma Linda, California, transplanted a baboon heart into a newborn baby. "Baby Fae" was born with a severe heart defect, and she lived for 20 days with the baboon's heart while doctors tried to find a suitable human donor for her. Sadly, no suitable heart was found in time. Blood group incompatibility may have contributed to the rejection of Baby Fae's baboon heart.

In 1992 and 1993 Dr. Thomas Starzl (University of Pittsburgh, Pennsylvania) used baboon livers on two patients with hepatitis B (baboon livers are resistant to hepatitis B). Both patients died of massive infections, but they survived longer than any other animal-liver recipients. Also in 1992, Astrid, the first transgenic pig, was born in Cambridgeshire, England. Astrid was created by scientists at Imutran, a British biotechnology firm. Transgenic pigs are considered the best hope to overcoming acute rejection problems in xenotransplants of solid organs.

Today, the shortage of human organs pushes research into xenotransplants forward. Researchers focus their efforts on cellular xenotransplants as potential cures for diseases such as Huntington's chorea and diabetes. Other researchers are trying to overcome the severe rejection processes that make solid organ xenotransplants currently impractical. In addition, efforts are ongoing to reduce the risk of a novel virus epidemic transmitted through xenotransplants as a result of exposure to donor animals.

—Adi Ferrara

Infectious Agents The danger of transmitting potentially lethal infections through transplants is not unique to xenotransplants. Hepatitis, HIV, cytomegalovirus (a virus that results in dangerous cell enlargement), and Epstein–Barr virus (associated with lymphoma and some cancers) are all examples of infections that have been transmitted through allotransplants. As we know, hepatitis and HIV can spread from the patient to other people. No one today is calling for a stop to allotransplants. Instead, better detection methods and screening procedures have been implemented, to try and minimize the danger of infections transmitted from donor to recipient. There are still cases, however, when the urgency of the situation overrides safety concerns, and the donor is not screened adequately, or is known to carry an infection and the organ is still used. In that respect, xenotransplant will have an advantage over conventional allotransplants. As the World Health Organization pointed out in a 1997 report: "Totally infectious agent–free human donors are not available." It is, however, possible to raise pigs in a specific pathogen–free environment (an environment free of specific disease–causing organisms). In such an environment (an "organ farm"), pigs can be raised free of most known and suspected infectious agents.

The chances of an infectious agent jumping species and establishing itself in a new host are greater the closer the two species are, genetically. This is a good argument against using non–human primates as organ donors, and indeed the current policy bans their use as donor animals. After all, the AIDS epidemic started in monkeys in Africa, when the Simian Immunodeficiency Virus (SIV) jumped species and established itself in humans as HIV. An argument could be made that by creating a transgenic pig, we are increasing the risk of an infectious agent (not necessarily PERV) jumping species. The argument has two valid points. By expressing human genes in pig organs, we are making pigs more genetically similar to humans. Additionally, the human genes that are currently expressed in transgenic pigs actually code for molecules that viruses use as receptors to gain access into the human cells. This concern is not to be taken lightly. However, even if a pig virus established itself in an immunosuppressed individual, it may not necessarily be able to spread beyond the recipient. Our immune system, when not compromised by drugs, can fight off a variety of viruses. In addition, not all viruses are created equal. Even in the case of the dreaded AIDS virus, not all the HIV virus groups spread through the population. Each of the two distinct types of HIV currently known is further divided into subgroups, some of which (HIV–2 C through F) have not spread at all, and some of which (HIV–1 O and N) only spread in a limited area in Africa. Variables in the environment, the host, and the virus itself can all combine to affect the pathogenicity of a virus—its ability to infect an individual and spread throughout the population.

Xenotransplant Research Must Continue

In a sense, the debate about xenotransplant is moot. The concept of xenotransplantation is not new. Xenotransplants have been done from a variety of animals since the beginning of the twentieth century. Pig–to–human xenotransplant has been taking place for a while, mostly on the cellular level in clinical trials. Parkinson's disease patients and stroke patients have been transplanted with pig neural cells. Diabetes patients have received pig islets cells. Pig livers have assisted several patients with liver failure. No one is arguing that there is no risk involved. As Dr. Weiss noted in a recent paper: "With biological products, as with crossing the street, there is no such thing as absolute safety." However, we must consider the potential benefits xenotransplant could bring to millions of people worldwide. The debate about the potential dangers is a valuable one. It supplies us with knowledge and helps us proceed with our eyes wide open. Strict safety regulations and patient monitoring are required because medicine offers no guarantees, and though the chances are small, we could face another AIDS. But when a vast number of people contracted AIDS as a result of tainted blood products, no one called for a stop to blood transfusions. Better screening procedures for blood and donors were developed, and a push was made to develop artificial blood, the latter already in clinical trials.

Weighed against a risk that is likely remote, and with few other options currently available, to abandon xenotransplant research and implementation would be wrong. Cloning, stem cell research, and artificial organs are other avenues that may offer us the same hope as xenotransplants. But until such a day that one of these options becomes a real solution, we cannot afford to turn our back on the others for fear of the unknown. —ADI FERRARA

Further Reading

Bryan, Jenny, and John Clare. *Organ Farm. Pig to Human Transplants: Medical Miracle or Genetic Time Bomb?* Carlton Books, 2001.

Cooper, David K. C., and Robert P. Lanza. *Xeno: The Promise of Transplanting Animal Organs into Humans.* New York: Oxford University Press, 2000.

Council of Europe. State of the Art Report on Xenotransplantation (2000). <http://www.social.coe.int/en/qoflife/publi/artreport/clinact.htm>.

Fishman, Jay. *British Medical Journal* 321 (September 23, 2000): 717–18.

Furlow, Bryant. "The Enemy Within." *New Scientist,* no. 2252 (August 19, 2000): 38–41.

Günzburg, Walter H., and Brian Salmons. "Xenotransplantation: Is the Risk of Viral Infection As Great As We Thought?" *Molecular Medicine Today* 6 (May 2000).

Hahn, Beatrice H., George M. Shaw, Kevin M. De Cock, and Paul M. Sharp. "AIDS as a Zoonosis: Scientific and Public Health Implications." *Science* 287 (January 28, 2000): 607–14.

Lower, Roswitha. "The Pathogenic Potential of Endogenous Retroviruses." *Trends in Microbiology* 7, no. 9 (September 1999): 350–56.

Olson, Lars. "Porcine Xenotransplants—Will They Fly?" *Nature Biotechnology* 18, no. 9 (September 2000): 925–27.

Organ Farm. A PBS Frontline Documentary on Xenotransplantation. <www.pbs.org/wgbh/pages/frontline/shows/organfarm/>.

Paradis, K., et al. "Search for Cross–Species Transmission of Porcine Endogenous Retrovirus in Patients Treated with Living Pig Tissue." *Science*, no. 20 (August 20, 1999): 1236–41.

Patience, C., Y. Takeuchi, and R. Weiss. "Infection of Human Cells by an Endogenous Retrovirus of Pigs." *Nature Medicine*, no. 3 (1997): 282–86.

"Report of the Cross–Species Infectivity and Pathogenesis Meeting." July 21 and 22, 1997. <www.niaid.nih.gov/dait/cross–species/final.pdf>.

Stolberg, Sheryl Gay. "Could This Pig Save Your Life?" *New York Times* (October 3, 1999).

Vanderpool, Harold Y. "Commentary: A Critique of Clark's Frightening Xenotransplantation Scenario." *The Journal of Law, Medicine and Ethics* 27, no. 2 (Summer 1999): 153–57.

Weiss, Robin A. "Adventitious Viral Genomes in Vaccines but Not in Vaccinees." *Emerging Infectious Diseases* 7, no. 1 (January/February 2001). <http://www.medscape.com/govmt/CDC/EID/2001/v07.n01/e0701.24.weis/e0701.24.weis.html>.

World Health Organization, Emerging and other Communicable Diseases, Surveillance and Control. "Xenotransplantation: Guidance on Infectious Disease Prevention and Management." <http://www.who.int/emc–documents/zoonoses/docs/whoemczoo981.pdf>.

Should the cloning of human beings be prohibited?

Viewpoint: Yes, because of the potential physical dangers and the profound ethical dilemmas it poses, the cloning of human beings should be prohibited.

Viewpoint: No, the cloning of human beings should not be prohibited because the potential for medical accidents or malfeasance is grossly overstated, and the ethical questions raised by detractors are not unique to cloning—indeed, ethical questions attend every scientific advancement.

Since the birth of Dolly, the cloned sheep, in 1997, several reproductive scientists, including Severino Antinori, Brigitte Boisselier, and Panayiotis Michael Zavos, have announced that they were ready to clone human beings. However, cloning mammals is still a highly experimental technique. Scientists involved in cloning various mammals have reported many technical problems. A large majority of the clones die during gestation or soon after birth. Placental malfunction seems to be a major cause of death. Many of the surviving clones are plagued with serious physiological and genetic problems. During embryological development, cloned sheep, cows, pigs, and mice tend to become unusually large. Clones are often born with a condition called "large offspring syndrome," as well as severe respiratory and circulatory defects, malformations of the brain or kidneys, or immune dysfunction. It is not yet known whether clones will develop and age normally, or whether subtle failures in genomic reprogramming or genetic imprinting might lead to various defects. Theoretically, tissues generated from cells cloned from a patient's own adult nucleus should not trigger an immune response, but it is possible that subtle differences caused by the foreign cytoplasm in the donor egg might cause a rejection response. Although scientists at Duke University suggested that human clones might not experience the problems encountered in cloned animals, the risks remains very high and quite unpredictable. Eventually animal research may indicate that human cloning can be accomplished with no greater risk than in vitro fertilization posed when Louise Brown, the first "test-tube baby" was born in 1978. However, scientists generally agree that human reproductive cloning should not be permitted before the scientific and technical issues have been clarified.

Many scientists believe that, at least in the near future, experiments in human cloning would involve many failures, miscarriages, stillbirths, and the birth of deformed babies. Some observers think that the reckless claims made by some scientists stimulated the passage of premature Congressional legislation that would ban all human cloning, both reproductive and therapeutic (non-reproductive). Similar reactions have occurred in other nations. For example, the French and German governments jointly asked the United Nations to call for a worldwide ban on human reproductive cloning.

After a heated debate about human cloning, on July 31, 2001, the U. S. House of Representatives voted 265–162 to institute a total federal ban on human cloning. The bill included penalties of up to 10 years in prison and a $1 million fine. The House rejected competing measures that would have banned cloning for reproductive purposes while allowing nonreproductive or therapeutic cloning for scientific research. The emotional nature of the debate, and the lack of understanding of the scientific aspects of the subject, is epitomized by House Majority Whip Tom Delay (R-Texas) who declared: "Human beings should not be cloned to stock a medical junkyard of spare parts." On the other hand, Rep. Jim Greenwood (R-Pennsylvania) lamented that the House had missed an opportunity to balance potential biomedical breakthroughs with ethical concerns.

The Human Cloning Prohibition Act outlaws the process known as somatic cell nuclear transfer (SCNT) using human cells. Although this process can be used for reproductive cloning, as in the case of Dolly the sheep, the technique can also be used for non-reproductive or therapeutic cloning, a process that could be used to create cells and tissues that would be immunologically compatible with the donor of the nuclear material. The potential uses of therapeutic cloning include cures and treatments for many diseases. Cloned cells could be used to create replacement tissue for diseased hearts, pancreatic cells for diabetics, treatments for neurodegenerative diseases, such as Parkinson's and Alzheimer's, nerve cells for victims of spinal cord injuries, and skin cells for burn victims. Researchers, as well as leaders of biotechnology and pharmaceutical companies, believe that therapeutic cloning will result in major medical breakthroughs. Therapeutic cloning could provide valuable new means of testing drugs for safety and efficacy, thus streamlining and improving the drug development process.

Another modification of the nuclear replacement technique known as oocyte nucleus transfer could help women with defective mitochondrial DNA. Defects in mitochondrial DNA are known to cause more than 50 inherited metabolic diseases. Theoretically, a child produced by this process would inherit its nuclear DNA from its mother and father and healthy mitochondrial DNA from a donor egg. This procedure would not constitute reproductive cloning because the child's genetic makeup would be as unique as that of a child produced by conventional sexual means.

Nonreproductive cloning is legal in the United Kingdom. Since the early 1990s British scientists have been allowed to create human embryos for research purposes and perform experiments in therapeutic cloning. The Human Fertilization and Embryology Act of 1990 established a system for regulating the creation and use of embryos. Research leading to reproductive cloning is banned, but therapeutic cloning in order to generate healthy replacements for diseased tissues and organs is permitted. Some ethicists and religious leaders object to all experiments on embryos, while others argue that even therapeutic cloning should be banned because it would eventually lead to reproductive cloning. While human cloning and human stem cell research are actually technically distinct, these issues have become virtually inseparable because both involve the use of human embryos.

President George W. Bush, who opposes cloning humans for research and reproductive purposes, immediately announced support for the anti-cloning bill passed by the House. President Bush has taken the position that: "like a snowflake, each of these embryos is unique with the unique genetic potential of an individual human being." Further confusion and controversy was, therefore, sparked by Bush's decision to allow limited federal financing for stem cell research on 60 cell lines that had already been established. Despite recognition of the potential value of the research, the president ruled out any future flexibility in his stem cell policy. Thus, Bush's decision angered groups on both sides of the argument: those who hoped that stem cell research would help people with serious diseases and disabilities and those who believe that the fertilized egg should be accorded full human status.

Less than a month after President Bush announced his decision on embryonic stem cell research observers noted that the substance of public debate had shifted from whether such research was ethical to a debate about whether essential biomedical research could be conducted in the United States. Despite confusion about scientific and technical issues, there is widespread public awareness of predictions that human cloning and stem cell research could provide treatments and cures for many diseases, if scientists are allowed the freedom to pursue all promising pathways. If patients and their advocacy groups think that cures are being withheld because of political and religious forces, there is little doubt that the nature of the public debate will become increasingly hostile.

The debate about the scientific and technical uncertainties of human cloning will presumably be settled in the not-too-distant future through experimentation in countries that allow this type of research to proceed. The situation is likely to be quite different where ethical issues are concerned. Those at one extreme believe that the use of embryos for research purposes is morally unacceptable on the grounds that an embryo should be accorded full human status from the moment of its creation. In contrast, some people believe that early embryos should not be considered human begins with special moral rights. Others consider the embryo a potential human being, but argue that the rights of the early embryo should be weighed against the potential benefits arising from research. The British Human Fertilisation and Embryology Act attempted to maintain an ethical "middle ground." Indeed, British authorities suggested that cloning embryos by the cell nuclear replacement technique that produced Dolly might be considered a form of transitional methodology that could eventually provide insights into genetic mechanisms for reprogramming adult cells.

Although many Americans believe that stem cell research and therapeutic cloning are morally justifiable because they offer the promise of curing disease and alleviating human suffering, others are unequivocally opposed to all forms of experimentation involving human embryos. Despite the potential benefits of cloning and stem cell research, most scientists acknowledge that there are still many technical and scientific problems. However, even if the scientific issues are resolved, ethical, emotional, and political issues will probably continue to dominate debates about human cloning. —LOIS N. MAGNER

Viewpoint:
Yes, because of the potential physical dangers and the profound ethical dilemmas it poses, the cloning of human beings should be prohibited.

On July 31, 2001, the U.S. House of Representatives passed legislation to prohibit the cloning of human beings (Human Cloning Prohibition Act of 2001, H.R. 2505). As defined in the bill, "human cloning means human asexual reproduction, accomplished by introducing nuclear material from one or more human somatic cells into a fertilized or unfertilized oocyte whose nuclear material has been removed or inactivated so as to produce a living organism (at any stage of development) that is genetically virtually identical to an existing or previously existing human organism." The bill makes it a federal crime with stiff penalties, including a fine of up to $1 million and imprisonment up to 10 years to attempt to or participate in an attempt to perform human cloning as defined by the act.

With this step the United States is now poised to join the international community. Many nations have already adopted the prohibition against human cloning. International scientific bodies with oversight for medicine, research, and health as well as biotechnology industry interest groups mostly support the prohibition. But the matter is not yet settled. Before the measure can take effect, it faces a vote in the U.S. Senate and it needs the president's signature.

What reasons or fears prompted U.S. lawmakers to take such strong measures? The sanctions in the act are more commonly associated with major felonies. Are there sufficient dangers to the nation and to its citizens to call for prohibition of a scientific technique whose applicability to humans seems as yet theoretical?

Since 1997, when Dolly the sheep, the first mammal ever derived from a cell (from the udder) of an adult mammal (sheep), was announced to the world, the possibility of human cloning has entered a new stage. After 276 unsuccessful attempts, the "cloned" embryo implanted in a ewe went through the normal gestation period and was born alive, surviving and healthy to the present. This was indeed a first and a major breakthrough. Since that event, the feat has been replicated in several types of livestock on different occasions. In 1998 mice were successfully cloned from adult cells. Humans bear certain genetic similarities to mice, and so often mice are the experimental model for what will be studied subsequently in humans. The technique of cloning by somatic cell nuclear transfer (SCNT) is steadily improving, suggesting a certain momentum.

The animal successes have inspired some groups and scientists to announce publicly their aim of embarking on the cloning of human beings through the same technique that gave us Dolly and the several other groups of mammals that followed.

Asexual Reproduction The technique itself raises fundamental questions. SCNT, as noted in the definition of the Human Cloning Prohibition Act, involves taking a cell—many types of cells seem to work—from a full-grown animal and removing the nucleus, which contains the complete genetic makeup, or DNA, of the animal. The nuclear DNA is then inserted into a separate unfertilized but enucleated ovum harvested from an adult mammal, even a mammal of a different species. After special preparation in a nutrient medium to allow the cell to reprogram itself and a small trigger of electric current to start mitosis, the "cloned" blastocyst, now with the nuclear DNA of an adult somatic cell in each of the multiplying cells, is implanted in the womb of a "gestational" mother, previously treated and made ready to accept the pregnancy. If the process is successful, the resultant birth yields a new individual animal, genetically identical to the animal that donated the nucleus yet at the same time different from any other animal: the DNA in the nucleus of its cells does not come from the mating of two animals, in which each parent's DNA combines to create a new individual. Rather the DNA is from one animal only.

Sexual reproduction with its commingling of the genetic endowment of two individuals has demonstrated its existential value. Over hundreds of millions of years, nature has used the

> ## KEY TERMS
>
> **BLASTOCYST:** Developing preimplantation embryo, consisting of a sphere of cells made up of outer support cells, a fluid-filled cavity, and an inner cell mass.
> **CHROMOSOME:** Composed chiefly of DNA, the carrier of hereditary information, these structures are contained in the nucleus of the cell. Normal human chromosomes contain 46 chromosomes, one half from each parent.
> **DNA:** Deoxyribonucleic acid; found primarily in the nucleus of the cell. It carries all the instructions for making the structures and materials necessary for bodily functions.
> **ENUCLEATED OVUM:** Ovum from which the nucleus has been removed.
> **FERTILIZATION:** Process by which sperm enters the ovum and initiates the process that ends with the formation of the zygote.
> **GENE:** Working subunit of DNA that carries the instructions for making a specific product essential to bodily functioning.
> **GENETIC IMPRINTING:** Process that determines which one of a pair of genes (mother's and father's) will be active in an individual.
> **GENOME:** Complete genetic makeup of a cell or organism.
> **IN VITRO FERTILIZATION:** Assisted reproduction technique in which fertilization occurs outside the body.
> **MEIOSIS:** Special type of cell division occurring in the germ cells by which each germ cell contains half the chromosomes of the parent cell. In humans this is 23 chromosomes.
> **MITOCHONDRION:** Cellular organelle that provides enrgy to the cell and contains maternal DNA.
> **MITOSIS:** Process of ordinary cell division, resulting in the formation of two cells identical genetically and identical to the parent cell.
> **NUCLEUS:** Cell structure that contains the chromosomes.
> **OOCYTE:** Mature female germ cell or egg.
> **SPERM:** Mature male reproductive cells.
> **ZYGOTE:** The single-celled fertilized egg.

process to great advantage to perpetuate and vary species, to bring new beings and new species into existence throughout the plant and animal kingdoms, to ensure the continued and enhanced adaptation of living things to changing environments, and to preserve life itself. It is also the process that in our experience gives the "stamp of individuality," producing new members of a species with a novel genetic endowment and, not rarely, new individuals with traits that are remarkable and prized as well as unique.

The Status of a Clone Perhaps a superior or functionally perfect breed of animal might offer special advantages—although a world of racing in which thousands of Secretariats challenged each other over and over might give oddsmakers and bettors nightmares. Then again, some rare or endangered species might be preserved. When we turn to humans, however, the technique raises many questions and challenges regarding our most important values and basic notions.

Intuitively, then, we ask, "What kind of human being would the individual be who has the nuclear DNA of another preexisting individual?" Would the "clone" be another individual human being or somehow not "truly" human? Is the "latest edition" of the cloned individual an individual in his or her own right? If he or she has the nuclear DNA from only one preexisting individual, if he or she was not the product of separate sperm and ovum, would such an individual fit the term "human," the way the rest of us do? What of the relationship to the individual supplying the DNA? How would we describe their relationship? Would the term "parent" fit the relationship? Could this clone of an individual ever be accepted by society? Or, much more profoundly, could the clone come to accept himself or herself as each one of us do: a unique and special person, with a nature and a future that is no one else's?

The ready response to questions like these is to recognize that the questions relate to imponderables. Of course, we wonder about them, but there is no evidence on which we can base an answer and no way of predicting the reactions of the clone or of society to the clone. We can only surmise that, just as now, if there were a world where clones existed, there would be tolerance and prejudice. Some individuals would display resilience and others would experience psychological challenges that test them sorely. People and the laws of society would adjust in time, just as we have for artificial insemination and in vitro fertilization. If anything, the imponderables

Matilda, a sheep cloned in Australia.
(© AFP/Corbis. Reproduced by permission.)

ought to give us pause, to prompt us to think before taking any fateful steps to clone a human being. Yet are the imponderables grounds enough to restrict the freedom of inquiry our society endorses? What does prohibiting experiments in human cloning—in fact, making it a crime—do for our right to pursue knowledge? Ironically, is it not the imponderables that intrigue us? They exert a pull on our minds to seek answers to precisely those questions we are on the verge of forbidding. Is this bringing us back to the message of the story of the garden of Eden of the Book of Genesis in the Bible: is there knowledge we are forbidden to seek?

A Threat to Free Inquiry? Is this a dilemma? A "catch-22"? Will prohibiting cloning of human beings cut us off from learning the answers to significant scientific information? Many argue that the research into the techniques of human cloning and the use of cloning technology are critical to advances that may lead to cures for many diseases. For example, the use of cloning may allow us to develop ways to culture replacement tissues and bodily fluids that can be used to treat individuals without the usual difficulties of HLA matching and tissue rejection. A somatic cell taken from a burn victim may be inserted into an oocyte and the process of cell division initiated to grow skin in tissue culture without the intent ever to create another human being. So too for bone marrow, so vital for treating forms of cancer and immune deficiency diseases. Nerve tissue can be developed in the same manner to be used to treat paralysis in accident or stroke victims without the problems of rejection of tissue and the complications of using immunosuppressive drugs.

It is not yet certain that such potential advances will actually occur. But it is questionable how feasible they will be in practical application. Obviously, at some early stage of life, even just after birth, somatic cells might be harvested and banked to be used if there was ever a therapeutic application for the individual whose cells would be preserved and reengineered in this way. But the system for accomplishing this is impractical. No matter where the somatic cells, however transformed and primed for therapeutic application, are stored, there are problems of availability at the point of need. Even more problematic is the management of a storage and centralized system of reference to manage the process. However, might the technology be used in an ad hoc way, case by case, to meet an individual patient's needs?

The prospect of cures like these is on the horizon. But they do not require human cloning by SCNT to accomplish. If the projections of scientists are correct, the techniques of embryonic and adult stem cell harvesting and transformation will yield a supply that can be universally available. It will not be necessary to clone an individual's own tissue via SCNT. SCNT is a more complicated process with many more foreseeable logistical and technical problems. The technology of stem cell development appears to present fewer problems and the prospect of more widespread uses and greater flexibility.

Potential Compassionate Applications
Perhaps these are reasons enough to deemphasize human cloning through SCNT. Even to abandon it. But the Human Cloning Prohibition Act of 2001 places an absolute ban on the procedure. Consequently, even humanitarian uses, out of compassion, are prohibited.

Advocates of human cloning through SCNT use the example of a couple whose child might be replaced through the use of SCNT. For example, a young child is so severely injured that recovery is impossible, and the child lingers a brief time before dying in a comatose state. A cell taken from the child could be inserted into the specially prepared oocyte of the mother and implanted in her. In this way, the child could be replicated as a replacement. The parents would have regained their lost child. A moving scenario, no doubt, even melodramatic, although we are very far from knowing how this situation would turn out. For one thing, it seems much more natural to have another child. The replacement may serve only to generate ambivalent feelings, and the child will bear a heavy burden, perhaps not loved for the distinct and separate individual he or she is.

Secondly, the harvesting of the oocytes is not straightforward. It takes time because there is not likely to be a supply of these on hand from the same mother. Thirdly, as we have learned from animal cloning, the technique has a minimal success rate. Disappointment and failure are likely, and at best it is a prolonged process. In making the decision to attempt cloning the lost child, the parents are acting under the strain of profound grief. Their loss is keenly felt and their judgment affected by the tragic circumstances. The doctor or research team required to carry out the procedure would need to divert their attention and resources to this set of tasks. Unless the parents had enormous wealth so they could assemble their own team, the scenario is more in the realm of fiction than a realistic exercise. Nor should we expect scientific research to channel itself to meet the rare occurrences of replacing children lost through tragedy.

However, cutting off research on human cloning through the technique of SCNT can cost us valuable knowledge of important benefit in another area: the treatment of infertility. Thousands of couples in the United States are unable to bear children because their reproductive systems function poorly or not at all. Assisted reproductive technology (ART) has been developed to remedy the condition. Artificial insemination, in vitro fertilization, drugs to stimulate ovulation, techniques of micro sperm extraction and direct injection into an ovum, and other techniques have been introduced successfully to treat human infertility.

Human cloning by means of SCNT offers a new method, giving hope to the thousands whose fundamental reproductive rights are thwarted by defects that cannot be treated by these accepted methods. For example, when the male is unable to produce sperm and the couple does not wish to use artificial insemination by a donor, cloning by SCNT would be an alternative. The insertion of the male spouse's nuclear DNA into the enucleated ovum of his wife allows the couple to have a child in a way that closely approximates a natural process of reproduction: the nuclear DNA, the ovum with the mitochondrial DNA of the woman, the implantation and natural process of pregnancy. Prohibition of human cloning cuts off this avenue for individuals to exercise their fundamental reproductive rights. A variant of this technique, in which the ovum of another woman is substituted for the ovum of the mother with a mitochondrial DNA defect and the nuclear DNA extracted from the fertilized embryo of the couple is inserted into the donated ovum, has been used with some success. Therefore, doesn't this prohibition interfere not just with the pursuit of knowledge for its own sake, but also run counter to the fundamental reproductive rights of individuals?

The Ethics of Human Experimentation The argument that prohibiting human cloning through SCNT interferes with freedom of inquiry or impedes gaining knowledge essential to treating conditions that interfere with fundamental human reproductive rights overlooks an important consideration. The quest for knowledge for its own sake or to benefit others is not unrestricted. Ethical rules and codes govern the kinds of experimentation allowed on humans. Harms are to be avoided; risks must be minimized. Worthy goals and hopes do not justify harmful procedures or those unlikely to produce reliable results. Good intentions or claims of widespread benefits do not automatically redeem procedures of uncertain outcome. In a phrase, the end does not justify the means. The rules and codes for experimentation on humans are precise. Those who propose to perform the experiments must provide evidence that the experiment can work; that any harms it may cause can be avoided or limited; that the expected outcomes are based on earlier work that allows an extrapolation to the proposed experiment, and that the experimental design is methodologically sound and well controlled.

It is not sufficient to cite lofty goals or pressing needs when experiments may entail costs of personal loss, distress, and harm to individuals—or even significant risks for these harms. Aircraft, vehicles, bridges, buildings, equipment are all designed with extra margins of safety and outfitted with safeguards as a fail safe to prevent harms. Experiments that involve humans must also meet rigorous standards, including review by experts and by nonscientists, to meet the so-called sniff test. No one contends that we know enough about SCNT as a technique in other mammalian species to try it out in humans. Livestock with specially prized traits, laboratory animals with unique characteristics, the endangered species whose existence is imperiled—if our cloning experiments fail or turn out to have unacceptable consequences, corrections are possible and unwanted outcomes can be disposed of. Society does not permit such liberties with human life. A breed of sheep or cattle that has unexpected flaws along with or instead of the sought after traits may still have other uses or can be euthanized and autopsied to understand the reasons for lack of success. These options do not exist with the cloning of human beings, and therefore the prohibition from even beginning the process is necessary until the highest standards of safety and effectiveness can be demonstrated.

The Risks of Cloning Some advance the argument that human cloning may be premature at this time, but scientific progress might reach the point of removing or offsetting the risk. However, prohibiting human cloning closes off the opportunity to ever reach the point of eliminating the possibility of failure or reducing it to acceptable levels. Humanity would never have succeeded in reaching outer space or landing on the moon if the terms were elimination of all risk. More to the point, there would be none of the cures, the vaccines, or the interventions of modern medicine if the advances needed to be free of risk. Life is full of risks. The achievements of humans could never have been made if the terms were "risk free."

This argument has force. But like many apparently persuasive arguments in favor of proceeding with human cloning, it obscures a feature of all the human achievements on which progress has been built. Of course there were risks. Those who accepted the challenge knew the risks and voluntarily accepted them. Experimental human cloning is different. The subjects in the cloning experiments are not only the healthy volunteers, for whom, after all, the risk is over once a cell is removed, once the oocytes are harvested and the pregnancy complete. The subjects in cloning experiments are not high-minded adventurers, heroic explorers, dedicated scientists, not even the desperate terminally ill or grievously afflicted or those showing their altruism by joining researchers in the quest to relieve suffering through biomedical experimentation. We recognize and appreciate the contributions and sacrifices that individuals make when they choose with a mind and voice of their own. The risks of human cloning experimentation fall disproportionately on the clones. And we who make the decisions on their behalf must think from what would be their point of view. When flaws and failures turn up later, we cannot point to the most fundamental precept for human experimentation: the well-informed and voluntary agreement of the individual to accept the risks, even those not currently foreseeable, and the freedom to quit the experiment when, in the individual's sole judgment, the risks no longer seem acceptable. This is the universal standard for human experimentation in the world today.

Who Decides? Human clones would of course have no opportunity to weigh the risks in advance. Moreover, the flaws or risks are now incorporated into their very existence. The inexorable outcome of the uncertainties or failures of the experiment determine their destiny. Hence there is no voluntary agreement to accept the risks as a price of coming into existence and no opting out in the event the burdens seem no longer acceptable.

It is true that no human gets to choose to be born or to whom or under what circumstances. We have no choice but to accept and make the most of the natural lottery of existence. This is part of what defines being human. For most individuals it is normal to honor and appreciate the choice our parents made for us in deciding to conceive and bring us into the world. The same rules and codes of human experimentation noted earlier recognize the special and pivotal role of parents whenever medical research calls for children to be the experimental subjects. Procedures are in place that reflect the universal assumption of law, policy, and common sense, that parents are best situated to make decisions in the best interests of their children. There are even provisions to have checks and balances, should the situation suggest the parent(s) may not be guided by normal parental instincts. In addition, by giving permission for experimentation on their children, parents are agreeing to accept the outcomes, even in the event that they are less than what was hoped for. Experiments always have a degree of uncertainty. Parents are free to withdraw their children from the research and children themselves may request that the experiment be terminated when they find they do not wish to continue.

Are There Reliable Safeguards? Inherent in the human cloning process is the unavoidable shifting of the burden of choice unequally to the clone, who is to be involuntarily recruited into the experiment with no chance to withdraw or to be withdrawn. Once begun, the experiment has passed the point of no return. Ideally, perhaps, an improved, even a fully perfected human cloning process ought to incorporate safeguards to detect potential flaws. Some of these safeguards already exist. Preimplantation genetic diagnosis (PGD) is a technique already available to screen embryos in vitro for a number of genetic diseases. With the rapid growth of genetic technology, it is possible that one day most, if not all, genetic defects could be screened for and so no anomalies would occur in the clone. Many other intrauterine techniques can detect developmental flaws during pregnancy. Serious congenital defects and damage as well as inherited metabolic deficiencies can be detected at birth or shortly thereafter. These problems can be managed in accordance with accepted standards of medical practice.

But so little is known at this time about the way the genome—each individual's genetic endowment—works developmentally and functionally before and after birth that many deficiencies or disorders will only emerge after many years. Evidence indicates that cells may have inherent in them a defined limit of passages: some reports claim Dolly the sheep is revealing indications of a more rapid cellular aging. Success rates in animal cloning through SCNT are poor: most clones never reach or complete the gestation process, or they display major abnormalities during gestation or at birth. The DNA of each individual has a complement of cellular mechanisms or switches that act to replicate cells at mitosis and to repair the damages to microscopic and molecular structures and processes essential to health, development, and survival. Some evidence of genetic instability has been identified in mice clones and in cellular processes with cognate human processes. More subtle differences in development or gene expression might only emerge from latency after years or in reaction to environmental influences, too late to detect or intercept.

Some scientists have offered the view that many of the more subtle effects in genetic structures may owe themselves to the sort of "imprinting" that takes place in the normal fertilization process. At that point, meiosis II, initiated by the single sperm that triggers a succession of biochemical events, reaches completion. Shortly thereafter, the events of embryonic development occur with the formation of the zygote and the beginning of cell division into daughter cells, each having the complement of DNA representing a component of each parent and with one set of the pair of parental genes becoming deactivated. It is possible that subcellular events at this stage, currently not well described or identified, may play a decisive role in normal development.

Recent reports have emerged surrounding a process of SCNT in a variant of the model we have been discussing so far. Scientists and clinicians have succeeded in overcoming defects in the mitochondrial DNA of women by the process of substituting a healthy oocyte from a donor for the oocyte with the deficiency. The donated oocyte has the nucleus removed or deactivated. Then the nucleus of an embryo, created in vitro with the sperm of the father and the mother's ovum, is inserted in the donor's enucleated ovum. The reconstituted embryo is implanted in the uterus of the wife for normal gestation. However, a number of the embryos thus created revealed signs of chromosomal defects, related to sex differentiation and mental retardation (Turner's syndrome). More subtler forms of deficiencies could only be ruled out by the process of trial and error, an unacceptable method for human experimentation. Thus the possibility of human cloning cannot be demonstrated without a huge and, in large part, unacceptable cost, both personal and financial, to the individuals involved—be they parents, the offspring, or the rest of society. To advocate investment of public research funds or expect any public endorsement for the dim and distant prospect of, at best, a highly limited success with slight benefit and significant risks is not justifiable.

The Public Interest Even so, the recent bill approved by the U.S. House of Representatives is not limited to use of public funds. It proscribes all efforts at human cloning by SCNT. This must surely intrude on an area of scientific inquiry and reproductive rights. Private funding and initiatives are banned. How can this be justified? The reason is clear: this is an area that lawmakers have seen as a matter of public policy and profound public interest. Over a decade ago, a federal law was passed, with no real controversy, prohibiting the sale and trafficking of human solid organs. It was determined that it was against public policy and not in the public interest to have individuals sell their organs, even if they as donor and the recipient would stand to benefit from what is a lifesaving procedure. The prospect of a commercialization of human organs was ruled to be against the public good. Although cloning of human beings under private auspices is very different, still the authority of the Congress and federal government to prohibit what is detrimental to the public interest is clear. Cloning of human beings is antithetical to the fundamental notions of what constitutes human conception. The rationale for prohibiting human cloning is that it entails outright commodification of human embryos derived from laboratory manipulations and puts society at risk for the untoward results of flawed and unethical experimentation on humans. It is not in the public interest to allow human cloning to go forward.

Proponents of human cloning rebut this view by arguing that it is based on the irrational fears of the so-called slippery slope argument. This argument uses the logic that without an absolute prohibition of some possible course of action, society will by degrees move down so far that precedent after precedent we will incrementally reach a point no one ever wanted or foresaw, and it will then be impossible to turn back, just as if one step on the icy slope makes the fall inevitable. The argument for prohibiting human cloning is not at all an instance of arguing based on slippery slope logic, however. It is not that the possibility of misuses of human cloning or abuses to create "novel human beings" or to replicate outstanding or favored types of human beings is what induces the prohibition. It is not just to rule out looming science fiction scenarios.

The prohibition of human cloning through the process of SCNT is based on a thorough understanding and appreciation of the methodology of scientific experimentation. The experimental method proceeds by a process of trial and error. A hypotheses is formed. Data are gathered that either support or fail to support the hypothesis. Based on better data and clearer objectives garnered through the first experiment, new hypotheses are developed and the steps repeated.

The result is greater accuracy and precision in the hypothesis or improved data collection. Or it leads to abandonment of the hypothesis as fundamentally flawed and not worth pursuing. The prohibition of human cloning is a declaration that efforts to engineer human embryos or beings in this way is off-limits to the method of trial and error. In effect, it is the rejection of a fundamentally flawed hypothesis.

Conclusion In experimentation on human beings, the basic principles of human society apply. There must be a reasonable likelihood of an unequivocal outcome, not some 1% to 3% limited success rate with failures and successes allowing incremental advances at intolerable cost. We can only proceed based on thorough preliminary research and sound scientific design, and at this point the evidence seems to be leading in a direction away from the attempt at human cloning. Harms are to be avoided and risks minimized. But it is not possible to foresee the harms or risks, and it is impossible to reverse them should we discover them later. There is no turning back if we are unsuccessful. There is no opportunity to quit the experiment if the subject so decides.

The test is not whether we should allow human cloning on the premise that useful information or some possible benefit may emerge. The test is to adhere to the standards for human experimentation that have been developed and refined over the past 50 years and codified in a succession of declarations, pronouncements, and laws by governmental and international bodies throughout the world. The burden of proof for meeting these standards falls on those who propose to clone human beings by SCNT.
—CHARLES MACKAY

Viewpoint:
No, the cloning of human beings should not be prohibited because the potential for medical accidents or malfeasance is grossly overstated, and the ethical questions raised by detractors are not unique to cloning—indeed, ethical questions attend every scientific advancement.

Every generation confronts moral, ethical, legal, and political problems regarding the appropriate use of new technology and the limits of scientific research. Which is to say, every generation confronts the problems that attend to the new knowledge and understanding about our world that such technology and research affords.

At the same time, however, societies continue to promote the advancement of the science and research that makes possible such new technologies and the societal advances they facilitate.

This essay attempts to demystify the cloning process and demonstrate how cloning, both in its reproductive and therapeutic capacities, provides much needed and sought after answers to otherwise intractable medical problems and conditions. As it turns out, the fears and ethical concerns regarding the cloning of humans (or even human embryos) that have moved many, including the U.S. Congress, to favor an absolute prohibition on human simply are unfounded. Indeed, many of the fears regarding cloning are not novel to human cloning; they are the same fears of the unknown that generally are raised more out of naive, knee-jerk reactions to new scientific procedures than well-thought-out, empirically supported assessments. On all fronts, the fears regarding human cloning can be allayed.

In light of the many benefits human cloning can provide, it would be foolish to prohibit by law the cloning of human beings or human embryos. This is not to say we should ignore the many ethical and legal issues, as well as other potential problems, that cloning humans may create, but rather, it simply is to recognize that human cloning per se is not the moral monstrosity that many are making it out to be.

There is, however, an overlap between the issues of human cloning and abortion, discussed in more detail later. Consequently, those who are morally opposed to abortion may also be morally opposed to human cloning, at least its therapeutic aspect. Many will continue to believe that human cloning is inherently morally unjustified, if not evil, for the same reasons they believe abortion is inherently immoral: both abortion and cloning of human embryos for medical research involves the destruction of (at least) potential life. Of course, whether human embryos are deserving of the same respect as a viable human fetus depends on personal definitions of what constitutes life and the scope of certain protections and this invariably is the point where moral intuitions clash.

Rather than attempting the futile enterprise of rearranging the moral intuitions of others, this essay explains first that cloning essentially is just another method of reproduction. Second, related to this first point, it demonstrates a common reoccurrence in the history of science and technology, namely, the presence of what could be called the "offensive factor," which tends to accompany new scientific and technological breakthroughs. Although initially the majority of people may view these breakthroughs with disdain, in time, that same majority comes to accept them.

New Technology and New Ideas: Overcoming the Offensive Factor As with virtually any new scientific advancement, there are a range of opinions on the issue of human cloning, from those that are morally opposed to it on religious grounds (e.g., the Catholic church), to, somewhat ironically, those that are in favor of it also on religious grounds (e.g., the Raelians, a religious order begun in the early 1970s in France that claims 55,000 members throughout the world; in 1997 the Raelians founded Clonaid, a company that conducts human cloning experiments).

The majority view, however, appears to be strongly opposed to the idea of cloning humans. That is not surprising. Indeed, throughout history, we can find many examples where a society initially condemned, or even was repulsed, by a new scientific practice or idea, although it would later come to accept and even embrace it. The performance of autopsies, organ donation, and the use of robots in manufacturing are but a few examples. Indeed, in the early 1970s, when in vitro fertilization (IVF) began on human beings, many in society—including medical professionals—opposed it vehemently on moral and ethical grounds. A doctor involved in one of the first IVF procedures was so appalled by the procedure that he deliberately destroyed the fertilized embryo because he thought the process was against nature. Now, of course, IVF clinics are so commonplace that the birth of so-called test-tube babies no longer are newsworthy events, and few, if any, contemporary medical professionals claim that IVF procedures are immoral, unethical, or should be prohibited.

For a more current example, witness the Internet. Although the Internet has greatly improved the exchange of information and ideas, it also has facilitated cybercrime and other economic crimes, and it has made it easier for individuals to engage in such illicit activities, which pose extreme threats to the security of nations and the stability of global markets. Does the fact that even a single individual could cause widespread damage mean we should dismantle the entire Internet? Few would agree, and the billions that benefit from and rely on its existence would undoubtedly fight such efforts. Indeed, in today's interconnected world, dismantling the Internet probably would be impossible.

Thus when new technology and new ideas spring forth, it often is difficult, if not impossible, to prevent their propagation. The "new" generally has both aspects of the "good" as well as the "bad." But the bad should not necessarily deny us the benefits of the good. Therefore, rather than futilely attempting to prevent human cloning because of the evils that might spring from it, we should, as we have done with other new technologies and ideas, embrace it for the benefits and new knowledge that will emanate from it, all the while doing the best we can to minimize abuses. Otherwise, we risk driving the cloning of humans underground, and that could be a far worse evil.

To be sure, many moral, ethical, and legal issues must be addressed with respect to cloning humans, not the least of which concerns the possibility that cloning humans will open a market for "designer babies," which could lead to eugenics. Again, the potential for abuse always accompanies the advent of new technology. Although human cloning now may be perceived as something bizarre and foreign to us, within a generation—perhaps even earlier—cloning humans will be as noncontroversial as IVF, surrogate motherhood, and organ transplantation. So while we grow accustomed to the new technology and knowledge that human cloning may bring, there simply is no good reason why human cloning should be prohibited, but there are many reasons—discussed later—why research in the area should continue. And while research continues, we can work toward addressing the ethical issues regarding human cloning by developing professional codes of conduct and governmental regulation where necessary in order to minimize potential abuse, just as we did regarding organ transplantation and initial fears that organ transplantation would create black markets in organs. The problem, then, is not the technology per se, but how we use that technology.

What Is Human Cloning? Perhaps the most significant impediment to new technology and ideas is that few people understand exactly what they are at first, and these misunderstandings breed fear and contempt. First and foremost, human cloning is not anything like photocopying oneself. For example, let's say Jackie A, a 30-year-old female, decides to clone herself. If she wants to "replicate" herself so there will be another 30-year-old female—we call this replicant Jackie B—that looks identical to her, Jackie A will be disappointed. Literally replicating oneself physically is impossible, and it is *not* what is meant by human cloning.

But Jackie A could do exactly what scientists in Scotland did with Dolly the sheep. Known as somatic cell nuclear transfer (SCNT), a doctor would first remove a mature but unfertilized egg from Jackie A or utilize one from a donor. The doctor then would remove the egg's nucleus, which only would have 23 chromosomes, half the necessary amount for human reproduction, and replace it with the nucleus of another cell—a somatic cell, which would have a full complement of 46 chromosomes. As we have assumed, if Jackie A wished to clone herself, the somatic cell simply would come from

IMPRINTING AND THE DISCOVERY OF IGF2R

On August 7, 2001, the United States National Academy of Sciences convened a panel of international experts in the scientific, medical, legal, and ethical aspects of human cloning. The debate centered on the potential hazards and current limitations of the cloning process in animals. Scientific and medical opinion weighed in heavily against the prospect of human cloning. Not enough was understood about the cloning process in animals to employ the current technology to clone humans. Scientists cited the inherent difficulties of inducing safe and successful gestation in animals as evidence of the high odds against achieving success in humans. The effects of culture media on embryos cloned by somatic cell nuclear transfer (SCNT) had not yet been examined enough to permit detection of possible flaws traceable to contamination by the nutrients in which the clones are maintained or to rule out cross-species pathogens. Other data pointed to possible interactions of the culture medium with the cloned embryo, which might trigger anomalies in the critical processes of gene expression, resulting in developmental failures or abnormalities.

More concrete parallels with problems encountered in animal cloning suggested that a technical barrier might exist as well. If so, the fundamental feasibility of human cloning was in question. Sexual reproduction involves a step called "imprinting." In this step one set of the pair of complementary genes from each parent becomes deactivated. The corresponding genes from the other parent ensure that the development and functions determined by the gene remain active. Because SCNT is asexual reproduction, the "imprinting" process does not occur. In the transferred genome critical parts of the genome are permanently deactivated. Consequently, genes critical to successful gestation and embryonic development do not function. If this proves to be the case, then we are up against an insurmountable barrier to human cloning.

On August 15, 2001, scientists startled the world with a new finding: humans and other primates receive two functioning copies of genes. In an article published in *Human Molecular Genetics,* scientists J. Keith Killian and Randy Jirtle and their collaborators identified a key gene called insulin-like growth factor II receptor (IGF2R) which they had traced in mammalian evolution. Sometime about 70 million years ago, the ancestors of primates evolved to possess two functional copies of this gene. The IGF2R gene is critical to normal fetal development. The absence of a working copy in the livestock and murine models used in cloning accounts for the failures, according to the scientists who published the article. Dr. Killian stated, "This is the first concrete genetic data showing that the cloning process could be less complicated in humans than in sheep." It is too soon to know whether other factors in the human genome also contribute to, or hinder, the technical possibility of cloning humans. But this key finding settles one question about the phenomenon of genomic "imprinting."

—Charles R. MacKay

Jackie A. The egg with the new nucleus then would be chemically treated so it would behave as if it had been fertilized. Once the egg begins to divide, it then would be transferred into Jackie A's uterus, initiating pregnancy.

Essentially, except for the nuclear transfer, the procedure is identical to now common IVF procedures. If all goes well, within approximately nine months, Jackie A will give birth to a daughter—an infant daughter, to be sure—who is genetically identical to Jackie A. Thus Jackie A has "cloned" herself. Jackie A's daughter, whom we have named Jackie B, is not a "replicant," but a completely separate individual—she will not share any of Jackie B's memories or consciousness, just as a child does not share any such traits of its parents. Jackie B, however, will grow up to look nearly identical to Jackie A inasmuch as she shares all of Jackie A's genetic identity. The "odd" part is that not only is Jackie B the daughter of Jackie A, but Jackie B also is Jackie A's genetic twin sister.

Some people may find this result a bit unsettling inasmuch as it appears to be quite unusual or "unnatural"—how can someone be both a daughter and a sister to her mother? First, whether we refer to someone as a sister, mother, brother, or father often turns not on genetic or biological relationships but on other social conditions and relationships. Adopted children, for example, do not share any of the genetic features of their adoptive parents, yet they still are recognized as the children and heirs of their parents. We speak of adopted children's birth parents as opposed to their (adoptive) parents to make such a distinction.

Similarly, in light of the phenomenon of surrogate motherhood, merely because someone is born from a particular woman does not necessarily mean that same woman is even the individual's biological mother: a surrogate mother could simply be a woman who has agreed to carry the fertilized egg of another woman. As a result, in such situations the child born to the surrogate mother shares none of the surrogate mother's genetic material. Thus familial relations no longer are simply a matter of biology and genetics, but more a matter of social constructions.

Second, sometimes even in nature we can witness so-called unnatural results. If your mother, for example, is an identical twin, then her twin, of course, is your aunt. But, interestingly enough, she also is your *genetic* mother in the sense that genetically speaking, she is identical to your mother. In such a situation, your aunt—genetically speaking—both is your aunt and your mother. Of course, we do not really consider your aunt to also be your mother merely because she is genetically identical to your mother. That would be silly. Your mother is your mother by virtue of the fact that she bore and reared you (although, as we discussed earlier, giving birth to someone does not necessarily mean the child and mother are genetically related). Likewise, it would be absurd to seriously consider Jackie A's daughter to also be her sister merely because she is genetically identical to Jackie A.

Why Should We Clone at All? As of this writing, it is unknown whether human cloning actually could be performed today. It also is unclear whether it already has occurred. What is evident, however, is that if human cloning is not technologically feasible now, it certainly is theoretically possible, and, once the procedures are refined, likely will occur soon. Indeed, in addition to the Raelians, Severino Antinori—a controversial fertility doctor from Italy—has announced publicly that he will be going forward with implanting cloned human embryos in volunteers shortly. Whether these experiments will be successful remains to be seen, but if they are, the first cloned human may be born in the summer of 2002.

In light of this (temporary) uncertainty with respect to the feasibility of successfully cloning a human, many opponents of human cloning argue in favor of a moratorium on human cloning. Such opponents point out that the process by which to clone mammals, let alone humans, is not yet perfected and could result in severely malformed and stillborn children. Certainly these are valid worries that cannot be ignored. Unfortunately, such worries cannot be resolved via a moratorium, for how can the necessary science advance if no research is being performed? Experimentation can be performed without impregnating human females of course. But just as there had to be a first heart transplant patient, there will have to be a first cloned embryo implanted into a woman if human cloning ever is to be achieved. There are always risks. The point of research is to minimize them, and a moratorium simply misses that point.

In any event, what follows with regard to human cloning assumes that the science will, if not now, eventually be perfected so these worries about malformations and stillbirths (both of which occur naturally anyway, lest we forget) can be greatly reduced. Assuming that human cloning will not present such dangers to either the child or the mother, there still remains the question of why we would want to clone humans at all.

Reproductive Cloning. As already mentioned, human cloning essentially is a method for asexual reproduction. As such, it can allow infertile couples or even single persons to reproduce offspring that are biologically related to them. A husband who no longer is able to produce sperm nevertheless still could have a biological heir by implanting his DNA into the egg of his wife (or, if the couple so choose, a surrogate). Likewise, via human cloning, a single woman, a single man, or even same-sex couples would be able to reproduce biologically related heirs. Adoption always is an option for such persons, but for some, the desire for biologically related children is a more attractive option. Essentially, then, human cloning makes available the possibility of reproduction to those who otherwise would be unable to reproduce.

The imagination can run wild with the possibilities that cloning humans for reproductive purposes provides. For instance, a couple's child dies in a horrific accident and they no longer are able to conceive a child. They are devastated. Would it be unethical of them to clone the child? Certainly, if they believe the cloned child will be just like their deceased child. They should be disabused of their misconception. The cloned child will be a completely new person even though it will look just like the deceased child. The parents cannot bring their deceased child back to life. Even if they understand that, is there still a problem with them wanting another child to replace the one they have lost?

Similarly, assume the same couple has a child dying of a terminal disease in need of a bone marrow transplant. No donors can be found. Some couples have, in fact, conceived children for the purpose of actually creating a possible donor for their dying child (which creates ethical issues of its own). Unfortunately, the new child may not necessarily be a good match for the dying child. Thus would it be unethical for the couple instead to clone their dying child in order to use the marrow from the clone—who necessarily will be a perfect match—to save their dying child? In effect, in order to save the child's life, the couple simply would be reproducing their child's twin.

Finally, cloning raises the specter of eugenics, which many associate with Nazi Germany and those who promote beliefs in racial supremacy. With cloning, we could "breed" superior humans who are smarter. Certainly, so-called designer babies would be possible with cloning. If that bothers us, however, we could impose specific restrictions on the cloning of humans and create a regulatory and enforcement body to oversee such procedures, just like we have done with organ donation. And just as importantly, we can inform those who are interested in cloning themselves, or a relative, or a friend, about what cloning actually is and clear up any misconceptions they might have about the procedure. In short, it is possible to define appropriate and inappropriate uses for cloning just as we have for a variety of other new medical procedures. Merely because some might abuse cloning should not preclude others from otherwise benefiting from the cloning procedure.

Therapeutic Cloning. Unlike reproductive cloning, therapeutic cloning does not involve reproduction, but the initial process is the same. A human embryo is cloned not to produce a child, but in order to retrieve its stem cells. Stem cells are unique in that they can be "grown" into virtually any type of cell in the body. According to some researchers, stem cell research may provide breakthroughs in the treatment of spinal cord injury and repair and a way to overcome the debilitating effects of many neurological disorders.

Because stem cells can be cultivated into the tissue of any organs, they also suggest a way for doctors to avoid the problems of organ transplantation. Often the patient's body receiving the organ rejects it. As a result, the patient must be given high doses of anti-rejection drugs. The same problems hold true for artificial organs. But anti-rejection drugs often have severe side effects. In addition to these problems, far more patients are in need of organs than are available. These patients often must wait for long periods of time, sometimes years, for an available organ and then hope the organ is not rejected by the body. Very often, such patients die just waiting for the organ.

Using stem cells to cultivate new organs would alleviate both these problems. First, because the organ simply would be cloned from the patient's DNA, the patient's body likely would not reject the organ for it essentially is the same organ from the perspective of the body. Second, there would be no waiting list for the patient. Whenever a new organ is needed, the patient's doctor would clone the needed organ from the patient.

So what possibly could be the problem with therapeutic cloning in light of these potential benefits? The retrieval of stem cells requires destroying the embryo. Note that stem cells can also be retrieved from adult cells, rather than embryos. However, whether such adult stem cells are as effective as embryonic stem cells still is an open question. Because of this, some religious groups are morally opposed to therapeutic cloning for the same reason they are morally opposed to abortion: they consider the embryo to be a human being (or, at least, with the potential for life), and therefore the destruction of the embryo is unjustified even if it will save the life of another.

Whether Americans consider an embryo to be a human being certainly is a matter of debate. According to an August 14, 2001, Gallup poll, only 36% of Americans believe that embryos deserve the same protection as human life. The same poll revealed that more than 70% of Americans think research on stem cells is necessary including, surprisingly, 31% who think that stem cell research is morally wrong. In light of this, it certainly appears that such moral objections do not dissuade most Americans from the importance of stem cell research. In contrast, a June 27, 2001, Gallup poll revealed that 89% of Americans are opposed to reproductive cloning. Evidently, therapeutic cloning is not as offensive as reproductive cloning.

Conclusion Human cloning essentially is human reproduction—nothing more, nothing less. How it is performed and the results that may follow, however, are new. This is where the offensive factor mentioned earlier comes in. As with other new technologies and new ideas that have been fostered by scientific research, we need to (and hopefully will), get beyond the repugnance we feel about human cloning. Recently, the U.S. House of Representatives passed the Human Cloning Prohibition Act of 2001 (H.R. 2505), and this legislation currently is being considered by the Senate. The act will make the cloning of human embryos a crime punishable by up to 10 years in prison and a civil penalty of at least $1 million.

Whether such legislation will become law is uncertain as of this writing, but undeniably the genie is out of the bottle—the technology already exists. Such legislation, most likely enacted out of political expediency in the face of the offensive factor, likely will do little to prevent determined researchers and advocates of human cloning from going forward with experimentation abroad.

Given that religious views often govern and guide our feelings toward such new technologies and new ideas, we return to the events surrounding what arguably was the first instance of cloning: God's creation of Eve from a rib of Adam. After their creation, both Adam and Eve were instructed by God not to eat the fruit of the Tree of Knowledge on pain of death. Tempted by a serpent, however, Eve disobeyed God by eating the fruit and offered it to Adam. By eating the fruit, both Adam and Eve suddenly became aware of their surroundings; they had acquired new knowledge. Afraid they would soon eat the fruit of the Tree of Life and thereby become immortal, God quickly banished them from the Garden of Eden "lest they become like one of Us."

Traditionally, the tale of Adam and Eve is thought to represent the first instance of sin, the "original sin," and the consequences that flowed from that sin. But there is another way to interpret their story. The serpent, after all, tempted Eve by informing her that should she eat of the Tree of Knowledge her eyes would be opened, and she would come to know right from wrong. In essence, it was this fact—a craving for new knowledge—that moved Eve to disobey God's command and risk death.

Although many take the moral of the story to concern the ramifications of disobeying the command of God, we also can see that it represents the ramifications of new knowledge. Eve, who essentially was the first scientist, sought out new knowledge, and like some of the millions of scientists who have followed her, also suffered some severe consequences. Thus new knowledge always comes at a price, but what is the alternative? Human cloning should not be prohibited because of the new knowledge it affords—its abuse, of course, should be. —MARK H. ALLENBAUGH

Further Reading

Andrews, Lori. *The Clone Age: Adventures in the New World of Reproductive Technology*. New York: Henry Holt, 2000.

Annas, George, and Michael Godein, eds. *The Nazi Doctors and the Nuremberg Code: Human Rights in Human Experimentation*. New York: Oxford University Press, 1992.

Human Cloning Prohibition Act of 2001. 107th Cong., 1st sess., H.R. 2505.

Jaenisch, Rudolf, and Ian Wilmut. "Don't Clone Humans!" *Science* 291 (March 2001): 2552.

Kass, Leon. "Preventing a Brave New World," *The New Republic*, May 21, 2001, p. 36.

———, and James Q. Wilson. *The Ethics of Human Cloning*. Washington, D.C.: American Enterprise Institute, 1998.

Kolata, Gina. *Clone: The Road to Dolly, and the Path Ahead*. New York: William Morrow, 1997.

National Bioethics Advisory Commission, *Cloning Human Beings: Report and Recommendations*. Rockville, Md.: June 1997. Also available at <http://bioethics.gov/pubs/cloning/>.

Silver, Lee. *Remaking Eden: Cloning and Beyond in a Brave New World*. New York: Avon, 1997.

———. *Remaking Eden: How Genetic Engineering and Cloning Will Transform the American Family*. New York: Avon, 1998.

U.S. House. *Report on Human Cloning Prohibition Act of 2001*, H.R. Rep. No. 107-170, 2001.

Historic Dispute: Did syphilis originate in the New World, from which it was brought to Europe by Christopher Columbus and his crew?

Viewpoint: Yes, syphilis originated in the New World and was brought to Europe by Christopher Columbus's crew.

Viewpoint: No, syphilis was a disease that had long been in the Old World; it was simply a coincidence that it flared up shortly after Columbus's return from the New World.

Since the emergence of HIV/AIDS (human immunodeficiency virus/acquired immune deficiency syndrome) in the 1970s, interest in the origin of syphilis has grown because of parallels between the history of syphilis and that of AIDS. Both diseases seemed to arise suddenly and spread rapidly, causing severe illness and death. Scholars have suggested that understanding the history of syphilis might provide a guide for AIDS policy. Today, AIDS, syphilis, and gonorrhea are considered the major venereal or sexually transmitted diseases, but other venereal diseases, such as genital herpes, trichomoniasis, nongonococcal urethritis, chlamydia, chancroid, lymphogranuloma venereum, and granuloma inguinale, can also cause serious complications. The term "venereal" has long served as a euphemism in matters pertaining to sex. In the late twentieth century, the phrase "sexually transmitted disease" officially replaced the older terminology. Although both terms have been applied to diseases that are incidentally transmitted by sexual contact, such as scabies and crab lice, a more restricted definition includes only those diseases that are never, or almost never, transmitted by any other means.

Syphilis and the Renaissance Syphilis has been called many things, including the "Scourge of the Renaissance." European art, culture, science, and scholarship were transformed during the period known as the Renaissance. This period is also remarkable for its changing patterns of epidemics and its vivid reports of apparently new and violent diseases. In part, the expanding medical literature is associated with the invention of the printing press in the early 1400s, one of the inventions that helped transform the medieval world. The printing press made it easier to disseminate reports of new diseases. Diseases previously rare, absent, or unrecognized—such as syphilis, typhus, smallpox, and influenza—became major public health threats. No disease raises more intriguing questions than syphilis, a disease that traces the hidden pathways of human contacts throughout the world, and the intimate links between social and medical concepts.

Despite the number of historical references to venereal diseases, many Renaissance physicians were convinced that syphilis was unknown in Europe until the end of the fifteenth century; others argued that there was only one "venereal scourge," with variable manifestations. The confusion is not surprising, given the natural history of the major venereal diseases.

The Symptoms of Syphilis Syphilis has been called the "great mimic" because its symptoms are similar to those of many other diseases. In fact, before the introduction of specific bacteriological and immunological tests, many physicians believed that "whoever knows all of syphilis knows all of medicine." Untreated, syphilis progresses through three stages of increasing severity. A small lesion known as a chancre is the first stage. The second stage involves generalized symptoms, such as fever, headache, sore throat, skin lesions, and swollen lymph nodes. Severe damage to major organ systems occurs in tertiary syphilis. Confusion between gonorrhea and syphilis has been a major theme in the history of venereal disease, but there is little doubt that gonorrhea is a very ancient and widespread disease. Fritz Richard Schaudinn (1871–1906) and Paul Erich Hoffmann (1868–1959) discovered the spirochete that causes syphilis in 1905. One year later August von Wassermann (1866–1925) discovered a specific blood test for the disease.

Early Descriptions When accounts of syphilis appeared in the sixteenth century, the disease was known by many names, including the "French disease," the "Neapolitan disease," the "Great Pox," and "lues venereum." Syphilis, the name used today, was invented by Girolamo Fracastoro (1478?–1553), an Italian physician, scientist, and poet. An acute observer of plague, typhus, and other contagious diseases, Fracastoro was a pioneer of epidemiology (the study of the spread of epidemic diseases) and an early advocate of the germ theory of disease. In *On Syphilis, or the French Disease* (1530), Fracastoro created the story of Syphilis the shepherd, whose blasphemy brought about the first outbreak of the scourge. Ever since Fracastoro described the disease, its natural history, mode of transmission, and contemporary remedies, medical historians have debated the same questions: when and where did syphilis originate?

The "Columbus Theory" According to sixteenth-century medical astrologers, a strange conjunction of Jupiter, Saturn, and Mars produced a toxic miasma, or poisonous gas, that brought a new epidemic disease to Europe. Even today, astrologers might argue that this theory has never been disproved. Another sixteenth-century theory was based on the idea that the New World was the source of new diseases, as well as new plants and animals. The expansion of commerce, travel, and warfare that characterized the fifteenth century transformed patterns of epidemic disease. Thus, many Renaissance physicians adopted the "Columbus Theory" as an answer to the origin of syphilis; that is, they assumed that Christopher Columbus and his crew had imported syphilis from the New World to the Old World.

Fracastoro recommended mercury as a remedy for syphilis, but many other physicians favored an expensive remedy known as "Holy Wood," which was made from the bark of trees indigenous to the New World. According to the ancient Doctrine of Signatures, if syphilis originated in the New World, the remedy should be found there. Therefore, physicians and merchants who profited from the use of Holy Wood were staunch advocates of the Columbus Theory.

The Columbus Theory appeared to explain many observations about syphilis, but critics argued that coincidence should not be confounded with causation. Some historians blamed the physical and mental deterioration of Columbus on syphilis, but other explanations are equally plausible. Rodrigo Ruiz Diaz de Isla (1462–1542) was one of the first physicians to assert that members of the crew of Columbus had imported syphilis to Europe. Although de Isla claimed that he had treated sailors with a new disease in 1493, he did not publish his observations until 1539.

The "French Gonorrhea" Circumstantial evidence supported the Columbus Theory, but Fracastoro argued that there was significant evidence that syphilis was not an imported disease. The physician and alchemist known as Paracelsus (1493?–1541) called the new venereal disease "French gonorrhea" and suggested that it arose through sexual intercourse between a leprous Frenchman and a prostitute with gonorrhea. Some physicians thought that the disease that was diagnosed as syphilis in the sixteenth century might have been misdiagnosed as leprosy in previous eras. Allusions to "venereal leprosy" and "congenital leprosy" before 1492 suggest that some medieval "lepers" might have been syphilitics. To determine the validity of this theory, scientists have looked for syphilitic lesions in bones found in leper cemeteries throughout Europe and the Middle East.

The "Unitarian Ancestral Treponematosis Theory" Two other interesting theories of the origin of syphilis require some knowledge of the causative agent of the disease as well as its natural history. These theories are known as the "African, or Yaws Theory" and the "Unitarian Ancestral Treponematosis Theory." If diseases were catalogued in terms of etiological, or causative, agents instead of means of transmission, syphilis would be described as a member of the treponematosis family. The treponematoses are diseases caused by members of the *Treponema* group of corkscrew-shaped bacteria known as spirochetes. The four clinically distinct human treponematoses are syphilis; yaws and pinta (contagious diseases commonly found in the Tropics); and bejel (or nonvenereal endemic syphilis). Some bacteriologists believe that these diseases are caused by variants of an ancestral spirochete that adapted to different patterns of climate and human behavior. Syphilis is caused by *Treponema pallidum*; bejel by *Treponema pallidum endemicum*; yaws by *Treponema pertenue*; and pinta by *Treponema carateum*.

The Yaws Theory According to the African, or yaws theory, syphilis was the result of the unprecedented mixing of the germ pools of Africa, Europe, and the Americas. With the Native American population decimated by smallpox and other Old World diseases, Europeans began importing African slaves into the New World within 20 years of the first contacts. If Africans taken to Europe and the New World were infected with yaws, changes in climate and clothing would have inhibited nonvenereal transmission of the spirochete. Under these conditions, yaws could only survive by becoming a venereal disease.

Skeletal Evidence The Columbus Theory requires, at the very least, conclusive proof of the existence of syphilis in the New World before 1492; unequivocal evidence of syphilis in Europe before the voyages of Columbus would disprove this theory. Human skeletal samples provide the bulk of the evidence for the presence of various diseases in pre-contact America and Europe. During the 1990s bones with diagnostic signs consistent with syphilis were discovered in some allegedly pre-Columbian skeletons in several European burial grounds. However, few diseases leave a characteristic mark in the bones, and different diseases may produce similar lesions. If the diagnosis and dating of skeletons in these European sites can be confirmed, it would mean that syphilis was present in both the Old World and the New World before the voyages of Columbus. However, given the difficulties inherent in paleopathology, many researchers argue that the diagnostic evidence for syphilis in pre-Columbian America and Europe remains problematic. Although partisans of various theories present many ingenious arguments, many experts believe the evidence is not yet compelling. Some historians argue that despite advances in understanding the treponematoses, medical detectives are no closer to a definitive account of the origin of syphilis than medical authorities are to eradicating sexually transmitted diseases. —LOIS N. MAGNER

Viewpoint:
Yes, syphilis originated in the New World and was brought to Europe by Christopher Columbus's crew.

The question of the origin of syphilis has been energetically debated since the disease first appeared in Europe in the late fifteenth century, ravaging the continent with an epidemic of great virulence. Charles VIII, the Valois king of France, invaded Italy in 1494 with an army of over 30,000 mercenary soldiers hired from all over Europe. Charles's army reached Naples, which was primarily defended by Spanish soldiers, in early 1495. After holding the city for a few months, Charles's forces retreated; by early summer King Charles had disbanded his army and returned to France. Unfortunately, his soldiers were infected with a mysterious disease, one so new that doctors had no name for it.

Initially called "the Neapolitan disease," over the next centuries the disease would be known by an extensive variety of terms, with "the French sickness" and "the great pox" being the most common. Although it was not until the nineteenth century that the name "syphilis" was universally accepted, to avoid confusion the term "syphilis" will be used throughout the essay. Regardless of the many names employed, as Charles's mercenaries returned to their native lands or moved elsewhere to wage war, they spread the disease throughout all of Europe in what quickly became a major epidemic. Since this was a period of European expansion, known as the Age of Exploration, syphilis was soon carried by European sailors into the non-European world, reaching India and China by 1504.

Symptoms of Syphilis Contemporaries of the initial epidemic agreed on two things: The first was that the disease was venereal; that is, it was spread by sexual intercourse. The first noticeable symptom of syphilis was (and still is) the development of a chancre or raised pustule on the genital organs of both male and female victims. The second factor agreed upon by all commentators was that syphilis had a frightening impact on the body. After some time, the first chancre would disappear, but new sores would develop over the body, especially on the face. This second period was coupled with great pain in the bones and often ended in death. Alexander Benedetto, a papal doctor, reported in 1497 that he had seen victims who had lost hands, feet, eyes, and noses because of the disease. Syphilis, he remarked, made "the entire body so repulsive to look at and causes such great suffering . . . that this sickness is even more horrifying than incurable leprosy, and it can be fatal." Europeans were terrified of syphilis. Scores of local rulers and municipal governments quickly issued decrees banning those with the disease from their territories.

These early symptoms of syphilis are important in determining the origin of the disease. After several decades, the severity of the symptoms began to abate. Documentary evidence establishes that syphilis was much more severe in the first half of the sixteenth century than it has ever been since then. This epidemiological pattern is consistent with the behavior of a disease in

SYPHILIS AND AIDS: SOME INTRIGUING SIMILARITIES

Although syphilis is caused by a bacterium while the causative agent of acquired immune deficiency syndrome (AIDS) is the human immunodeficiency virus (HIV), there are intriguing similarities in the ways these diseases have afflicted humans. Both erupted suddenly in Western societies: syphilis in Europe in the 1490s; AIDS in the United States in the early 1980s. In both instances, medical experts were baffled over the cause of the new disease. Within a decade of their initial appearance, both diseases became pandemic, spreading rapidly around the world creating a global calamity of massive proportions. In each case, this dispersion was facilitated by improvements in transportation. The expansion of Europe in the sixteenth century expedited the spread of syphilis, while growing international air travel, urbanization, and the construction of roads has accelerated the transmission of HIV in Africa, Asia, and Latin America. Both diseases are spread primarily by heterosexual intercourse. It is true that AIDS was first identified in male homosexuals and intravenous drug users. Nevertheless, today the primary cause of the worldwide spread of HIV is heterosexual intercourse. Unfortunately, both syphilis and HIV can also be contracted congenitally, as the fetus develops in the uterus.

As with syphilis, the sudden emergence of AIDS triggered intense debate over the origin of the disease. Some early theories viewed AIDS as a punishment, either natural or divine, for deviant behavior such as homosexuality and drug abuse. Other theories have postulated that HIV was developed in government germ warfare laboratories and had somehow accidentally "escaped." Those attracted to conspiracy theories agreed that the virus was the product of a secret biological warfare project, but they argued that its release was deliberate, not unintentional. Still others have theorized that HIV was the inadvertent result of a medical procedure. Medical historian Edward Hooper, for example, has suggested that the virus was spread to humans through an African oral polio vaccine developed in the late 1950s. This vaccine was grown in fluid containing kidneys of chimpanzees, some of which might have been infected with an animal form of HIV. Throughout the 1990s, however, an increasing number of researchers adopted the hypothesis that HIV was one of those diseases, such as dengue fever (a tropical disease) and the liver disease hepatitis B, that originated in other species. Today, based especially on the work of Beatrice Hahn, the scientific community widely accepts the view that HIV developed first in west-central Africa where humans, when butchering chimpanzees for meat, were infected with an animal version of the virus which closely resembles the HIV now plaguing us.

—Robert Hendrick

---

an area that has never experienced it and whose inhabitants thus have no immunity to it. The disease loses its virulence over time as people's bodies begin developing some protective immunity to it. This indicates that until 1495, Europeans had never suffered from syphilis. The bone pains are of even greater significance in determining the disease's origin. Syphilis is one of several diseases that modifies bone structures by causing skull and bone lesions. As will be seen, the existence of these bone lesions will ultimately settle the debate over the origin of syphilis.

Oviedo's New World Theory This debate had divided doctors and other learned authorities since the onset of the disease. Some adopted the usual bromide that syphilis had been inflicted upon humans as punishment for their sins, in particular for their sexual transgressions, given the venereal nature of the disease. Others relied on an astrological explanation, holding that epidemics were the result of unfortunate alignments of planets. However, a new theory of the disease's origin appeared in 1526 with the publication of Fernández de Oviedo y Valdés's *Summary of the Natural History of the Indies,* in which Oviedo, who spent over a decade in Hispaniola as a Spanish administrator, wrote: "[You] may take it as certain that this malady comes from the Indies" and that it was brought to Europe by Columbus's sailors. Rodrigo Ruy Diaz de Isla made exactly the same claim in his *Treatise on the Serpentine Malady* (1539). Together these two publications initially proposed the New World, or Columbian, theory of the origin of syphilis.

Opponents of this thesis have attacked it on the basis of the timing of its first appearance. Why, they question, did over 30 years pass after

KEY TERMS

CHANCRE: A dull, red, insensitive ulcer of sore, usually at the site at which the infectious organism, in this case, the syphilis-causing bacterium, enters the body.

CONGENITAL: A disease or condition developed by a fetus in the uterus rather than being acquired through heredity. Syphilis can be acquired congenitally.

LEPROSY: A chronic and contagious infection caused by a bacillus (rod-shaped) bacterium; the symptoms of this infection include skin and bone lesions or sores.

LITHOPEDION: An extremely rare and fatal occurrence related to the extrauterine formation of a fetus in the abdominal cavity. Following the fetus's death, severe calcification results in lesions on its skeleton.

PALEOPATHOLOGY: A branch of anthropology and medicine dealing with the analysis of skeletal remains, concentrating on the structural changes produced by diseases.

PINTA: A tropical skin disease caused by a spirochete which is indistinguishable from the treponema responsible for syphilis and yaws. It usually develops in childhood and is spread by skin contact, and not through sexual intercourse.

PLAGUE: A contagious and often fatal epidemic infection caused by a bacterium; it is spread from rats to humans through flea bites.

SPIROCHETE: A general term for any of the slender, spiral–shaped, and mobile bacteria of the order Spirochaetales.

YAWS: A tropical disease caused by a spirochete which is indistinguishable from the treponema responsible for syphilis and pinta. It usually develops in childhood and is spread by skin contact, and not through sexual intercourse. Like syphilis, the disease results in bone lesions.

Columbus's first voyage before this hypothesis was proposed? There is strong evidence that de Isla, a Spanish doctor, wrote his account in 1506 or earlier, even though it was not published until more than three decades later. Hence, his claim to have treated men who had sailed on Columbus's first 1492–1493 voyage and who had developed syphilis upon their return was written relatively soon after the event. He also claimed to have treated individuals suffering with syphilis in Spain before Charles VIII's invasion of Italy, implying that it was Spanish soldiers who carried the disease to Naples. While this is not incontrovertible evidence for the New World theory, it is difficult to imagine why Diaz de Isla would fabricate his account.

Other critics of the New World theory have pointed out that Oviedo had purchased considerable amounts of land in Hispaniola. In his writings Oviedo stated, correctly, that the natives of the island, the Taino, believed that wood from the guaiac trees which grew there could cure syphilis. Oviedo subsequently formed a partnership with the Fuggers of Augsburg (a family of prominent German financiers and merchants) and obtained a monopoly to import guaiac wood into the Holy Roman Empire. Critics claim that Oviedo was exploiting the popular belief that for every harmful affliction, God had placed a cure nearby. If Oviedo could establish that syphilis came from the New World then, according to this assumption, the cure would come from there as well. He stood to make a fortune importing the wood into Europe. However, these critics have failed to question the origin of the idea that guaiac wood cured syphilis. It would only be prevalent among the Taino people if syphilis, or something like it, also existed there. Nor do these critics take into account the fact that a German author named Nicholas Poll claimed in 1517 that guaiac wood from Hispaniola reportedly cured the disease.

The Old World Hypothesis Even before the New World theory was advanced, a number of authors claimed that syphilis had been present in Europe since the time of the ancient world. Those holding this thesis of an Old World origin used passages from the Bible, Galen's writings, medieval texts and works of literature, and Arabic sources to demonstrate that the "new" venereal disease had always been present in Europe and the Mediterranean area. Essentially, their argument was that the descriptions of leprosy and other afflictions in these ancient sources were really misdiagnoses of syphilis. They combed all sorts of literary sources for evidence, much of which even some present-day defenders of the Old World hypothesis admit is often forced and unconvincing.

In the rapidly developing polemic over origin—there were 58 books on syphilis published by 1566—neither side could offer irrefutable evidence to win the debate. But the "American," or Columbian, theory was certainly more persuasive than the labored literary interpretations of the Old World adherents, and by the eighteenth century, belief in the New World origin of syphilis was very widely held. Both Montesquieu and Voltaire accepted the Columbian theory, and for good reason: In their 1988 exhaustive review of the controversy, modern scholars Baker and Armelagos conclude that, with regard to reliance on the documentary evidence of written sources, "the case for pre-Columbian syphilis in the Old World rests solely on vague and ambiguous disease descriptions and must, therefore, be rejected."

The controversy was far from over, however. In the second half of the nineteenth century, microbiology emerged as a new scientific field. With the formulation of the germ theory, with Robert Koch's demonstration that specific diseases were caused by different bacteria, and with the development of improved microscopes, research into syphilis yielded valuable information. By 1905, it had been discovered that the spirochete bacterium *Treponema pallidum* caused syphilis and could only be contracted either through sexual intercourse or congenitally. It was soon learned that other treponemata were the agents that produced the diseases yaws and pinta, both of which were endemic to the American tropics and both of which were present on Hispaniola at the time of Columbus's first voyage. These bacteria are so similar that even under today's high-powered microscopes they are still indistinguishable. Evidence also accumulated that under changed environments, treponemata could mutate from one form to another. So *Treponema pertenue*, the bacterium that caused yaws in Hispaniola, might mutate to *Treponema pallidum* in Europe's climate, thus causing syphilis there.

Archaeological Evidence As indicated earlier, syphilis is one of the diseases that causes abnormal changes in bone structure. Its lesions leave skeletal evidence of its presence, as do diseases such as yaws and conditions such as lithopedion. Although the documentary evidence indicated a New World origin of syphilis, adherents of the Old World origin argued that this evidence was not absolutely conclusive. This being the case, from the 1890s to the present the focus of the debate shifted from a search for literary sources to the field of archaeology, the study of ancient history through the examination of physical remains. As archaeology developed as an independent science in the late nineteenth century, its practitioners learned that skeletal remains could reveal a great deal about past events and provide information that written sources could not.

If bones and skulls were found in the Americas with lesions indicating the presence of syphilis before 1492, but could not be found in the Old World prior to that date, then the Columbian origin of syphilis would be conclusively proven. If, however, the reverse were the case, then the Old World theory would hold sway. That is, if pre-1492 skeletal evidence was found in the Old World with syphilitic lesions but not in the Americas, then obviously the disease could not have come from Hispaniola. If pre-Columbian skeletal evidence of syphilis was found in both areas, the Old World case would be strengthened to the point that it could be demonstrated that while syphilis may not have originated in Europe, it was not a fifteenth-century New World import.

The results of this century-long search of archaeological sites in both Europe and the Americas have demonstrated the validity of the New World thesis of syphilis's origin. Numerous sites in the Americas stretching from New York to Peru have provided an abundance of skeletal evidence indicating the pre-1492 existence of the treponemata diseases of yaws and syphilis. On the other hand, no pre-Columbian skeletons bearing syphilitic lesions have been found in Europe. For example, in an exhaustive study made in the 1970s and 1980s in Czechoslovakia of over 10,000 skeletons found in medieval cemeteries, no evidence of syphilis was found. These findings conclusively demonstrate the New World origin of the disease.

Occasionally a news item will appear claiming that pre-Columbian bones with lesions indicating that the individual suffered from syphilis have been found in Europe, and the debate will momentarily flare up. However, over time it invariably turns out that the lesions were caused by some other disease or condition. In 1996, for instance, it was announced that the skeleton of a seven-month-old fetus dating from the fourth century was found in Costebelle, France, bearing lesions from congenital syphilis. If true, this would have certainly revived the Old World hypothesis. However, a leading paleopathologist examined the skeleton and convincingly demonstrated that the bones were deformed by a case of lithopedion, an extremely rare and fatal occurrence related to the extrauterine formation of a fetus in the abdominal cavity; the fetus was not a victim of congenital syphilis. More recently, in August 2000, English paleopathologists announced that lesions of skeletons excavated at a friary in Hull proved the existence of syphilis in Europe by about 1475. In this case, however, the dating is uncertain, and some scientists believe the disease involved may have been bejel, which is prevalent in North Africa. Hence, while this discovery has refocused interest in the debate, it does not appear to be a viable threat to the New World thesis.

There is still, however, some debate among the adherents of the New World theory. A small group of New World theorists advocate what is sometimes referred to as the "unitary" hypothesis. This theory maintains that following their sexual assaults on the Taino women, Columbus's men were infected with yaws, which is spread by skin contact. When these sailors returned to Europe, environmental changes caused the bacterium to mutate into the form that causes venereal syphilis, for which the Europeans had no immunity. Baker and Armelagos hold this position, concluding that "nonvenereal treponemal infection is a New World disease that spread to the Old World and became a venereal disease following European contact." While somewhat modifying the New World theory, this "unitary" position does not vitiate it because it holds that without contact with the diseases of the American tropics, there would have been no initial epidemic of syphilis in Europe.

However, leading paleopathologists, such as Bruce and Christine Rothschild, argue that the causal agent of yaws and/or pinta had undoubtedly mutated and was causing syphilis in the West Indies long before Columbus arrived. Their examination of almost 700 skeletons from various New World sites convinced them that venereal syphilis was present in the tropical Americas at least 800 (and possibly as many as 1,600) years ago. They, and many other archaeological medical detectives, strongly claim that syphilis originated as a New World disease, most plausibly from a mutation of yaws. Thus Columbus's crews, after their brutal mistreatment of Taino women, proved to be the vector for the introduction of syphilis into Europe, where the disease had never been encountered. From there it was carried around the world. —ROBERT HENDRICK

Viewpoint:
No, syphilis was a disease that had long been in the Old World; it was simply a coincidence that it flared up shortly after Columbus's return from the New World.

Controversies often occur over scientific issues where there is evidence on both sides of the question, but the evidence is not overwhelming on either side. In the case of medical controversies that involve the history of diseases such as syphilis, it is frequently difficult to obtain compelling evidence because descriptions of the diseases written in the distant past are usually vague since writers lacked the precise vocabulary to describe symptoms that is available to medical writers today. Thus it is difficult to make a confident diagnosis on the basis of these early writings. In addition, these descriptions are frequently based on ideas about disease processes that are often different from those we accept today, when evil spirits and punishment for sin—often cited as causes of syphilis—are not seen as valid topics for discussion in medical analyses. Also, at times the symptoms of more than one condition may have been described as signs of a single disease. For example, the symptoms of the two sexually transmitted diseases syphilis and gonorrhea were often confused and combined. This is not surprising since they both can be contracted through sexual intercourse, so it is very possible that the same person could be suffering from the two infections simultaneously. It was not until the creation of improved microscopes in the nineteenth century and the development of chemical tests for these diseases in the twentieth century that doctors were able to clearly differentiate between these two bacterial infections.

What Is Syphilis? Despite the problems involved in diagnosing the diseases of the past, there is a great deal of convincing evidence that syphilis was a disease that had been present in Europe long before Columbus's time, thus making it a disease of the Old World rather than of the New. In order to present this case, some background on the symptoms and course of the infection is useful. Syphilis is caused by a spiral-shaped bacterium, a spirochete, of the species *Treponema pallidum pallidum*. The corkscrew shape may explain why *T. pallidum pallidum*, unlike most other bacteria, can pass through the placenta into a developing fetus and cause abnormal development and even miscarriage. This bacterium is a delicate organism that can only survive on warm, moist tissue, which is why it is spread through sexual intercourse. The infection it causes is a slow one, with a series of stages during which the host's immune system attempts to fight off the bacteria, though the bacteria often survive in the body to flare up later. Until the antibiotic penicillin became available in the mid-1940s, there was no effective treatment for syphilis, no effective way to stop the bacterium from causing its later and more damaging effects. Mercury compounds could slow the infection somewhat, but mercury is itself a poison that can damage the nervous system, so this is definitely a case where the cure was, if not worse than the disease, at least as harmful.

Though the course of syphilis varies widely from individual to individual and not everyone experiences all the symptoms, the normal course of the disease begins with a chancre or raised pustule that usually develops a few weeks after exposure at the site where the bacteria entered the body—in other words, the chancre usually

Columbus arriving in the New World.
(Columbus, Christopher, painting. Corbis. Reproduced by permission.)

appears on the genitals. Even if the disease is left untreated, the chancre disappears after several days. Three or four months later a rash and low-grade fever may develop, sometimes accompanied by inflammation of the joints. These are the symptoms of secondary syphilis and are signs that the spirochete has moved from the skin into the blood stream and that the immune system is now mounting a whole-body rather than a localized attack against the bacteria.

One third of those with untreated syphilis seem to suffer no further problems after secondary syphilis and show no further evidence of the disease. Another third continue to have what is called a latent infection, which means that while there are no symptoms, they still test positive in blood tests for the infection. Months or years after secondary syphilis, the third group of patients develops tertiary syphilis, in which the immune response to the continuing presence of these spirochete bacteria can lead to destruction of the bones, the nervous system, and the heart and blood vessels, leading to disfigurement and crippling, paralysis and blindness, and ultimately to death.

One other point is important to keep in mind about syphilis when weighing evidence as to its origins: *T. pallidum pallidum* is very similar to the organism *T. pallidum pertenue,* which causes a non–sexually transmitted disease called yaws. The names of these two organisms indicate just how similar they are: they are two varieties of the same species, *T. pallidum*. Yaws is now common only in the tropics as a skin disease in children, though, like syphilis, it can have long-term effects and can cause damage to the bones.

Not a "New" Disease The major point in the argument of those who see syphilis as originating in the New World and reaching Europe through the men who had accompanied Columbus on his voyage of discovery is that beginning in 1495 and extending into the first decades of the 1500s, syphilis spread throughout Europe, and at least some observers of the time considered it a new disease. But there are problems with this view, including the fact that it was not until this new wave of syphilis had existed for at least 15 years that observers began to speculate that it was a new disease carried back to Europe by Columbus. In addition, the debate over ancient versus Colombian origins for European syphilis did not intensify until the seventeenth century. This seems odd; if it was in fact so different from the diseases of the past, why was this not noted immediately? Also, could a disease that is relatively slow in the development of its symptoms have spread so fast throughout Europe, to be widespread by the late 1490s when Columbus had only returned in 1493?

What everyone does agree on is that the form of syphilis which spread at the time was much more dangerous and deadly than it had been in the past or was to become in the future. Spread by sexual contact, it was highly contagious and caused pustules, pain, and itching of the skin, often spreading all over the body. These symptoms were followed by intense pains and a deterioration of the bones. This stage of the disease often ended in death. A possible explanation for these symptoms is not that they were caused by a new disease, but by a more virulent or deadly form of a long-occurring organism. This is not an uncommon phenomenon among bacterial infections, with a modern-day example being that of *Staphylococcus aureus*, in most cases a relatively harmless organism commonly found on the skin. But there is a deadly form of *S. aureus* responsible for toxic shock syndrome, an often-fatal infection that surfaced in the 1980s.

When a more virulent form of *T. pallidum pallidum* did arise in Europe at the end of the fifteenth century, the political conditions were ideal for its spread. There were large armies amassing in several areas and large-scale troop movements. Wherever there are armies there are likely to be illicit sexual activities that spread sexually transmitted diseases; these, coupled with troop movements, hastened the spread of syphilis. Descriptions recorded by firsthand observers at the time tell of patients with many pustules, rather than the single pustule usually seen today, and with the symptoms of rash and painful swellings that are now associated with secondary syphilis occurring much more rapidly than they do today. After about 50 years, this particularly virulent form of syphilis seems to have subsided and was replaced by the more slowly progressing form that we see today.

Biblical and Other Ancient Evidence But is there evidence of syphilis in any form present in Europe and other parts of the Old World before the end of the fifteenth century? Those who support the hypothesis that it was the return of Columbus's expedition that triggered the outbreak of syphilis claim that there is no such evidence, but this view ignores a great deal of information pointing to the presence of this disease in Europe, the Middle East, and elsewhere in the Old World for centuries and maybe even longer. There are many reports in the Bible and other writings of a disease in which there are pustules and swellings, quite reminiscent of syphilis. In the recent study *Microbes and Morals,* Theodore Rosebury argues that while there are many passages in the Bible that can be more accurately seen as descriptions of syphilis than of any other infection we know of today, the descriptions of the plagues in Egypt and in Moab are the most detailed and convincing passages on syphilis. These plagues are described as causing scabs and itching, perhaps as a result of chancres and rashes. There are also descriptions of madness, blindness, and "astonishment" of the heart, which could be interpreted as heart failure—all three of these symptoms are related to tertiary syphilis. In describing the plague of Moab, there are many references to loose sexual practices leading to disease, another indication that the infection involved was syphilis.

Those who support the Colombian origin view argue that many of the passages cited by the anti-Colombian advocates describe leprosy, not syphilis. Admittedly, it is difficult to make a specific diagnosis one way or the other on the basis of sketchy descriptions, but a few details are telling. One is that in some cases, there are Medieval and biblical reports of the disease being transmitted to unborn children. This is never the case with leprosy, but it is possible with syphilis since the spirochete can penetrate the placenta. Also, a number of historians have amassed a large collection of evidence that syphilis was known in ancient China and Japan, and was familiar to Hindus and to ancient Greeks and Romans. This evidence includes comments on the disease's relationship to sexuality and to birth defects, both hallmarks of syphilis as opposed to other diseases with which it might be confused, including leprosy and the plague. Such reports are likely to be documenting cases of syphilis in the ancient world and throughout Asia.

Skeletal Remains There is also more solid evidence rather than just verbal reports. In 2000, British researchers unearthed 245 skeletons from a cemetery in northeastern England. Three of these skeletons showed clear signs of the type of damage associated with syphilis, with 100 more showing slight indications of such damage. Radiocarbon dating was used on the skeleton showing evidence of the most severe disease and revealed that the man died sometime between 1300 and 1450, well before Columbus's voyage. Some argue that the bone lesions could be the result of yaws rather than syphilis. But the same argument can be used against a favorite piece of evidence presented by pro-Colombian advocates: the discovery of syphilis-like skeletal lesions in bones dated to pre-Columbian times and found in various parts of South, Central, and North America. Advocates contend that this proves that syphilis was present in the New World before the time of Columbus, a view that some anti-Colombians argue against. So it would seem that the physical evidence from bone lesions is less than convincing on either side of the dispute.

Of course, the presence of syphilis in England in the fourteenth and fifteenth centuries does not necessarily mean that the infection was present in Europe from ancient times. Still another hypothesis, somewhat of a compromise between the Colombian and anti-Colombian views, is that it was not Columbus but the Vikings who brought syphilis back from the New World. There is evidence that the Vikings had reached Canada's eastern shores hundreds of years before Columbus's voyage, and they began trading in northeastern England around 1300, which was about the time the recently discovered skeletons started to show signs of the disease.

There may eventually be a resolution of this controversy thanks to molecular genetics. In 1998, the genome, or genetic makeup, of *T. pallidum pallidum* was completely deciphered, so biologists now have a record of all the information in the genes for the organism that causes syphilis. The same information is now being accumulated for *T. pallidum pertenue*, the cause of yaws. When this sequence is deciphered, researchers can hunt for bacterial DNA (deoxyribonucleic acid—the nucleic acid found in the nuclei of all cells) in the bones that have been unearthed in England and at other sites and determine which sequence most closely matches the bacterial DNA from the bones.

In one sense it would be disappointing if such information finally settled this centuries-old disagreement. As R. R. Wilcox, who argued against the Colombian hypothesis, wrote over a half a century ago: "It would be a great pity if someone did produce irrefutable evidence for either side and thus prevent any further such interesting speculation." —MAURA FLANNERY

Further Reading

Baker, Brenda J., and George J. Armelagos. "The Origin and Antiquity of Syphilis: Paleopathological Diagnosis and Interpretation." *Current Anthropology* 29, no. 5 (1988): 703–737. Reprinted in Kenneth F. Kiple and Stephen V. Beck, eds. *Biological Consequences of the European Expansion, 1450–1800.* Aldershot, Great Britain: Ashgate, 1997, pp. 1–35.

Crosby, Alfred W. *The Columbian Exchange: Biological and Cultural Consequences of 1492.* Westport, Conn.: Greenwood Press, 1972.

Guerra, Francisco. "The Dispute over Syphilis: Europe versus America." *Clio Medica* 13 (1978): 39–62.

Hackett, C. J. "On the Origin of the Human Treponematoses." *Bulletin of the World Health Organization* 29, no. 7 (1963): 7–41.

Holcomb, Richmond C. "The Antiquity of Syphilis." *Bulletin of the History of Medicine* 10 (1941): 148–77.

Kiple, Kenneth, and Stephen Beck, eds. *Biological Consequences of the European Expansion, 1450–1800*. Brookfield, Vt.: Ashgate, 1997.

Malakoff, David. "Columbus, Syphilis, and English Monks." *Science* 289 (August 4, 2000): 723.

Origins of Syphilis. <http: www.archaeology.org/9701/newsbriefs/syphilis>.

Quétel, Claude. *History of Syphilis*. Trans. Judith Braddock and Brian Pike. Baltimore: Johns Hopkins University Press, 1990.

Rosebury, Theodore. *Microbes and Morals: The Strange Story of Venereal Disease*. New York: Ballantine Books, 1973.

Salyers, Abigail, and Dixie Whitt. *Bacterial Pathogenesis: A Molecular Approach*. Washington, D.C.: ASM Press, 1994.

Watts, Sheldon. *Epidemics and History: Disease, Power and Imperialism*. New Haven, Conn.: Yale University Press, 1997.

Weisman, Abner I. "Syphilis: Was It Endemic in Pre–Columbian America or Was It Brought Here from Europe?" *Bulletin of the New York Academy of Medicine* 42, no. 4 (1966): 284–300.

Zimmer, Carl. "Can Genes Solve the Syphilis Mystery?" *Science* 292 (May 11, 2001): 1091.

PHYSICAL SCIENCE

Is the cost of high-energy laboratories justified?

Viewpoint: Yes, research in the area of high-energy physics has already led to numerous advances that have provided great public benefit, and the search for the ultimate building blocks of the universe is a fundamental part of human nature and an expression of human civilization.

Viewpoint: No, the enormous public money spent on high-energy laboratories would be better spent on scientific work with more relevance and clear applications.

The field of particle physics revolves around finding a simple, orderly pattern to the behavior of atomic and subatomic matter. Among its tools are particle accelerators that act like giant microscopes, peering into the atom, past electrons and into the proton- and neutron-populated nuclear core.

The modern view of the structure of matter is based upon thousands of experiments that showed over time how all matter could be broken into particles called atoms. More than 100 kinds of atoms occur in nature and combine to make simple molecules like water (H_2O—two hydrogen atoms, one oxygen atom) or complex million-atom molecules like DNA (deoxyribonucleic acid), the building block of life on Earth.

An atom is mainly empty space, occupied by cloudlike swarms of electrons, each with a tiny negative electrical charge. The atom's core—the nucleus—is 100,000 times smaller than the atom and is composed of neutrons and protons. To see the nucleus takes particle accelerators—devices in which charged particles are accelerated to extremely high speeds—that push the limits of microscopic sight.

In the 1950s and 1960s, new, powerful accelerators smashed atomic particles together at increasingly higher energies, producing a gaggle of new particles. Experiments in the 1970s showed that protons and neutrons were not the fundamental, indivisible particles physicists thought—instead they were made of smaller, pointlike particles called quarks. Quarks come in flavors (or kinds) that signify their symmetry properties, and with three strong charges called colors. The universe itself, in its early stages, existed as a soup of quarks and leptons (both fundamental kinds of particles). According to a theory called the Standard Model, there are 12 fundamental particle types and their antiparticles. In addition, there are gluons, photons, and W and Z bosons, the force-carrier particles responsible for strong, electromagnetic, and weak interactions. The Standard Model was the triumph of 1970s particle physics and is now a well-established theory that applies over a range of conditions. Today, experiments in particle physics seek to extend the Standard Model in order to discover what has been called the Theory of Everything—the grand unification of many theories into an ultimate theory of how the universe works.

Those who say the cost of high-energy labs is justified base their belief on extraordinary spinoffs from particle physics research—advances in cancer treatment, medical scanners, nuclear waste transmutation, microchip manufacturing techniques, and other industrial applications—and on the notion that the search for the ultimate building blocks of the universe is a fundamental part of human nature and an expression of human civilization.

Atomic and subatomic physics research has led to many useful and important innovations. The renowned English physicist J. J. Thomson discovered the electron in 1897 by using a very simple kind of particle accelerator—a cathode-ray tube. Today cathode-ray tubes are the basic component of television screens, computer monitors, and medical and scientific instruments. Using a basic particle beam, English physicist Ernest Rutherford, often referred to as the founder of nuclear physics, revealed the nature of the atom and ushered in the atomic age. The quest to understand the atom led to a major scientific revolution in the 1920s—a new understanding of nature in the form of quantum theory. And this led to laser technology, solar cells, and electronic transistors—the basis of modern electronics. In the 1930s, American physicist E. O. Lawrence created the first working circular particle accelerator (cyclotron) and used it to produce radioactive isotopes for cancer treatment. Accelerated beams of neutrons, protons, and electrons have been used to treat cancer in thousands of hospitals. Computed tomography (CT scan) owes its development to particle detection methods used in high-energy physics. Positron emission tomography (PET scan) uses components developed in particle physics research. Magnetic resonance imaging's (MRI) noninvasive diagnostics is a particle physics spin-off.

Even research problems—like synchrotron radiation—have led to useful applications. Tools that use synchrotron radiation now are used in medical imaging, environmental science, materials science and engineering, surface chemistry, biotechnology, and advanced microchip manufacturing.

High-energy accelerators allowed more accurate measurements of the particles that make up the atomic nucleus. In the process, new theories of the nature and structure of the matter were developed and tested, leading to a better understanding of the origin and evolution of the universe.

Technical knowledge gained in scientific research is also used in broader research and development. Vacuum technology, developed for accelerators, has many commercial uses. Superconducting magnets are being developed for commercial use in power transmission, energy storage, conservation, medicine, and high-speed magnetically levitated trains called maglevs—literally "flying trains" that use magnets chilled in helium to a achieve a superconducting state.

The most practical development of high-energy physics was the World Wide Web, created by CERN, the European high-energy physics lab, as a way for high-energy physicists to exchange information. It transformed information sharing and revolutionized communications. It did no less than change the world, making possible many collaborations in science, medicine, entertainment, and business.

Scientific research may seem removed from everyday life, but in the long term esoteric knowledge often enters the mainstream with startling consequences.

Those who cannot justify the cost of high-energy labs claim that even though budgets for high-energy physics experiments can run into the billions of dollars, physicists never know exactly what they will discover about the nature and behavior of matter or how the knowledge will pay off. Public money invested in such experiments, opponents argue, would be better spent on scientific work with more relevance and clear applications, like medical research, programs that directly improve people's lives, research that leads to feeding the world's population, and research into energy alternatives for nonrenewable resources. Private funding, some say, might be a better way to pay for particle physics experiments.

One example is the Superconducting Supercollider (SSC), an $8 billion paticle accelerator proposed to be built in Texas in the late 1980s and early 1990s. Congress canceled the SSC in 1993 after legislators refused to continue funding the project. Because the SSC caused such controversy, cost so much money, and changed the course of science funding, one physicist called it the "Vietnam of physics." In 1993 critics argued that no high-energy lab project could stick to its budget. The price of particle physics research remains high, with duplicated efforts in several labs in the United States and Europe.

Particle physics scientists are still unable to explain the relevance of and applications for their work when asked why their projects should be publicly funded. Their best argument for doing the research is that if the United States does not fund the expensive projects, the country will not be in a position of leadership in the field. Americans should not worry about not being at the forefront of particle physics. Once the knowledge is gained—whether in Europe, Asia, or North America—it can be applied anywhere.

Medical research has an impact on everyday life. It is unlikely that any knowledge from high-energy physics will have a clear impact on the wealth and health of Americans. —CHERYL PELLERIN

Viewpoint:

Yes, research in the area of high-energy physics has already led to numerous advances that have provided great public benefit, and the search for the ultimate building blocks of the universe is a fundamental part of human nature and an expression of human civilization.

Billions of dollars each year are spent around the world on the continuing investigation into the structure of matter, funding giant particle accelerators and international research groups. Some have questioned the value of such research, which they see as esoteric and of little use to the average person. However, research into high-energy physics has had many extraordinary spin-offs in the last century, including cancer treatment, medical scanners, nuclear waste transmutation, microchip manufacturing techniques, and other industrial applications. Aside from the obvious material benefits it can also be argued that the search for the ultimate building blocks of the universe is a fundamental part of human nature and an expression of human civilization. Perhaps a better question is whether high-energy physics is receiving enough funding considering its proven practical value and its potential to unlock the secrets of the structure of the universe.

New Tools for Human Development Scientific innovation and research is an important aspect of modern life. The world we live in has been revolutionized by technological and scientific breakthroughs time and time again. Innovation in one field can have great impact in other seemingly unrelated areas, and it can change the way we think about, and interact with, the universe itself. Developments in timekeeping led to better sea navigation, materials developed for spacecraft led to nonstick cookware, and research into the components of the atom has given us a host of medical and industrial machines. Time and time again seemingly esoteric scientific research has provided new tools for human development. Basic scientific research always seems to pay off in the long term, giving back far more than it consumes in terms of resources.

Yet the value and utility of pure scientific research is often questioned on grounds of cost. In a world that cries out for more food, more medicine, more everything, can we afford to spend billions of dollars on research that appears to have no practical values or goals? The study of particle physics is often singled out by would-be cost cutters. It is an expensive field of study, using gigantic accelerators that can be many miles in length. Inside these devices small highly focused bundles of particles are sped up to near the speed of light and then aimed at particle targets. The subatomic debris from the moment of collision is then studied. Accelerators are costly to build, to maintain, and to run, and high-energy physics is often regarded as a field that does not offer useful applications, yet nothing could be further from the truth.

Applications of High-Energy Physics
Research into atomic and subatomic physics has already led to many useful and important innovations, as well as giving us a clearer picture of the nature of reality and the building blocks of the universe. The discovery of the electron, by J. J. Thomson in 1897, was made using a very simple kind of particle accelerator, a cathode-ray tube. The cathode-ray tube was an important early tool for probing the structure of the atom, and over the years it was improved on, refined, and modified. Today the cathode-ray tube is the fundamental component of television screens, computer monitors, and medical and scientific instruments.

Using a basic particle beam, Ernest Rutherford revealed the unexpected nature of the atom, a tiny compact nucleus surrounded by an insubstantial cloud of electrons. His discovery was to have unforeseen consequences. "Anyone who expects a source of power from the transformation of these atoms is talking moonshine," said Rutherford, yet his discovery ushered in the atomic age. Although human foibles and flaws led some to use the power within atoms for destruction, it has also become a valuable source of power. And although nuclear fission has unwanted products and problems, the hope is

---

> **KEY TERMS**
>
> **PARTICLE ACCELERATOR:** Device designed to accelerate charged particles to high energies. Often used in scientific research to produce collisions of particles to observe the subatomic debris that results.
>
> **PARTICLE PHYSICS:** Branch of physics concerned with subatomic particles, especially with the many unstable particles found by the use of particle accelerators and with the study of high-energy collisions.
>
> **PURE SCIENCE RESEARCH** (as opposed to applied science): Research that is conducted to advance science for its own sake. It is not undertaken to meet commercial or other immediate needs.

worldwide. The CT scan (computed tomography) owes its development to particle detection methods used in high-energy physics. The PET scan (positron emission tomography) is yet another medical machine that contains components initially developed in particle physics research. The MRI (magnetic resonance imaging) is also a spin-off from particle physics research that gives doctors a powerful diagnostic tool for noninvasive imaging of the body. The MRI relies on the esoteric quantum property of "spin" in the atomic nucleus and the discovery by chemists that the energy associated with the flip of this spin depends on the chemical environment of the nucleus.

Even the problems that have beset researchers of high-energy physics have led to useful applications. Accelerated charged particles lose energy as they travel, in the form of light. This problematic phenomenon, called synchrotron radiation, is one of the reasons why accelerating particles requires so much power and such large-scale constructions. However, although synchrotron radiation makes research into particle physics more difficult, it has been harnessed as a source of useful radiation. Tools using synchrotron radiation are used in medical imaging, environmental science, materials science and engineering, surface chemistry, biotechnology, and the manufacture of advanced microchips. The scientific uses of synchrotron radiation include research into hemoglobin (the oxygen carrier in blood), and it has been important in the quest to find a cure for ALS (Lou Gehrig's disease) and the AIDS virus.

High-energy accelerators allowed for more accurate measurements of the size of the particles that make up the nucleus of the atom, the neutron and the proton. However, they also revealed the unexpected result that these particles are themselves made from yet smaller particles. The strangeness of the subatomic world of quarks was discovered, and as particle accelerators have improved and developed, more and more has been revealed about the internal structure of what were once considered elementary particles. Six quarks have been found and a host of other subatomic particles. In the process new theories of the nature and structure of the matter have been created and tested, and a better understanding of the origin and development of the universe has been gained.

Commercial Uses In order to build more powerful accelerators to probe ever deeper into the subatomic world, great leaps in technology and engineering have been made. The technical knowledge gained in the name of pure scientific research has then been used in broader research and development. Sophisticated vacuum technology, developed for accelerators, has found

E. O. Lawrence
(© Corbis. Reproduced by permission.)

that soon nuclear fusion will usher in an age of safe and plentiful energy.

Using Quantum Physics The quest to understand the atom led to a major scientific revolution in the 1920s. A tide of experimental results cast doubt on the underlying principles of classical physics and gave rise to a new understanding of nature in the form of quantum theory. Understanding of the quantum reality of the universe has led to laser technology, solar cells, and the discovery of the electronic transistor, which underlies all of modern electronics. Quantum theory may seem unreal or contrary to common sense, but all around us are devices that rely on quantum effects and have become integral parts of our lives.

Many other practical benefits have been discovered in the process of exploring the subatomic particles and forces that shape the physical universe. In the 1930s, the first working circular particle accelerator (cyclotron) was used by its creator, E. O. Lawrence, to produce radioactive isotopes for cancer treatment. After the day's physics experiments had been run, Lawrence would keep the device on through the night in order to produce isotopes for local California hospitals. Since then there have been many developments in the use of accelerators for medical use. Accelerated beams of neutrons, protons, or electrons have been used to treat cancer successfully in thousands of hospitals many commercial uses. The boundaries of superconducting magnet technology have been pushed by high-energy physics research, and such technology is now being developed for commercial use in power transmission, energy storage, conservation, medicine, and high-speed magnetically levitated trains (maglevs).

The people who build, maintain, and use the huge accelerators of high-energy physics gain valuable knowledge that has application beyond particle physics. Many of the medical and industrial tools mentioned here were designed and built by people trained in the research environment but who moved into the commercial workplace. Only one in five Ph.D. students in the field of high-energy physics remains in basic research. Many of the others go into industrial research and development.

High-energy physics is a field rich in international collaboration. Many countries share information, reproduce and confirm experimental results, and share the costs involved. This collaboration led to one of the most unexpected practical developments of high-energy physics: the origin of the World Wide Web. The Web was created by the European high-energy physics laboratory CERN as a tool to enable high-energy physicists to exchange information instantaneously and make collaborations among them easier. The Web has unexpectedly grown to include millions of users and a new global economy. It has transformed information sharing and has revolutionized how we communicate and the way we view the world and its people. It has also enabled many other collaborations in science, medicine, entertainment, and business.

Conclusion Although a few billion dollars a year for scientific research may seem like a huge amount of money, in the larger scheme of things it is a small price to pay. The benefits that society has already gained from basic research are so enormous as to be incalculable. High-energy physics is just one field that competes for the few billions of public funding given to science, but it is one of the more expensive. The costs of running a single high-energy complex can be over $300 million a year, and construction costs can be even larger. However, compared to the trillions the world spends on military weapons and the vast fortunes accumulated by companies and individuals in business, the money spent on scientific research is far more beneficial to humankind. Opponents of scientific funding may argue that the money could be spent to help feed the starving or house the homeless, but it must be asked how many more people would be starving or homeless (or worse) if not for the benefits basic research has given the world. Indeed, given the track record of pure scientific research to improve the human condition and the possible benefits that may come from such research, we should ask whether science receives enough public funding around the world.

An American politician questioning the funding of high-energy physics once said, "We've got enough quarks already. What do we need another one for?" Indeed, the discovery of quarks has no obvious benefits to the average person and will not in itself make the world a better place to live in. However, the knowledge gained in studying particle physics does help make the universe a more understandable place. High-energy physics helps us understand how the universe is constructed and how it works. That in itself is a goal worth pursuing. It is part of the human condition to seek answers about ourselves and the universe we live in.

The search for the fundamental truth about the nature of nature has led to many important and valuable discoveries over the millennia that humans have pondered such questions. Many tools and products we take for granted today, in medicine, telecommunications, and in the home, owe their existence to the pure science research of the past. In the future the secrets unlocked inside particle accelerators may lead to many different applications and new fields of technology that are as yet undreamed of.

Research into high-energy physics has led to many concrete benefits, from medical scanners and treatments, to industrial superconductivity technology, and even the World Wide Web. History has shown us that although much pure scientific research may seem wholly removed from everyday life, in the longterm esoteric knowledge often enters the mainstream with startling consequences. High-energy physics is the quest to understand the fundamental building blocks of nature and how they fit together. The knowledge we have gained has already given the world a variety of important discoveries and in the process many unexpected benefits and tools. Although we have little idea at present what particular advances further basic research will give us, we have every historical reason to expect that the benefits will greatly outweigh our investment. —DAVID TULLOCH

Viewpoint:

No, the enormous public money spent on high-energy laboratories would be better spent on scientific work with more relevance and clear applications.

When physicists work in high-energy laboratories doing experiments in elementary particle physics, they use powerful and expensive equipment to learn more about the universe and, in particular, to gain better understandings of matter and its behavior in its smallest form. Some experiments in particle physics look at matter as small as a billionth of a billionth of a meter (one meter is approximately 39.37 inches). This requires producing particle beams at the highest possible energies and the use of very expensive equipment. The most costly aspects of their experimental equipment is the accelerator, the complex of high-energy magnets and the construction of underground tunnels—miles in length—in which the experiments take place. An accelerator is like a super microscope (although miles long) designed to afford physicists their closest look at matter in its subatomic state.

Overpriced and Unpredictable Research
Although the goals of high-energy lab experiments are clear, the practical applications of the information physicists might discover are not. Budgets for high-energy labs run into the billions of dollars, but physicists do not know exactly what they will discover about the nature and behavior of matter in such small forms. They do not know how that knowledge will pay off. They cannot even predict how this knowledge will be relevant to our world, nor can they predict how this knowledge will benefit humankind. For example, physicists who work with expensive accelerators hope to find Higgs boson—a particle that is extremely small with no mass of its own, yet it gives matter its mass. Billions of dollars are being spent in search of Higgs boson, with no indication that it will be found.

Because of the great expense associated with high-energy labs and the lack of relevance and direct application of knowledge that will be gained from high-energy lab experiments, these labs are not worth the expense. If public money is to be invested in such experiments, that money would be better spent on scientific work with more relevance and clear application, such as medical research, programs to directly improve people's lives, research that leads to feeding the world's population, and seeking energy alternatives of the earth's nonrenewable resources.

Private funding may be a better way to foot the bill for elementary particle physics experiments. Private funding gives scientists more freedom from legislators politicizing their work, and it also provides a smoother vehicle for making commercial use of their finds through patents.

SSC Controversy To better explain why high-energy labs are not a prudent expenditure of science dollars drawn from taxpayers and how high-energy experiments carried out at public expense become political footballs, we must go back not too far into the history books—to the early 1990s—and look at the experience with the Superconducting Supercollider (SSC). The SSC project was a particle collider, an "atom smasher" of sorts, aimed at finding a better understanding of matter. As mentioned earlier, accelerators are very expensive and SSC was no exception. Its budget was $8 billion, and, had it been completed, it was certain to go way over budget.

The U.S. Congress canceled SSC in 1993 after legislators refused to continue funding the project. It has always been difficult for physicists to explain the value of such research to humanity and justify the spending in terms of direct benefits, and legislators perhaps feared a public backlash from the expenditure, although the economy was strong in the early 1990s.

Because the SSC caused so much controversy, cost so much money, and became a political quagmire that changed the course of science funding, it has been called "the Vietnam of the physics world" by one physicist.

For example, an August 1992 article in *Physics Today*, a magazine that serves the physics community, was entitled "What's Gone Wrong with the SSC? It's Political, Not Technological." The best argument for private funding for expensive science is that the work cannot be so easily politicized. SSC became even more politicized when some Texas politicians rallied behind the project. With SSC based in Texas, they felt it was good for Texas. Politicians from other states were less willing to keep the dollars coming. One of the more outspoken congressional foes of SSC back in 1992 was Dale Bumpers (D-Ark.), who used the argument that SSC was irrelevant research.

Once more, critics including Bumpers argued in 1993—and it is still true—that no high-energy lab project can stick to its budget. The Department of Energy, they said, cannot control "cost overruns." Exceeding a budget is an increasing and unwelcome burden on taxpayers.

Work Abroad Since 1993 other accelerator projects started, however, and many American physicists began working abroad, still attempting to unravel the secrets of matter. The price of this work remains high with duplicated efforts spread out through several labs in the United States and in Europe.

At SLAC (Stanford Linear Accelerator Center), at the CERN lab in Geneva, Switzerland, and at the Fermilab National Accelerator Lab in Chicago, scientists are working with accelerators. At CERN, scientists are constructing the Large Hadron Collider (LHC) with a $6 billion price tag. The Fermilab collider is the highest energy collider in the world—for now—but will be surpassed over the next few years when the European-based CERN LHC starts producing data.

Many physicists are concerned that the United States won't contribute. "The U.S. won't be at the energy frontier then," lamented Johns Hopkins University physicist Jonathan Bagger, professor and co-chair of a U.S. Department of Defense and National Science Foundation subpanel. "Do we want to host another facility at the energy frontier, or is our country content with sending our physicists abroad to use there facilities elsewhere?" Bagger added, "We are going through a cyclic period of reassessment of our goals and long-term plans in high-energy physics."

"Reassessment" means thinking about new accelerator projects. New efforts have always been on the drawing board since the demise of SSC, and physicists are lobbying to begin anew. For example, in July 2001, physicists from around the world met at Snowmass, Colorado, to discuss the future of particle physics and seek international cooperation on a $6 billion 20-mile-long new particle accelerator so large, so powerful, and so expensive, it could only be built if they all cooperated.

At Snowmass, American physicists from the high-energy Fermilab worried aloud that lack of U.S. funding might kill the project. Michael Holland, from the White House Office of Management and Budget, advised scientists that to get funding, they needed to demonstrate the device would be important to science in general.

This is precisely the point. Because elementary particle physicists have never been able to explain what practical applications—if any—might come from their work, high-energy labs are too expensive to be funded with taxpayer dollars. Simply put, particle physics scientists are still unable to withstand the relevancy and applications test when confronted with questions about why their projects should be publicly funded.

Physicists on the Offensive Their best argument for doing the research is the one stated earlier: if we do not fund the expensive projects, the United States will no longer be in a leadership role in particle physics. Their second argument is that if we do not have funding, workers and physicists will lose their jobs. Neither argument is strong enough to warrant such high-cost public funding.

Physicists seem to realize this, and they are already on the offensive. At a meeting at Johns Hopkins University in March 2001, prior to the June particle physics meeting at Snowmass, physicists expressed concern that their research would not get funding and, as a result, go the way of SSC.

"That decision (SSC) was a real blow to the high-energy physics community, and ended up forcing many of us to go for extended periods of time to collider facilities in Europe," said Morris Swartz, professor of physics at Johns Hopkins.

The U.S. Congress and taxpayers are not stingy when it comes to funding science; however, they prefer funding medical research and for good reasons. Medical research has been shown to have an impact on everyday life. Finding Higgs boson will likely not. It is unlikely that any knowledge from high-energy physics will have a clear impact on the wealth and health of Americans.

Although expensive research, it was prudent in the 1990s to fund work such as the Human Genome Project. With the human genetic system now mapped, scientists are daily gaining insight into which genes may be responsible for diseases and disability. No such practical and beneficial outcomes can result from expensive experiments in particle physics.

Letting Go of a Cold War Mentality Americans should not be concerned that the nation is not at the forefront of elementary particle physics. Once the knowledge is gained—whether in Europe, Asia, or North America—it can be applied anywhere. People in other countries will not have a monopoly on the information gained through this research. Just as people the world over, not just Americans, have benefited from American medical research (and even from American outer space research) so will Americans benefit from physics research, if there is truly a benefit to be had.

That the United States should lead the world in science, no matter what science, no matter what the cost, is a Cold War mentality that seeks to one-up the rest of the world. Being one up is expensive and unnecessary in the twenty-first century. Even particle physicists seem to have realized this as they talked about international cooperation at Snowmass.

In a letter to *Physics Today* during the SSC debate, Mark Phillips, a researcher in radiation physics at the University of Washington Medical Center in Seattle, said the aims of not only SSC but big funding for particle physics research is suspect. "Not every scientist, nor every citizen, for that matter, believes that spending tens of billions of dollars on this project [SSC and high-energy physics] is the sine qua non of American scientific commitment. In essence, every man, woman and child in the US is being asked to donate $30 (for construction alone) of the SSC. . . . I hope that every physicist is equally generous with his or her own money when environmental, religious or lobbying organizations come to the door asking for donations." His argument is still a good one today.

Also writing in *Physics Today* at the time of the SSC debate was scientist and entrepreneur Lon Hocker. His arguments against SSC then are also relevant to the argument against spending money on particle physics today.

"The SSC [was] a huge opportunity for pork barrel spending," wrote Hocker, "as well as an uninspired extension of a remarkably useless science.... High-energy physics shows no conceivable benefit to anyone other than a few politicians, contractors and scientists.... The $10 billion planned for SSC could be spent on $10,000 educational grants to a million physics students, or ten thousand $1,000,000 research grants."

Other funding questions then and now for so-called big science ask by which criteria do we decide that one kind of science is more important to fund than another. Small science and not terribly expensive science can also be productive.

Conclusion Currently, physicists are lobbying the Bush administration for increased funding for the Department of Energy (DoE), the agency under which high-energy labs would receive funding and do their work. The Bush budget for 2003 gives the DoE $19 billion, $700 million less than it had in 2001. By contrast, the National Institutes of Health will get $23 billion in 2003 and the Defense Department $310.5 billion. The portion of the DoE funds that would go toward particle physics is rather small by comparison. These figures mean the government thinks the tangible benefits from elementary particle research are not worth increased funding.

The cost of high-energy labs is too great for public funding, and scientists should seek out private funding. The DoE also has small science projects in materials engineering, computing, and chemistry, all of which could have greater relevance to life on Earth. More emphasis should be given to alternatives to big science research money. —RANDOLPH FILLMORE

Further Reading

CERN (European Organization for Nuclear Research). <http://welcom.cern.ch>.

Freeman, Chris, and Luc Soete. *The Economics of Industrial Innovation*. Cambridge: MIT Press, 1997.

Kursunoglu, Behram, and Arnold Perlmutter, eds. *Impact of Basic Research on Technology*. New York and London: Plenum Press, 1973.

Ne'Eman, Yuval, and Yoram Kirsh. *The Particle Hunters*. Cambridge: Cambridge University Press, 1983.

Snowmass 2001. <www.snowmass2001.org>.

Weinburg, Steven. *The Discovery of Subatomic Particles*. London: Penguin, 1990.

Historic Dispute:
Do neutrinos have mass?

Viewpoint: Yes, the Japanese–U.S. research team called the Super-Kamiokande Collaboration announced experiment results in 1998 that proved that neutrinos do indeed have mass.

Viewpoint: No, the experiments of earlier twentieth-century scientists repeatedly indicated that neutrinos did not have mass.

In 1931 Wolfgang Pauli first predicted the existence of a subatomic particle that Enrico Fermi would later name the neutrino. The neutrino must exist, Pauli reasoned, because otherwise the atomic process known as beta decay would violate the physical laws of conservation of energy and conservation of angular momentum. Neutrinos had never been detected, so Pauli concluded that they didn't interact with most other particles or forces. This implied they were extremely small particles with no charge or mass.

The only force that noticeably affects neutrinos is the "weak" force, a subatomic force that is not as strong as the force that holds the atomic nucleus together, but that likewise operates only at very short range. Because of their extremely limited interactions with other particles and forces, neutrinos can travel huge distances unimpeded by anything in their path. Neutrinos arising from the nuclear reactions in the Sun stream through Earth and all its inhabitants without having any effect whatsoever. For this reason they are extremely difficult to detect, and their existence was not confirmed until a quarter century after Pauli's prediction.

Today's neutrino detectors, kept deep underground to avoid stray particles on Earth's surface, may contain thousands of tons of fluid. While trillions of neutrinos pass through the fluid every day, only a few dozen are likely to be detected.

Scientists have discovered that there are three types of neutrinos, each associated with a different charged particle for which it is named. Thus they are called the electron neutrino, muon neutrino, and tau neutrino. The first type of neutrino to be discovered was the electron neutrino, in 1959. The muon neutrino was discovered in 1962. The tau neutrino has yet to be directly observed. It was inferred from the existence of the tau particle itself, which was discovered in 1978. The tau particle is involved in decay reactions with the same imbalance that Pauli solved for beta decay by postulating the electron neutrino.

One ongoing issue in neutrino research is called the "solar neutrino problem." This refers to the detection of fewer electron neutrinos than expected, given the known energy output of the Sun. One possible explanation for this phenomenon could be "oscillation" between the different neutrino types. That is, electron neutrinos could change into muon or tau neutrinos, which are even more difficult to detect. Similarly, scientists have observed a deficit in the number of muon neutrinos they would expect to see coming from cosmic rays.

Neutrino oscillations, if they exist, are a quantum mechanical phenomenon dependent on the difference in the masses of the two types of particles. That means that if researchers could prove that neutrino oscillation occurs, at least one of the neutrino types involved must have a non-zero mass. In 1998 a Japanese–U.S. research team called the Super-Kamiokande Collaboration announced that they had discovered evidence of oscillation between muon neutrinos and either the tau neutrino or a new, unknown type. —SHERRI CHASIN CALVO

Viewpoint:
Yes, the Japanese–U.S. research team called the Super-Kamiokande Collaboration announced experiment results in 1998 that proved that neutrinos do indeed have mass.

Wolfgang Pauli inferred the existence of the neutrino in 1930 from the discovery that a small amount of energy was missing in the decay products in certain types of radioactivity (beta decay). In this type of decay, a neutron in an atomic nucleus is converted into a proton, with emission of an electron and an antineutrino, a neutral particle that could not be detected directly. Pauli also suspected that the mass of this particle would be zero because it could not be detected in the decay experiments. The mass difference between the initial and final nuclei produced during this decay should correspond to the highest energy of the emitted particles. Any difference would then correspond to the mass of the neutrino itself, according to Einstein's mass-energy relationship. This energy difference, however, proved too small to be measured.

Neutrinos escaped direct detection until 1956 when Frederick Reines and Clyde Cowan Jr. captured neutrinos produced in a nuclear reactor. In subsequent experiments, the speed of neutrinos could not be distinguished from that of light. Also, there was no spread in the arrival time of neutrinos produced by the supernova SN observed in 1987. According to relativity, particles such as photons that travel at the speed of light have zero masses. This reinforced the idea that neutrinos might have zero mass.

The first hint that something might be wrong with this picture came in 1967 when scientists tried to detect neutrinos created by nuclear reactions in the sun's core. In this and subsequent experiments, scientists consistently observed a shortfall of about half of the neutrinos that were predicted by theory.

However, already in 1957, long before the detection of solar neutrinos proved to be a problem, the Italian physicist Bruno Pontecorvo, then working in the Soviet Union, proposed that neutrinos could change continuously from one type into another. Now we know that three types, or "flavors," of neutrinos exist: tau, muon, and electron neutrinos, named after the particles they are associated with during their creation. The continual change from one type into another type of neutrino is called "oscillation." This phenomenon is a quantum mechanical effect (superposition of several quantum states) and is also observed in other elementary particles, such as K0 mesons. For example, if we detect a muon neutrino, it is possible that it just oscillated from being an electron neutrino before arriving at the detector. But according to quantum mechanics, oscillations are only permitted if neutrinos have mass.

The Standard Model of particle physics, a global theory that describes elementary particles and their interactions, developed since the 1970s, predicted neutrinos with zero mass. Therefore, at first the fact that neutrinos may have mass was viewed as a possible crack in the Standard Model. But theorists are now extending the Standard Model so it can incorporate neutrino masses.

The Atmospheric Neutrino Anomaly The neutrino deficit shown by the solar neutrino experiments was the first indication that the Standard Model did not describe these particles accurately: some changed flavor and thus escaped detection. Another so-called neutrino anomaly became apparent around the mid-1980s. Both the Kamiokande detector in the Japanese Alps and the Irvine-Michigan-Brookhaven (IMB) detector in the Morton salt mine in Ohio started observing neutrinos produced by cosmic-ray particles colliding with atomic nuclei in the earth's atmosphere. According to theory, these interactions should produce twice as many muon neutrinos as electron neutrinos, but the detectors discovered roughly equal amounts. Scientists explained this discrepancy by the possibility that muon neutrinos change into either electron or tau neutrinos before reaching the detectors.

The successor to the Kamiokande is the Super-Kamiokande, the world's largest neutrino detector. It consists of a huge tank containing 50,000 tons of purified water, and it is protected from cosmic rays by a 1,000-m (c. 1,094 yd) rock layer. Eleven thousand photomultiplier tubes line the tank and track the light flashes (Cerenkov radiation) caused by muons or elec-

NEUTRINOS AND DARK MATTER

According to current theory, the universe is "flat"; in other words, it contains just the right amount of matter so that it is at a point exactly in between collapse and infinite expansion. However, astronomers are only able to observe a fraction of the matter that is required to make the universe flat and therefore believe that most of the matter in the universe is unobservable. In fact, the edges of many galaxies rotate faster than they should if they contained only the visible matter. The fast motion of galaxies in clusters indicates that they are gravitationally held together by the presence of mass that cannot be observed.

According to estimates, about 90% of the mass of the universe consists of dark matter—matter that does not radiate light and therefore is invisible to us. A part of the dark matter consists of "normal," or "baryonic" matter, matter made up of electrons, protons, and neutrons, the constituents of atoms with which we are familiar. This matter would be found in the invisible "brown dwarfs," stars that are very cool and radiate little, cool intergalactic gas, and the so-called MACHOs (massive compact halo objects).

A large part of dark matter should also consist of non-baryonic particles—exotic particles that do not make up normal matter. Several of these that still would have to be discovered have been proposed over the years: WIMPs—weakly interacting massive particles, such as gravitinos, axions, and neutralinos. Up until now none of these particles has been detected. However, astrophysicists know that huge quantities of neutrinos were produced during the Big Bang. The discovery that neutrinos have mass has far-reaching implications for cosmology. The density of neutrinos is so high—more than 300 neutrinos per cubic centimeter—that even if the mass of neutrinos is very tiny (millions of times less than that of electrons) they should account for a sizeable portion of dark matter, perhaps up to 20%. Because neutrinos move at velocities close to that of light, they would make up the "hot" dark matter in the universe. Because of their speed they would "erase" smaller structures, such as galaxies. The existence of these galaxies is viewed as an argument that "cold" dark matter, in the form of normal matter or the slow-moving massive WIMPs, must coexist with hot dark matter.

—Alexander Hellemans

---

trons produced by neutrinos interacting with nuclei that travel through the water at velocities higher than the velocity of light in water.

The photodetectors also allow the tracking of the direction of the incoming neutrinos. In 1998 Super-Kamiokande results showed that about half of the muon neutrinos produced in the earth's atmosphere that had traveled through the earth over a distance of 13,000 km (c. 8,078 mi) had disappeared, as compared to those produced in the atmosphere just overhead of the detector.

Accelerator-Based Experiments One of the weaker points in the neutrino experiments with solar or atmospheric neutrinos is the impossibility of controlling the neutrino source: there was always the possibility that theory simply predicted wrong neutrino fluxes. Therefore researchers tried to use neutrinos created in nuclear reactors or particle accelerators. At CERN, the European Laboratory for Nuclear Physics, a proton beam from the Super Proton Synchrotron (SPS) was aimed at a beryllium target for the production of muon neutrinos. These muon neutrinos were aimed at two detectors, NOMAD and CHORUS, placed at a 900-m (c. 984 yd) distance from the neutrino source that tried to pick up tau neutrinos that would have been formed by the muon to tau oscillation. This type of search is called an "appearance" search, and an appearance of a different type of neutrino than those that created in the accelerator would be a much stronger proof that neutrinos oscillate than a "disappearance" experiment. But data taken with both detectors have not revealed a measurable neutrino oscillation. Several similar experiments, using nuclear reactors for neutrino sources, such as CHOOZ in the French Ardennes, also could not confirm neutrino oscillations.

In 1996 researchers using the Liquid Scintillating Neutrino Detector (LSND) at Los Alamos announced the detection of electron antineutrinos produced by oscillating muon antineutrinos. The muon antineutrinos were produced by a proton beam hitting a target in an accelerator. However, some researchers doubt the results because the obtained oscilla-

KEY TERMS

ATOM: Smallest unit of matter that can take part in a chemical reaction and cannot be broken down chemically into anything simpler.

BETA DECAY: Transformation of a radioactive nucleus whereby its number of protons is increased by one through the conversion of an neutron into a proton by the emission of an electron (and an antineutrino).

CHERENKOV RADIATION: Named for Soviet physicist Pavel Cherenkov, who discovered this effect in 1934. It occurs as a bluish light when charged atomic particles pass through water or other media at a speed greater than the speed of light.

COSMIC RAYS: Highly energetic particles, mainly electrons and protons, that reach the earth from all directions.

DARK MATTER: Any matter in the universe that gives off no light of its own and does not interact with light the way typical matter does.

EINSTEIN'S MASS-ENERGY EQUIVALENCY PRINCIPLE: Principle stating that the total energy of a body is equal to its rest mass times the square of the speed of light.

ELECTRON: Stable, negatively charged elementary particle that is a constituent of all atoms and a member of the class of particles called leptons.

FLAVOR: Property that distinguishes different types of particles. Three flavors of neutrinos exist: tau, muon, and electron neutrinos.

LAW OF CONSERVATION OF ENERGY: Law that says energy can be converted from one form to another, but the total quantity of energy stays the same.

MUON: Fundamental charged particle (lepton) comparable to the electron in that it is not constituted of smaller particles, but with a mass approximately 200 times that of the electron.

NEUTRON: One of the three main subatomic particles. A composite particle made up of three quarks. Neutrons have about the same mass as protons, but no electric charge, and occur in the nuclei of all atoms except hydrogen.

PROTON: Positively charged elementary particle and a constituent of the nucleus of all atoms. It belongs to the baryon group of hadrons. Its mass is almost 1,836 times that of an electron.

RADIOACTIVE DECAY: Process of continuous disintegration by nuclei of radioactive elements like radium and isotopes of uranium. This changes an element's atomic number, turning one element into another, and is accompanied by emission of radiation.

STANDARD MODEL OF PARTICLE PHYSICS (not to be confused with the "Standard Model" of the Sun): General theory unifying the electric, weak, and strong force that predicts the fundamental particles. All predicted have been observed except for the Higgs particle.

SUPERNOVA: The explosive death of a star, which temporarily becomes as bright as 100 million or more suns for a few days or weeks. The name "supernova" was coined by U.S. astronomers Fritz Zwicky and Walter Baade in 1934.

SYNCHROTRON: Circular particle accelerator in which the applied magnetic field increases to keep the orbit radius of the particles constant when their speed increases.

TAU LEPTON: Fundamental charged particle comparable to the electron with a mass about 3500 times that of the electron.

---

tion rates differ from oscillation rates measured with other experiments. Others believe that a fourth kind of neutrino, a so-called sterile neutrino that only interacts very little with matter, may explain the discrepancy in results.

Long Baseline Experiments A new generation of long baseline experiments will allow physicists to pin down neutrino oscillations much more accurately. Because the neutrinos travel over much longer distances, it is more likely they would undergo oscillations. A first successful experiment, dubbed K2K, was announced in Japan in July 2000. Neutrinos produced in an accelerator at the Japanese National Accelerator Facility (KEK) in Tsukuba were aimed at the Super-Kamiokande. The neutrinos traveled over a distance of 250 km (155 mi). During an experimental run that lasted from June 1999 to June 2000, the Super-Kamiokande only detected 27 muon neutrinos from the 40 muon neutrinos it would have detected if neutrinos would not oscillate. In July 2001 K2K researchers announced that a total of

44 muon neutrinos have been detected from 64 that should have arrived if neutrinos did not oscillate.

The U.S. Department of Energy's Fermilab near Chicago is now planning to set up a long baseline terrestrial experiment called MINOS (Main Injector Neutrino Oscillation). Muon neutrinos, produced by a proton accelerator at Fermilab, will be aimed at a detector placed 800 m (c. 875 yd) underground in the Soudan iron mine in Minnesota at a distance of 730 km (438 mi) from Fermilab. Neutrinos will be detected by electronic particle detectors placed in between stacks of steel plates. The energy range of the neutrinos will be similar to the atmospheric neutrinos detected by Super-Kamiokande, and therefore MINOS would be an independent check on the Super-Kamiokande results.

In Europe, a long baseline experiment is planned for 2005, when CERN will aim pulses of neutrinos at the Gran Sasso National Laboratory, the world's largest underground laboratory excavated deep under a mountain in the Appenines in central Italy. At CERN, a proton beam from the SPS (Super Proton Synchrotron) will hit a graphite target and produce muon neutrinos. While traveling over a distance of 730 km (438 mi) to Gran Sasso, some of the muon neutrinos will change into tau neutrinos. Two detectors, OPERA and ICANOE, will look for these tau neutrinos. Unlike the other long baseline experiments, which look for the disappearance of certain types of neutrinos, the CERN-Gran Sasso experiment will look for the appearance of neutrinos that have undergone oscillations, and therefore the results will be much more reliable.

Longer baselines will also allow a higher accuracy, and scientists are already studying the possibilities of building "neutrino factories," muon storage rings that would produce very intense beams of neutrinos that could be aimed at detectors several thousands of kilometers away. —ALEXANDER HELLEMANS

Viewpoint:
No, the experiments of earlier twentieth-century scientists repeatedly indicated that neutrinos did not have mass.

From stone circles arranged by the ancient Celts to space-based telescopes, astronomers have always kept their eyes on the skies—unless they are looking for neutrinos, not galaxies or planets, but subatomic particles that have had scientists pounding the theoretical pavement since 1931.

It all started around 1900 when a series of experiments and some good luck showed physicists that the atom was not a featureless ball. Rather, it had an internal structure. The atomic age was born. It was radiation that got their attention—atoms of elements like uranium emitted some kind of ray.

Discovery of Neutrinos By the 1920s they knew the atom had a core, which they named a nucleus, with electrons moving around it. Inside the nucleus were protons and neutrons. Some of the nuclear radiation turned out to be beta particles (speeding electrons) that the nucleus emitted when heavier elements decayed into lighter ones. But atoms that emitted beta particles did so with less energy than expected. Somehow energy was being destroyed, and it was starting to look like physicists might have to abandon the law of conservation of energy.

Then, in 1931, physicist Wolfgang Pauli suggested that when an atom emits a beta particle it also emits another small particle—one without a charge and maybe without mass—that carries off the missing energy.

In 1932 physicist Enrico Fermi developed a comprehensive theory of radioactive decays that included Pauli's hypothetical particle. Fermi called it a *neutrino,* Italian for "little neutral one." With the neutrino, Fermi's theory explained a number of experimentally observed results. But it took another 25 years or so to prove neutrinos existed.

American physicists Frederick Reines and Clyde Cowan Jr. conducted an elaborate experiment in 1956 at the Savannah River nuclear reactor. They set up a detection system that focused on one reaction a neutrino might cause and detected the resulting gamma rays produced at just the right energies and time intervals. In 1959 they announced their results, and they later shared the 1995 Nobel Prize in physics for their contribution to the discovery. This neutrino was later determined to be an electron neutrino. Their findings confirmed Pauli's theory, but that was not the end of questions about neutrinos.

Further Experiments In 1961 physical chemist Ray Davis and theoretical physicist John Bahcall started wondering if there was a direct way to test the theory of how stars shine. They wanted to find a way to observe neutrinos that were supposed to be produced deep inside the Sun as hydrogen burned to helium. They knew it was a long shot, because anything that could escape from the center of the Sun would be very hard to detect with a reasonable-sized experiment on Earth.

The next year, in 1962, experiments at Brookhaven National Laboratory and CERN, the European Laboratory for Nuclear Physics, discovered that neutrinos produced in association with particles called muons did not behave like those produced in association with electrons. They had discovered a second neutrino flavor, the muon neutrino.

Two years later, in 1964, Bahcall and Davis proposed the feasibility of measuring neutrinos from the Sun, and the next year—in a gold mine in South Africa—Reines and colleagues observed the first natural neutrinos, and so did researchers Menon and colleagues in India.

In 1968 Ray Davis and colleagues started the first radiochemical solar neutrino experiment using 100,000 gallons of cleaning fluid—carbon tetrachloride—a mile underground in the Homestake gold mine in South Dakota. The chlorine in the cleaning fluid absorbed neutrinos and changed chlorine atoms into detectable radioactive argon. But the Homestake experiment captured two or three times fewer solar neutrino interactions than Bahcall calculated on the basis of standard particle and solar physics. Physicists called it the solar neutrino problem.

Several theorists suggested the missing electron neutrinos had oscillated—turned into another kind of neutrino that the Homestake experiment could not detect—but most physicists thought it was more likely that the solar model used to calculate the expected number of neutrinos was flawed. Besides, neutrino oscillation would be possible only if neutrinos had mass, and current theories said neutrinos had no mass.

Debate over Neutrino Mass But the discrepancy between standard calculation and experimental detection was telling physicists about something new in particle physics—that something unexpected happened to neutrinos after they were created deep inside the Sun.

In 1985 a Russian team reported measuring neutrino mass—10,000 times lighter than an electron's mass—but no one else managed to reproduce the measurement.

The Kamiokande detector went online in 1986 in the Kamioka Mozumi mine, 186 miles northwest of Tokyo in the Japanese Alps. For 1,200 years the mine had given up silver, then lead and zinc. Now, down among its 620 miles of tunnels, was a plastic tank, bigger than an Olympic swimming pool, full of ultraclean water and thousands of photomultiplier tubes that detected tiny flashes of light in the dark mine water—spontaneous proton decay. When a tube detected a flash it sent an electronic signal to a computer that recorded time and location.

There were no flashes at Kamiokande until February 23, 1987, at 7:35 A.M., Greenwich mean time. That morning, the Kamioka detector recorded 11 events, each made up of 30 photomultiplier flashes in a certain time sequence and pattern. Something had entered the detector and interacted. It all took 10 seconds.

When a massive star explodes as a supernova, the blast unleashes a thousand times more neutrinos than the Sun will produce in its 10-billion-year lifetime. More than 20 years earlier, theoretical astrophysicists said supernovas should release huge numbers of neutrinos. They had come from the core of the exploding star, escaping into space. Hours or days later, a shock wave from the main explosion would reach the surface, producing a blast of light as the star blew apart. And that is what happened at Kamiokande.

Several hours after the neutrinos hit the Kamioka mine, Supernova 1987A became the first exploding star in 384 years to be seen by the naked eye. Two years later, in 1989, Kamiokande became the second experiment to detect neutrinos from the Sun and confirmed the long-standing solar neutrino problem—finding about a third of the expected neutrinos.

Exploring Oscillation For years there were experimental hints for neutrino oscillation, mainly from the smaller than expected number of solar electron neutrinos. Other experiments hinted at oscillations by muon neutrinos produced in the upper atmosphere, in a decay chain that yielded two muon neutrinos for every electron neutrino.

But early experiments at Kamiokande and at the Irvine-Michigan-Brookhaven detector near Cleveland suggested the muon-to-electron-neutrino ratio was one, not two. If that was true, half the muon neutrinos were missing. Physicists needed proof to show the cause was neutrino oscillation. To answer that question, in 1996 another detector went online at the Kamioka mine. Super-Kamiokande was a $130 million neutrino detector built to find out whether neutrinos had mass.

This detector is a big tank of clean water 1 km (0.6 mi) underground—a 50,000-ton cylinder, 132 ft around and high, with 11,146 photomultiplier tubes lining the tank. The tubes are each sensitive to illumination by a single photon of light—a level about equal to the light visible on Earth from a candle at the distance of the Moon.

Any charged particle moving near the speed of light in water produces a blue Cerenkov light, sort of a cone-shaped optical shock wave. When an incoming neutrino collides with an electron, blue light hits the detector wall as a ring of light. A photomultiplier tube sees the light and amplifies it, measuring how much arrived and when, and the computer registers a neutrino hit.

NOMAD (Neutrino Oscillation Magnetic Detector)
(Courtesy of European Organization for Nuclear Research. Reproduced by permission.)

The tube array also samples the projection of the distinctive ring pattern to determine a particle's direction. Details of the ring pattern—especially whether it has a muon's sharp edges or an electron's fuzzy, blurred edges—is used to identify muon-neutrino and electron-neutrino interactions.

In 1997 Super-Kamiokande reported a deficit of cosmic-ray muon neutrinos and solar electron neutrinos at rates that agreed with measurements by earlier experiments. And in 1998, after analyzing 535 days of data, the Super-Kamiokande team reported finding oscillations—and so mass—in muon neutrinos.

Scientists Explore Neutrino Properties
This was strong evidence that electron neutrinos turned into muon and tau neutrinos as they streamed away from the Sun, but astrophysicists needed something more. The Sudbury Neutrino Observatory (SNO), 2 km (1.2 mi) underground in INCO's Creighton mine near Sudbury, Ontario, went online in November 1999 to determine whether solar neutrinos oscillate on their trip from the Sun's core to Earth and to answer other questions about neutrino properties and solar energy generation.

The SNO detector is the size of a 10-story building. Its 12-m-diameter (c. 13.1 yd) plastic tank contains 1,000 tons of ultrapure heavy water and is surrounded by ultrapure ordinary water in a giant 22-m-diameter by 34-m-high cavity. Outside the tank is a 17-m-diameter geodesic sphere that holds nearly 9,500 photomultiplier tubes that detect light flashes emitted as neutrinos stop or scatter in the heavy water.

The detector measures neutrinos from the Sun in two ways—one spots a neutrino as it bounces off an electron (any of the three neutrino flavors cause a recoil and are detected); the other detects an electron neutrino when it hits a neutron in SNO's 1000-ton sphere of heavy water. Only an electron neutrino can make the neutron emit an electron and trigger the detector. The two methods, along with results from Super-Kamiokande, were designed to show how many neutrinos come from the Sun and what proportion are muon or tau neutrinos.

On June 18, 2001, SNO's Canadian, American, and British scientific team announced they had spotted neutrinos that had been missing for 30 years. SNO confirmed what several experiments, especially Super-Kamiokande in Japan, had already shown—the missing electron neutrinos from the Sun had changed to muon and tau neutrinos and escaped detection.

The transformation also confirmed earlier observations that neutrinos had mass, and the SNO measurements agreed with first-principles calculations of the number of solar neutrinos created by the Sun. The solar neutrino problem was solved, according to a team member, with a 99% confidence level. The answer is oscillations.

The Standard Model But does the solution to the solar neutrino problem create more problems for the Standard Model of particle physics? Neutrinos are massless in the Standard Model, but the model could be extended to include massive neutrinos through the Higgs mechanism—a phenomenon that physicists believe gives other particles mass.

But most particle theorists do not want to extend the model. They prefer using another version of the Higgs mechanism called the see-saw mechanism. This includes neutrino interactions with a very massive hypothetical particle. For the range of parameters indicated by data from Super-Kamiokande, the heavy mass would be within a few orders of magnitude of the scale where physicists believe strong and electroweak forces unify.

Massive neutrinos also could contribute to dark matter. Dark matter, like missing mass, is a concept used to explain puzzling astronomical observations. In observing far-off galaxies, astrophysicists see more gravitational attraction between nearby galaxies and between inner and outer parts of individual galaxies than visible objects—like stars that make up the galaxies—should account for.

Because gravity comes from the attraction between masses, it seems like some unseen, or missing, mass is adding to the gravitational force. The mass emits no light, so it is also called dark matter.

Scientists have known for a long time that visible matter is only a small fraction of the mass of the universe; the rest is a kind of matter that does not radiate light. Neutrinos with the kind of mass difference measured by Super-Kamiokande could make up a big part of that dark matter if their mass is much larger than the tiny splitting between flavors. —CHERYL PELLERIN

Further Reading

Bahcall, John N. *Neutrino Astrophysics.* New York: Cambridge University Press, 1989.

Chang, Kenneth. "The Case of the Ghostly Neutrinos." *New York Times* (June 21, 2001).

Feynman, Richard. *Surely You're Joking, Mr. Feynman! Adventures of a Curious Character.* New York: Norton, 1985.

Lederman, Leon. *The GOD Particle: If the Universe Is the Answer, What Is the Question?* New York: Dell, 1993.

———, and David Schramm. *From Quarks to the Cosmos: Tools of Discovery.* New York: Scientific American Library, 1989.

"Mass for a Massless Particle? (Neutrinos)" *Sky & Telescope* (September 1, 1998).

Shklovskii, Iosif. *Stars: Their Birth, Life and Death.* New York: W. H. Freeman, 1978.

Turner, Michael S. "More Than Meets the Eye." *The Sciences* (November 2000).

Weinberg, Steven. *The Discovery of Subatomic Particles.* New York: Scientific American Library, 1989.

Winters, Jeffrey. "Much Ado about Neutrinos." *Discover* (January 1, 1999).

Zukav, Gary. *The Dancing Wu Li Masters: An Overview of the New Physics.* New York: William Morrow, 1979.

Do hidden variables exist for quantum systems?

Viewpoint: Yes, hidden variables are necessary to explain the contradictions and paradoxes inherent in quantum theory.

Viewpoint: No, experimental evidence, including the work of John Bell, Alain Aspect, and Nicolas Gisin, has continually shown that hidden variables do not exist.

Quantum physics is a daunting subject that often seems to be beyond comprehension. Nobel prize–winning quantum physicist Richard Feynman once said, "I think I can safely say that nobody understands quantum mechanics," and there would be few, if any, who would disagree with him.

Quantum theory contains many ideas that defy common sense. The popular concept of the atom is that of a tiny planetary system, with a nucleus "sun," and electron "planets" orbiting. However, quantum theory describes atoms and particles as having wavelike properties and avoids talking about specific positions and energies for particles, using instead ideas of probability. In quantum theory quantities such as energy can only exist in specific values, which contradicts the generally held notion that quantities have a continuous range and that any value in that range is possible.

Nobel prize–winning physicist Albert Einstein vehemently disliked many aspects of quantum physics, particularly the seemingly random and probabilistic nature of reality that the discipline implies, which he dismissed with his famous quote "God does not play dice." In 1932 Einstein, along with two colleagues, Boris Podolsky and Nathan Rosen, published a paper directly challenging some of the fundamental aspects of quantum theory. The EPR paper, as it came to be known, uses a thought experiment—an experiment that cannot be physically attempted, only imagined—to prove that quantum physics is an incomplete theory.

The three scientists argued that the missing bits that made quantum theory incomplete were "hidden variables" that would enable a more deterministic description of reality. Essentially, these scientists, and others, worried that quantum theory contains a number of scientific and philosophical problems and paradoxes. Examples include the infamous Schrödinger's Cat paradox, another thought experiment, in which quantum theory predicts that a cat exists in both dead and alive states until observed, or the two–slit experiment, which appears to break down the barriers of logic when single particles are used.

The Copenhagen interpretation of quantum physics, as formulated by Danish physicist Niels Bohr, German physicist Werner Heisenberg, and others, took the view that reality at the quantum level does not exist until it is measured. For example, a particle such as an electron orbiting an atomic nucleus could be in many different locations at a particular point in time. Quantum mechanics allows one to calculate the probabilities of each viable location of the electron as a wave function. However, the theory goes further, saying that until the electron is observed, it is in all possible positions, until the wave function that describes it is collapsed to a specific location by an observation. This creates some interesting philosophical problems and has been seen by some as implying that human beings create reality. Hidden variables, the EPR papers argue, would overcome these problems and allow for reality to be described with the same certainty that applies in Newtonian physics.

Hidden variables would also remove the need for "spooky" forces, as Einstein termed them—forces that act instantaneously at great distances, thereby breaking the most cherished rule of relativity theory, that nothing can travel faster than the speed of light. For example, quantum theory implies that measurement of one particle can instantaneously change another particle that may be light years away, if the particles are an entangled pair. Entangled particles are identical entities that share a common origin and have the same properties. Somehow, according to quantum theory, these particles remain in instantaneous contact with each other, no matter how far apart they separate. Hidden variables would allow two entangled particles to have specific values upon creation, thereby doing away with the need for them to be in communication with each other in some mysterious way.

The EPR paper caused many concerns for quantum theorists, but as the experiments it describes cannot be performed, the paper presented more of a philosophical problem than a scientific one. However, the later work of John Bell implied that there were specific tests that could be applied to determine whether the "spooky" forces were real or not, and therefore whether there are hidden variables after all.

In the 1980s the first such experiment was performed, and many more have been done since. The results imply that "spooky" actions–at–a–distance do indeed exist. Some scientists have challenged the validity of these experiments, and there is still some room for debate. These experiments only mean that "local" hidden variables do not exist, but would still allow "non–local" hidden variables. In this case, local effects are those that occur at or below the speed of light. You can think of the locality of an object as a sphere around it that expands at the speed of light. Outside of this sphere only non–local effects can take place, as nothing can travel faster than the speed of light. Non–local hidden variables, therefore, would have the same spookiness that the EPR paper was trying to avoid.

The debate over "hidden variables" is in some sense an argument over the completeness of quantum theory. Newton's laws once seemed to describe all motion, from particles to planets. However, the laws were found to be incomplete and were replaced by relativity, with regards to planets and other large–scale objects such as humans, and by quantum physics, with regards to particles and other very small–scale objects. It seems likely that one day relativity and quantum physics will also be replaced by other theories, if only because the two of them, while explaining their respective areas extremely well, are not compatible with one another.

In another sense, the hidden variables debate is a philosophical argument over whether the universe is deterministic and concrete, or merely probabilistic and somewhat spooky. Einstein and others have argued that reality must, on some deeper level, be fixed and solid. The majority of physicists, however, see no need for this desire for physical determinism, arguing that quantum mechanics can currently explain the world of the small–scale very well without the need to add in extras such as "hidden variables." —DAVID TULLOCH

Viewpoint:
Yes, hidden variables are necessary to explain the contradictions and paradoxes inherent in quantum theory.

The modern understanding of the nature and behavior of particles is most thoroughly explained by quantum theory. The description of particles as quantum mechanical waves replaces the age–old notion of particles as "balls" or "bullets" in motion. With important limitations or uncertainties, the quantum wave interpretation of nature, and the mathematical description of the wave attributes of particles, allow accurate predictions about the state (e.g., attributes such as velocity and position) and behavior of particles. Yet, Albert Einstein and others have asserted that the quantum mechanical system is an incomplete description of nature and that there must be undiscovered internal variables, to explain what Einstein termed "spooky" forces that, in contradiction to special relativity, seemingly act instantly over great distances. Without hidden variables, quantum theory also presents a paradox of prediction because the sought–after attributes of a particle can, in fact, determine the collapse of the quantum wave itself.

The Quantum Theory This quantum mechanical view of nature is counter–intuitive, and stretches the language used to describe theory itself. Essentially, reality, as it relates to the existence and attributes of particles, is, according to quantum theory, dependent upon whether an

KEY TERMS

DETERMINISM (DETERMINISTIC): Causes precede effects—and there is a clear chain of causality that can be used to explain events. In essence, if a set of conditions is completely known, then an accurate prediction can be made of future events. Correspondingly, behavior of particles could be explained if all of the variables or mechanisms (causes) influencing the behavior of a particle were completely known.

NON-LOCALITY: The phenomena wherein a force, act (observation), or event in one place simultaneously (instantly) influences an event or particle in another place, even if there is a vast (e.g., light years) distance between them.

PROBABILITY FUNCTION: In quantum theory, not all possible states of matter—attributes such as position, velocity, spin, etc.—have equal probabilities. Although states are undetermined until measured, some are more likely than others. Quantum theory allows predictions of states based upon the probabilities represented in the quantum wave function.

QUANTUM ENTANGLEMENT: The ability to link the states of two particles. Entanglement can be produced by random or probability-based processes (e.g., under special conditions two photons with correlated states can sometimes be produced by passing one photon through a crystal). Quantum entanglement is essentially a test for non-locality. Recently NIST researchers were able to entangle an ion's internal spin to its external motion and then entangle (correlate) those states to the motion and spin of another atom. The concept of quantum entanglement holds great promise for quantum computing.

QUANTUM WAVE: The properties of matter can be described in terms of waves and particles. De Broglie waves describe the wave properties of matter related to momentum. Waves can also be described as a function of probability density. The quantum wave represents all states and all potentialities. The differential equation for quantum waves is the Schrödinger equation (also termed the quantum wave function) that treats time, energy, and position.

SUPERLUMINAL: Faster-than-light transmission or motion.

SUPERPOSITION: A concept related to quantum entanglement. For example, if one of two particles with opposite spins, that in a combined state would have zero spin, is measured and determined to be spinning in a particular direction, the spin of the other particle must be equal in magnitude but in the opposite direction. Superposition allows a particle to exist in all possible states of spin simultaneously, and the spin of a particle is not determined until measured.

---

event is observed. Unlike other measurable waveforms, however, the quantum wave is not easily measured as a discrete entity. The very act of measurement disturbs quantum systems. The attempted observation or measurement of a quantum wave changes the wave in often mathematically indeterminate and, therefore, unpredictable ways. In fact, the act of measurement leads to the collapse of the quantum wave into traditional observations of velocity and position.

Despite this counter-intuitive nature of quantum mechanics, it is undoubtedly successful in accurately predicting the behavior of particles. Well-tested, highly verified quantum concepts serve as a cornerstone of modern physics. Although highly successful at predicting the observed properties of atomic line spectra and the results of various interference experiments, there remains, however, problems with simply adopting the irreducibility of quantum mechanisms and the philosophical acceptance of an inherently statistical interpretation of nature that must exist if there are no hidden variables in quantum systems.

The EPR Argument Einstein, Boris Podolsky, and Nathan Rosen treated the problems of quantum mechanics with great detail in their 1935 classic paper titled "Can Quantum-Mechanical Description of Physical Reality Be Considered Complete?" Eventually their arguments became known as the Einstein–Podolsky–Rosen (EPR) paradox. At the heart of the argument advanced by the three was an attempt to set forth a definition of reality. The EPR definition of reality is well grounded in both classical and relativistic physics (descriptions reconciled with relativity theory) and asserts that physical reality exists (as measured by physical quantities such as velocity and position) when, without disturbing a system, there is a certainty in the ability to predict the value of the physical quantity in question. Although this definition of reality is intuitive (makes sense with our common understandings based upon experience), it then required Einstein, Podolsky, and Rosen to set forth a method by which one could observe a system without disturbing that system.

The EPR paper created a thought experiment to meet that challenge. In the EPR example, two bound particles were at rest relative to the observer in a closed system. If the particles were then to suddenly separate and begin moving in opposite directions, the total momentum of the closed system must, in accordance with the law of conservation of momentum, be conserved. Given that the two particles in their bound–together state were at rest relative to the observer, the initial momentum of the system was zero. Accordingly, as the particles move in different directions, their momentum must be equal and opposite so that the sum of the particle momenta always remains zero. Because the particles move in opposite directions, it is possible that they carry the same magnitude of momentum cut with differing signs (positive or negative) related to the coordinate systems in use to describe the particle motion (i.e., one particle moves in the positive direction as the other particle moves in the negative direction). If the sum of the two particles' momentum were to exceed zero, this condition would violate the law of conservation of momentum.

Because the sum of the momenta of the particles must equal zero, Einstein, Podolsky, and Rosen argued that by measuring the momentum of one particle, the momentum of the other particle can be determined with absolute certainty. If the velocity of one particle is known, the velocity of the other particle can be exactly determined without uncertainty. Correspondingly, if the position of one particle is known, the other can also be exactly determined. Given observation of one particle, no interaction on the part of the observer with the second particle is required to know with certainty the state or relevant physical quantity of the particle. In essence, in opposition to fundamental quantum theory assertions, no observation is necessary to determine the state of the particle (e.g., either the particle's velocity or position).

In accord with the uncertainty principle, it remains impossible to simultaneously determine both the velocity and the position of the second particle because the measurement of the first particle's velocity would alter that particle's velocity, and then subject it to different conditions than the second particle—essentially rupturing the special bound relationship of the two particles in which the sum of their respective momenta must remain zero.

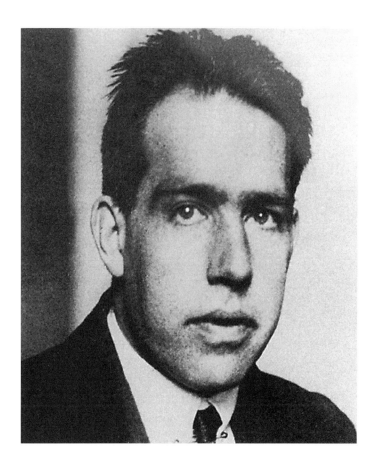

Niels Bohr
(Bohr, Niels, photograph. The Library of Congress.)

In the EPR experiment, the fact that the velocity and position of the second particle can be specified imparts a physical reality to the second particle. More importantly, that reality (the physical states of the second particle) is determined apart from any influence of the observer. These factors directly challenge and stand in contrast to the inability of quantum theory to provide descriptions of the state of the second particle. Quantum theory can only describe these attributes in terms of the quantum wave. The EPR paradox then directly challenges this inability of quantum theory by asserting that some unknown variables must exist, unaccounted for by quantum theory, that allow for the determination of the second particle's state.

Some quantum theorists respond with the tautology (circular reasoning) that the observation of the first particle somehow determines the state of the second particle—without accounting for a lack of observer interaction or other mechanism of determination. Hidden variable proponents, however, counter that that argument only strengthens the assertion that hidden variables, unaccounted for by quantum theory, must operate to determine the state of the second particle. An attack on the EPR premise and definition that physical reality exists when physical states are independent of observation is an inadequate response to the EPR paradox because it simply leaves open the definition of reality without providing a testable alternative.

More importantly, if, as quantum theory dictates, the observation of the first particle serves to determine the state of the second particle, there is no accounting for the distance between the particles and the fact that the determination of state in the second particle must be instantaneous with any change in the state of the first particle. Given the speed of light limitations of relativity theory, any transmission over any distance that is simultaneous (i.e., requires zero time of transmission) violates relativity theory.

Hidden variable interpretations of quantum theory accept the validity and utility of quantum predictions while maintaining that the theory remains an incomplete description of nature. In accord with deterministic physical theory, these hidden variables lie inside the so–called black box of quantum theory and are determinant to the states currently described only in terms of statistical probability or the quantum wave. Moreover, the sum influence of these quantum hidden variables becomes the quantum wave.

The Limitations of Quantum Theory
Although quantum theory is mathematically complete, the assertion that no hidden variables exist leaves an inherently non–deterministic, probability–based explanation of the physical world. Hidden variable proponents, while granting that quantum mechanics represents the best and most useful model for predicting the behavior of particles, assert that only the discovery and explanation of the hidden variables in quantum theory will allow a complete and deterministic (where known causes lead to known effects) account of particle behavior that will remove the statistical uncertainties that lie at the heart of modern quantum theory.

The reliance on an indeterminate probability–based foundation for quantum theory rests heavily on the work of physicist and mathematician John von Neumann's 1932 elegant mathematical proof that deterministic mechanisms are not compatible with quantum theory. Other physicists and mathematicians, however, were able to construct and assert models based upon deterministic hidden variables that also completely explained the empirical results. David Bohm in the 1950s argued that von Neumann's assumptions, upon which his proof rested, may not have been entirely correct and that hidden variables could exist—but only under certain conditions not empirically demonstrated. Although subsequently John Bell's theorem and experiments testing Bell's inequalities are often touted as definitive proof that hidden variables do not exist, Bell's inequalities test only for local hidden variables and are, therefore, more properly only a test of locality.

Bell went on to revisit the EPR paradox and compare it to several popular hidden variable models. Bell's work demonstrated that for certain experiments, classical (i.e., deterministic) hidden variable theories predicted different results than those predicted by standard quantum theory. Although Bell's results were heralded as decisive in favor of quantum theory, without the need for hidden variables, they did not completely explain quantum entanglements, nor did they rule out non–local hidden variables. As a result, Bell's findings properly assert only that if hidden variables exist, they must be non–local (i.e., an effect in one reference frame that has the ability to simultaneously influence an event in another reference frame, even of the two reference frames are light years apart).

The acceptance of the argument that there are no hidden variables also entails the acceptance of quantum entanglement and superposition wherein particles may exist in a number of different states at the same time. These "Schrödinger's cat" arguments (e.g., that under a given set of circumstances a cat could be both dead and alive) when applied to particle behavior mean that particles can, for example with regard to radioactive decay, exist simultaneously in a decayed and non–decayed state. Moreover, the particle can also exist in innumerable superpositioned states where it exists in all possible states or decay. Although investigation of quantum entanglement holds great promise for communication systems and advances in thermodynamics, the exact extent to which these entangled states can be used or manipulated remains unknown. Although the EPR paradox powerfully argues that quantum entanglement means that quantum theory is incomplete and that hidden variables must exist, the fact that these variables must violate special relativity assertions is an admittedly powerful reason for modern physicists to assert that hidden variables do not exist.

Despite the weight of empirical evidence against the existence of hidden variables, it is philosophically important to remember that both relativity theory and quantum theory must be fully correct to assert that there are no undiscovered hidden variables. Without hidden variables, quantum theory remains a statistical, probability–based description of particle theory without the completeness of classical deterministic physics. —BRENDA WILMOTH LERNER

Viewpoint:

No, experimental evidence, including the work of John Bell, Alain Aspect, and Nicolas Gisin,

has continually shown that hidden variables do not exist.

Although the standard model of quantum physics offers a theoretically and mathematically sound model of particle behavior consistent with experiment, the possible existence of hidden variables in quantum theory remained a subject of serious scientific debate during the twentieth century.

Based upon our everyday experience, well explained by the deterministic concepts of classical physics, it is intuitive that there be hidden variables to determine quantum states. Nature is not, however, obliged to act in accord with what is convenient or easy to understand. Although the existence and understanding of heretofore hidden variables might seemingly explain Albert Einstein's "spooky" forces, the existence of such variables would simply provide the need to determine whether they, too, included their own hidden variables. Quantum theory breaks this never-ending chain of causality by asserting (with substantial empirical evidence) that there are no hidden variables. Moreover, quantum theory replaces the need for a deterministic evaluation of natural phenomena with an understanding of particles and particle behavior based upon statistical probabilities. Although some philosophers and philosophically minded physicists would like to keep the hidden variable argument alive, the experimental evidence is persuasive, compelling, and conclusive that such hidden variables do not exist.

The EPR Paradox The classic 1935 paper written by Einstein, Boris Podolsky, and Nathan Rosen (EPR) and titled "Can Quantum-Mechanical Description of Physical Reality Be Considered Complete?" presented a *Gedankenexperiment* (German for "thought experiment") that seemingly mandates hidden variables. What eventually became known as the EPR paradox struck at the ability of particles to remain correlated in entangled states even though those particles might be separated by a great distance. Quantum entanglement is a concept of quantum theory that relies on the superposition of possible states for particles. In a two-particle entangled system, the act of measuring one of the entangled particles causes that particle's quantum wave to collapse to a definite state (e.g., a defined velocity or position). Simultaneous with the collapse of the first particle's wave state, the quantum wave of the second particle also collapses to a definite state. Such correlations must be instantaneous, and EPR argued that if there were any distance between the particles, any force acting between the particles would have to exceed the speed of light. Einstein called these forces "spooky actions at a distance."

EPR specifically identified three main problems with the standard interpretations of quantum mechanics that did not allow for the existence of hidden variables. Because of the limitations of special relativity, EPR argued that there could be no transacting force that instantaneously determines the state of the second particle in a two-particle system where the particles were separated and moving in opposite directions. EPR also challenged the uncertainty limitations found in quantum systems wherein the measurement of one state (e.g., velocity) makes impossible the exact determination of a second state (e.g., position). Most importantly, the EPR paper challenged the quantum view of nature as, at the quantum level, a universe explained only by probability rather than classical deterministic predictability where known causes produce known results. Einstein in particular objected to the inherent fundamental randomness of quantum theory (explaining his often quoted "God does not play dice!" challenge to Niels Bohr and other quantum theorists) and argued that for all its empirical usefulness in predicting line spectra and other physical phenomena, quantum theory was incomplete and that the discovery of hidden variables would eventually force modifications to the theory that would bring it into accord with relativity theory (especially concerning the absolute limitation of the speed of light).

Quantum theory, highly dependent on mathematical descriptions, depicts the wave nature of matter with a wave function (quantum waves). The wave function is used to calculate

Albert Einstein
(Einstein, Albert, photograph. The Library of Congress.)

probabilities associated with finding a particle in a given state (e.g., position or velocity). When an observer interacts with a particle by attempting to measure a particular state, the wave function collapses and the particle takes on a determinable state that can be measured with a high degree of accuracy. If a fundamental particle such as an electron is depicted as a quantum wave, then it has a certain probability of being at any two points at the same time. If, however, an observer attempts to determine the location of the particle and determines it to be at a certain point, then the wave function has collapsed in that the probability of finding the electron at any other location is, in this measured state, zero.

The EPR paradox seemingly demands that for the wave function to collapse at the second point, some signal must be, in violation of special relativity, instantaneously transmitted from the point of measurement (i.e., the point of interaction between the observer and the particle) to any other point, no matter how far away that point may be, so that at that point the wave function collapses to zero.

David Bohm's subsequent support of EPR through a reconciliation of quantum theory with relativity theory was based upon the existence of local hidden variables. Bohm's hypothesis, however, suffering from a lack of empirical validation, smoldered on the back burners of theoretical physics until John Bell's inequalities provided a mechanism to empirically test the hidden variable hypothesis versus the standard interpretation of quantum mechanics.

John Bell's Inequalities Bell's theorem (a set of inequalities) and work dispelled the idea that there are undiscovered hidden variables in quantum theory that determine particle states. Bell's inequalities, verified by subsequent studies of photon behavior, predicted testable differences between entangled photon pairs that were in superposition and entangled photons whose subsequent states were determined by local hidden variables.

Most importantly, Bell provided a very specific mechanism, based upon the polarization of photons, to test Bohm's local hidden variable hypothesis. Polarized photons are created by passing photons through optical filters or prisms that allow the transmission of light polarized in one direction (a particular orientation of the planes of the perpendicular electromagnetic wave) while blocking differently oriented photons. Most useful to tests of the EPR assertions are polarized photons produced by atomic cascades. Such photons are produced as electron decay from higher energy orbitals toward their ground state via a series of quantum jumps from one allowable orbital level to another. The law of the conservation of energy dictates that as electrons instantaneously transition from one orbital level to another they must give off a photon of light with exactly the same amount of energy as the difference between the two orbitals. An electron moving toward the ground state that makes that transition through two discreet orbital jumps (e.g., from the fourth orbital to the third and then from the third to the first) must produce two photons with energy (frequency and wavelength differences) directly related to the differences in total energy of the various oribitals. Of particular interest to EPR studies, however, is the fact that in cascades where there is no net rotational motion, the photons produced are quantum–entangled photons with regard to the fact that they must have specifically correlated polarizations. If the polarization of one photon can be determined, the other can be exactly known without any need for measurement.

Although the details of the measurement process, based upon the angles of various filters and measurement of arrival times of polarized photon pairs taking different paths, are beyond the scope of this article, the most critical aspect is that the outcomes predicted by standard quantum theory are different than the outcomes predicted if hidden variables exist. This difference in predicted outcomes makes it possible to test Bell's inequalities, and, in fact, a number of experiments have been performed to exactly test for these differences. In every experiment to date, the results are consistent with the predictions made by the standard interpretation of quantum mechanics and inconsistent for the existence of any local hidden variables as proposed by Bohm.

In 1982, the French physicist Alain Aspect, along with others, performed a series of experiments that demonstrated that between photons separated by short distances there was "action at a distance."

In 1997, Nicolas Gisin and colleagues at the University of Geneva extended the distances between entangled photons to a few kilometers. Measurements of particle states at the two laboratory sites showed that the photons adopted the correct state faster than light could have possibly traveled between the two laboratories.

Any Empirical Evidence? In modern physics, Einstein's "spooky" actions underpin the concept of non–locality. Local, in this context, means forces that operate within the photons. Although Bell's inequality does not rule out the existence of non–local hidden variables that could act instantaneously over even great distances, such non–local hidden variables or forces would have a seemingly impossible theoretical and empirical barrier to surmount. If such non–local hidden variables exist, they must act or move faster than the speed of light and this, of course, would violate one of the fundamental assertions of special relativity. Just as quantum theory is well supported by empirical evidence, so too is relativity theory. Accordingly, for hidden variables to exist, both quantum and relativity theories would need to be rewritten. Granting that quantum and relativity theories are incompatible and that both may become components of a unified theory at some future date, this is certainly not tantamount to evidence for hidden variables.

The only hope for hidden variable proponents is if the hidden variables can act non–locally, or if particles have a way to predict their future state and make the needed transformations as appropriate. Such transactional interpretations of quantum theory use a reverse–causality argument to allow the existence of hidden variables that does not violate Bell's inequality. Other "many worlds" interpretations transform the act of measurement into the selection of a physical reality among a myriad of possibilities. Not only is there no empirical evidence to support this hypothesis, but also it severely strains Ockham's razor (the idea that given equal alternative explanations, the simpler is usually correct).

In common, hidden variable proponents essentially argue that particles are of unknown rather than undefined state when in apparent superposition. Although the hidden variable, transactional, or "many worlds" interpretations of quantum theory would make the quantum world more understandable in terms of conventional experience and philosophical understanding, there is simply no experimental evidence that such interpretations of quantum theory have any basis or validity. The mere possibility that any argument may be true does not in any way provide evidence that a particular argument is true.

In contrast to the apparent EPR paradox, it is a mistake to assume that quantum theory demands or postulates faster–than–light forces or signals (superluminal signals). Both quantum theory and relativity theory preclude the possibility of superluminal transmission, and, to this extent, quantum theory is normalized with relativity theory. For example, the instantaneous transformation of electrons from one allowed orbital (energy state) to another are most properly understood in terms of wave collapse rather than through some faster–than–light travel. The proper mathematical interpretation of the wave collapse completely explains quantum leaps, without any need for faster–than–light forces or signal transmission. Instead of a physical form or independent reality, the waveform is best understood as the state of an observer's knowledge about the state of a particle or system.

Most importantly, although current quantum theory does not completely rule out the existence of hidden variables under every set of conceivable circumstances, the mere possibility that hidden variables might exist under such special circumstances is in no way proof that hidden variables do exist. There is simply no empirical evidence that such hidden variables exist.

More importantly, quantum theory makes no claim to impart any form of knowing or consciousness on the behavior of particles. Although it is trendy to borrow selected concepts from quantum theory to prop up many New Age interpretations of nature, quantum theory does not provide for any mystical mechanisms. The fact that quantum theory makes accurate depictions and predictions of particle behavior does not mean that the mathematical constructs of quantum theory depict the actual physical reality of the quantum wave. Simply put, there is no demand that the universe present us with easy–to–understand mechanisms of action. —K. LEE LERNER

Further Reading

Bell, J. "On the Einstein Podolsky Rosen Paradox." *Physics* 1, no. 3 (1964): 195–200.

Bohr, N. "Quantum Mechanics and Physical Reality." *Nature* 136 (1935): 1024–26.

Cushing, J. T., and E. McMullin. *Philosophical Consequences of Quantum Theory.* Notre Dame: University of Notre Dame Press, 1989.

Einstein, E., B. Podolski, and N. Rosen. "Can Quantum Mechanical Description of Physical Reality Be Considered Complete?" *Physical Review* 47 (1935): 776–80.

Heisenberg, W. *The Physical Principles of the Quantum Theory.* Trans. C. Eckart and F. C. Hoyt. New York: Dover, 1930.

Popper, K. *Quantum Theory and the Schism in Physics.* London: Hutchinson, 1982.

Schrödinger, E. "Discussion of Probability Relations Between Separated Systems" *Proceedings of the Cambridge Philosophical Society* 31 (1935a): 555–62.

Von Neumann, J. *Mathematical Foundations of Quantum Mechanics.* Trans. R. Beyer. Princeton: Princeton University Press, 1955.

Wheeler, J. A. and W. H. Zurek, eds. *Quantum Theory and Measurement.* Princeton: Princeton University Press, 1983.

Does cold fusion exist?

Viewpoint: Yes, the experiments of Martin Fleischmann and Stanley Pons offered legitimate proof that cold fusion exists, and subsequent efforts by other scientists have supported their claim.

Viewpoint: No, Fleischmann and Pons did not utilize proper control experiments, failed to address errors, and incorrectly used significant figures. Properly conducted experiments by other scientists have not supported their claims.

Nuclear fusion is the coming together of two smaller atomic nuclei to form a larger one, with the release of energy. Nuclei, which consist of protons and neutrons, are positively charged. They tend to repel one other, and it is difficult to get a fusion reaction started.

If the nuclei can be forced very close together, the attractive nuclear force, which is effective only at close range, overcomes the repulsive electrostatic force. This is what happens in the hot core of a star, where the force of gravity pulls the nuclei into a densely packed condition in which fusion can take place.

To an industrialized world with dwindling supplies of fossil fuels, being able to harness the energy that powers the stars is a very attractive prospect. However, here on Earth, forcing the nuclei close together over the opposition of the electrostatic force requires slamming them into each other at very high speeds. This is done using particle accelerators or complicated reactors in which extremely high temperatures and densities are achieved. Nuclear fusion is not a practical energy source given current technologies.

For decades, scientists have wondered whether it was possible for nuclear fusion to occur at or near room temperature. Such "cold fusion" research has concentrated on dissolving deuterium, a form of hydrogen with an extra neutron, in a solid such as palladium metal. The theory is that the structure of the solid would confine the deuterium nuclei, and the negatively charged electrons in the metal would help to counteract the electrostatic repulsion. The deuterium can in fact dissolve to high concentrations under such conditions, with the nuclei getting even closer together than they would in a pure solid form. However, most scientists point to theoretical calculations indicating that the nuclei would not be anywhere near close enough for detectable fusion to result.

In the late 1980s University of Utah electrochemists Martin Fleischmann and Stanley Pons were working on a cold fusion experiment involving an electrolytic cell, in which current was passed between a palladium cathode and a platinum anode, immersed in a solution containing deuterium. The reaction causes deuterium to enter the palladium cathode. Pons and Fleischmann built their cell into a calorimeter, or heat-measuring device. At some points they measured 10% more power than they were using to run the cell. This excess heat, they reasoned, indicated that fusion was taking place.

Pons and Fleischmann knew that Steven Jones, a researcher at nearby Brigham Young University, was also working on cold fusion. Concerned about getting their results out first, they took an unusual step that severely hampered their credibility in the physics community. Instead of subjecting their work to the peer review inherent in the normal scientific publication process, they held a press conference on March 23, 1989. The potential economic implications of cold fusion sparked a media circus, and serious scientific debate became difficult.

Many scientific teams attempted to reproduce the results. But none saw the gamma radiation expected to be produced in fusion reactions, and both the larger nuclei that were supposed to be produced and the excess heat were detected only sporadically. Most researchers concluded that the Fleischmann-Pons results were due to experimental error. The excess heat could have come from the experimental equipment. The helium nuclei that were occasionally detected, which could have been fusion products, could also have come from the helium naturally present in the air. And no theory adequately accounted for the absence of the gamma radiation.

On May 1, 1989, at a meeting of the American Physical Society, strong presentations refuting cold fusion were made by several respected physicists. As a result, papers on the subject are rarely accepted by peer-reviewed journals or presented at mainstream scientific conferences in the United States. Some research in the field is still ongoing, in the United States, Japan, Italy, and elsewhere. But the absence of peer review makes it difficult to distinguish the bearers of new ideas from ordinary "crackpots." All in all, it is clear that the story of cold fusion research did not unfold in a way that was well suited to separating new physics from experimental error. —SHERRI CHASIN CALVO

Viewpoint:
Yes, the experiments of Martin Fleischmann and Stanley Pons offered legitimate proof that cold fusion exists, and subsequent efforts by other scientists have supported their claim.

Cold fusion is a label loosely applied to nuclear reactions that occur in materials without the usual application of very high energy. Much attention has been applied to making these reactions occur in palladium that has been saturated with deuterium. However, a wide range of materials appears to have this ability, from superconducting oxides to gold and nickel. The extent of this phenomenon is just beginning to be understood.

The first systematic attempt to initiate such reactions was undertaken in 1927 by Fritz Paneth and Kurt Peters, who thought they had succeeded in making helium from hydrogen; instead they discovered how to flush helium out of quartz using hydrogen. They retracted their cold fusion claim, thereby setting the standard some people hoped would be applied when cold fusion was rediscovered six decades later.

The Fleischmann-Pons Research In 1949, Martin Fleischmann began to wonder whether deuterium in palladium might undergo a fusion reaction. He was unable to test this notion until he and Stanley Pons started on a self-financed project in 1983. They were blessed from the start because, if the hypothesis was correct and they were able to detect any extra energy, the neutron emission would have killed them, or so it was thought at the time.

Fleischmann and Pons, who were affiliated, respectively, with Great Britain's Southampton University and the University of Utah, considered using four methods of loading palladium with deuterium: electrodiffusion, electrochemical charging, highly reducing or super basic media, and super acid media. Because they are electrochemists, they finally settled on electrochemical charging, which they understood in detail, although it is an extremely difficult technique.

Since 1989 many other methods of loading palladium have been used successfully to produce the cold fusion effect, including gas loading, ion beam loading, electrodiffusion, transient cavitation bubbles, sparking, and high temperature electrochemical glow discharge. These other methods are considered easier than the original technique, and their success rate is often higher. In this regard, Fleischmann and Pons were not so lucky.

Several years into their project, they found what they were seeking. In a few test runs, the palladium deuteride produced measurable levels of anomalous excess heat. That is to say, the palladium cathode in the electrochemical cell grew hotter than it should have, given the amount of electricity being fed into electrolysis. It was producing a fraction of a watt more heat than the power being fed in. Chemical sources of heat were ruled out. The cell contains only water and a few grams of metal, which are chemically inert. (Water does not burn at room temperature.) Not only was there no chemical fuel in the cell, no chemical ash was found after the reaction, and the heat continued for days at a time, pro-

Martin Fleischmann
(Fleischmann, Martin, photograph by Barbara DelloRusso. Infinite Energy Magazine. Reproduced by permission.)

ducing in the aggregate far more energy than any chemical fuel. The small palladium wire was apparently putting out thousands of times more energy than any match, candle, rocket fuel, or battery of the same mass.

Only one source of this much concentrated heat energy is known to science: a nuclear reaction. However, it was obvious that this reaction was not like any known form of nuclear fusion or fission. A conventional reaction that produces a fraction of a watt of heat will also generate deadly neutron radiation so intense that it would kill an observer in a few minutes, yet the alarms showed nothing dangerous, and Fleischmann, Pons, and their colleagues safely observed the cell for hours. They had discovered a whole new field of safe nuclear interactions.

In 1989, they were still at work, planning to keep their results secret for at least five more years, while they investigated and confirmed their findings, but news of the research leaked out. A worldwide mass-media brouhaha broke out. Many peculiar and mistaken ideas about the research began circulating, and many circulate even today. Some scientists attacked the research because they thought the results were theoretically impossible. They made statements that violate scientific ethics and common sense. To this day, the spokesman for the American Physics Society (APS) brags that he has never read a single experimental paper on cold fusion but he knows it must be wrong.

Scientists Replicate the Research Science works its wonders by examining nature rather than by attacking other scientists. To learn the truth, a scientist skilled in the technique does the experiment again, independently, using the same materials and methods, to see whether the same results occur. Changes are then made and new patterns of behavior are sought. Gradually, a picture emerges and the phenomenon is understood. Everyone who examined the problem knew such an understanding would be difficult to achieve. Pons and Fleischmann were well-known experts, and they warned it had taken them years to master the technique. Distinguished electrochemist R. Oriani, who successfully replicated some of the results, says it was the most difficult experiment he encountered in his 50-year career.

Despite the difficulties, within a few years several hundred skilled scientists replicated and published more than 1,000 positive papers in peer-reviewed journals and conference proceedings. Many of these scientists worked in prestigious laboratories, including the Los Alamos and Oak Ridge National Laboratories, the Naval Research Laboratory, the Bhabha Atomic Research Centre in Bombay, and many others. A skilled researcher who pays close attention to the literature will find it easier to replicate cold fusion today than in 1989, although the task remains challenging.

The first challenge is to get the cell to produce energy, and to be sure that the energy is real. In early replications, the experiment was performed very much the way Pons and Fleischmann did it. That was not good enough. Many other variations were called for, using entirely different instruments and techniques. Other kinds of nuclear evidence had to be confirmed. Pons and Fleischmann had only $100,000, barely enough to do a rudimentary experiment with simple instruments, looking for a few signs of a nuclear reaction. If everyone had performed the same experiment using the same type calorimeter they used, researchers would not have learned much more than Pons and Fleischmann already knew. What is worse, it is conceivable that dozens of scientists worldwide might all make the same systematic error.

To ensure against this remote possibility, in the years following the announcement, many different instrument types were employed. Heat measurement, for example, was first performed using single-wall isoperibolic calorimeters. Later, many errors were removed by using the double-wall type. Mass flow calorimeters and a variety of electronic Seebeck calorimeters were also used. Sometimes these techniques were used simultaneously; heat from the same sample was measured with different methods. Although many experiments produce only marginal heat, or no heat, every type of instrument has seen

examples of strong positive results, or what is called a large signal-to-noise ratio.

The same principle applies to the measurement of nuclear effects. Autoradiographs with ordinary X-ray film often show that a cathode has become mildly radioactive during an experiment. But suppose the sealed film is somehow affected by the minute quantity of deuterium gas escaping from the cathode? Very well, devise a plastic holder that keeps the film a millimeter away from the metal surface, and put the cathode and the film together into a vacuum chamber, so the outgasing hydrogen never reaches the sealed X-ray film. When that works, try again with an electronic detector. The outgasing deuterium cannot have the same effect with all three methods, yet the shadow of radioactivity from the hot spots in the cathode surface shows up every time. Every conventional type of spectrometer has been harnessed to detect tritium and metal transmutations, time after time, in experiment after experiment. A researcher will examine a specimen with SIMS, EPMA and EDX spectroscopy, to be sure he sees the same isotope shifts with different instruments. He will probably examine the specimen before and after the experiment, and compare it to an unused sample.

This is the cheap and easy technique, but researchers at Mitsubishi were not satisfied with it. They developed highly sensitive on-line spectroscopy that takes data as the reaction is occurring. Such equipment costs tens of millions of dollars, but the results are highly reliable, and they tell us far more about the nature of the reaction than static results taken after the nuclear process has ceased. For the past six years Mitsubishi has been able to perform the experiments several times per year, with complete success each time. They observe excess heat, transmutations and gamma rays.

Dozens of parameters can be changed in a cold fusion experiment, but only one reliably predicts the outcome: the metallurgical and physical characteristics of the active material. For example, researchers at the National Cold Fusion Institute in Utah did twenty different types of experiment. In the final and best version, they achieved 100 percent reproducible tritium production, proof that an unusual nuclear reaction had occurred. Four out of four heavy water experiments produced significant tritium, while none of the light water controls did. The final report said, "tritium enhancements up to a factor of 52 were observed," meaning there was 52 times more than the minimum amount their instruments could detect. In these tests, 150 blank (unused) samples of palladium were tested extensively, by dissolving in acid and by other exhaustive methods. None of the blank samples had any measurable level of tritium.

KEY TERMS

CALORIMETER: An instrument that measures heat energy, of which there are five basic types. In one type, the sample is surrounded by flowing water. By combining the rise in water temperature with the flow rate and the heat capacity of water, the amount of heat being generated can be determined.

DEUTERIUM, DEUTERON, DEUTERIDE: *Deuterium* is heavy hydrogen, that is, a hydrogen atom with one additional neutron. The individual nucleus is called a *deuteron*. A *deuteride* is a metal combined with deuterium.

ELECTROCHEMISTRY: When electric current passes from one electrode to another through a liquid, electrolysis breaks apart the molecules of liquid into ions. The positively charged ions are attracted to the negative electrode (the cathode), and the negative ions to the positive electrode (the anode).

HEAVY WATER: Water composed of oxygen and deuterium, instead of hydrogen. Ordinary water contains one part in 6,000 heavy water. Heavy water is 10 percent heavier than ordinary water.

NEUTRON: A subatomic particle with no charge, and roughly the same mass as a proton. Atoms heavier than hydrogen are composed of electrically negative electrons, surrounding a nucleus composed of positive protons and neutrons with no charge.

TRITIUM: Super-heavy hydrogen, with two neutrons per atom. Hydrogen with one neutron is called deuterium. Tritium is radioactive, with a half-life of 12.3 years.

What Conclusions Can Be Drawn? Because myriad different instrument types have detected some cold fusion effects time after time, in experiment after experiment, in hundreds of different laboratories independently throughout the world, often at a high signal-to-noise level, we can now be sure the effect is real. To be sure, some of the nuclear effects have been seen only by a handful of laboratories, mainly because it costs millions of dollars to detect them, and only a few Japanese corporate researchers have this kind of funding. Nevertheless, most of the important nuclear reactions have been observed in several independent laboratories.

We could not have known that cold fusion is real until hundreds of man years of research were completed and reported in the peer-reviewed scientific journals. In 1989, no one could have predicted this is how things would turn out, yet the possibility was generally rejected. In experimental science, there can be no assurance at the outset, and no rush to judgment. Once a result has

Julian Schwinger
(Schwinger, Julian, photograph. Courtesy of Infinite Energy Archives. Reproduced by permission.)

been confirmed in hundreds of independent, quality replications, it has to be real. There is no other standard to judge reality. To reject cold fusion at this point would be tantamount to rejecting the experimental method itself.

Finally, it should be noted that some critics demand "an explanation" or a theory before they will believe that cold fusion phenomena exist. Never, in the history of science, has this been held as a standard for belief. If it had been, biologists would have denied that cells reproduce until 1953, when the nature of DNA was unraveled. Physicists would not believe that high temperature superconductors exist. Needless to say, there are an infinite number of other natural phenomena beyond our present understanding. Textbook theory can only be an imperfect guide, not the absolute standard.

A person might reasonably ask why the claims have been so hard to duplicate. The environment in which the nuclear reactions occur is rare and difficult to produce. If this were not the case, we would have seen this effect long ago. In 1989, most people who attempted to duplicate the claims unknowingly used defective palladium. Gradually researchers discovered how to identify useful palladium that does not crack when it loads, and how to treat the metal so that it would become active. As a result, the effect can now be duplicated at will. Such understanding of how the properties of a material affect a phenomenon is slow to develop because a large number of variables have an influence. Research in semiconductors and catalysis was hampered for many years by similar problems. Skepticism must be tempered by patience.

Overreaction by the Nay-sayers The reader may wonder why so few research papers on this subject are available in mainstream science journals in the United States, and why no news of it appears in the mass media. More than 500 positive replications of cold fusion were published in the five years after the 1989 announcement, mainly by senior scientists who had enough clout to perform controversial research. This led to a crackdown. In the United States, most of the scientists who published positive results were ordered to stop, demoted to menial jobs, forced into early retirement, or in some cases summarily fired. As a result, the field is largely ignored in the United States. Fortunately, it is tolerated in Japan and Europe. In Italy the government funds modest programs at the National Physics Laboratory.

Some skeptics point to the dwindling number of publications as proof that the subject is moribund, but cold fusion researchers believe that many more papers would be published if traditional academic freedom was respected. Nobel laureate and cold fusion theorist Julian Schwinger protested the crackdown by resigning from the APS. He wrote: "The pressure for conformity is enormous. I have experienced it in editors' rejection of submitted papers, based on venomous criticism of anonymous referees. The replacement of impartial reviewing by censorship will be the death of science." One might well ask what this attitude means to science in general in the United States. —JED ROTHWELL AND EDMUND STORMS

Viewpoint:

No, Fleischmann and Pons did not utilize proper control experiments, failed to address errors, and incorrectly used significant figures. Properly conducted experiments by other scientists have not supported their claims.

"Cold fusion" is a term that has been applied to a number of phenomena from the natural fusing of atoms that occurs randomly due to cosmic rays to the supposed deliberate fusing of heavy hydrogen as a fantastic source of

THE N-RAY AFFAIR

René Blondlot "discovered" n rays in 1903 and named them after the University of Nancy in France, where he did his research. He had been experimenting with x rays, found in 1895 by Wilhelm Conrad Roentgen. X rays had been big news and were considered one of the most important scientific discoveries of the time. Blondlot hoped his n rays would be just as significant.

Many strange claims were made for n rays. Supposedly, they could penetrate many inches of aluminum but were stopped by thin foils of iron; they could be conducted along wires, like electricity; and when n rays were directed at an object there was a slight increase of brightness. Blondlot admitted, however, that a great deal of skill was needed to see these effects. A large number of other, mainly French, physicists confirmed and extended his work, and many scientific papers were published on n rays.

The American physicist R. W. Wood had been unable to reproduce the n-ray work, and so he visited Blondlot's laboratory to observe the experiments firsthand. Blondlot demonstrated to Wood that he could measure the refractive indexes of n rays as they passed through a prism. Wood was startled by the degree of accuracy Blondlot claimed to observe, and he was told that n rays do not "follow the ordinary laws of science." Wood himself could see nothing and was very sceptical of Blondlot's claims. The story goes that Wood managed to quietly pocket the prism in the experiment without anyone noticing, and then he asked for the measurements to be repeated.

Blondlot obtained exactly the same results, down to a tenth of a millimeter, confirming Wood's suspicion that there was no visible effect at all, merely self-deception on Blondlot's part and all those who had confirmed his results.

Wood wrote an account criticizing n rays, published in *Nature,* and this effectively ended n-ray research outside of France. However, the French continued to support Blondlot's work for some years, and many other French physicists reconfirmed his results. National pride seems to have fueled a defiance of foreign scientific opinion. Eventually the nonexistence of n rays was accepted in France, and Blondlot's career was ruined, leading to madness and death.

—David Tulloch

power with almost no side effects. It is this last definition that is generally associated with the term in the public imagination. However, cold fusion as a source of power does not exist. It is a myth, and one so seductive that people around the world are still chasing it.

Early Research in Cold Fusion The first attempts to fuse hydrogen at "cold" temperatures (as opposed to "hot fusion," which is an attempt to mimic the processes that occur inside the sun) took place in the 1920s. Two German chemists, Fritz Paneth and Kurt Peters, were attempting to manufacture helium from hydrogen, to fill airships. They used the properties of the metal palladium, which can absorb a great deal of hydrogen and compress it to high pressures within the atomic lattice of the metal, in the hope that such pressure would force the hydrogen to fuse. Initially, their experiments seemed to produce helium, but they were fooled by the ability of glass to absorb helium from the air and release it when heated. Their basic premise was also flawed, because they were unaware that normal hydrogen atoms, which are merely a single proton with one orbiting electron, do not contain enough atomic material to form helium by fusion, as they lack neutrons. (Helium atoms consist of two protons, and one or more neutrons.)

Neutrons were not discovered until 1932. This achievement prompted a Swedish scientist, John Tanberg, to revisit the Paneth and Peters experiment. He had already attempted to duplicate it earlier with the added method of electrolysis, passing an electric current through water to break some of it into hydrogen and oxygen. Tanberg used palladium as the negatively charged electrode, so that hydrogen atoms lacking an electron (and therefore positively charged) collected inside it. Tanberg then modified the experiment to use heavy water. Heavy water is made from hydrogen and oxygen, just like normal water, except that the hydrogen used contains an extra atomic ingredient, a neutron. Now, Tanberg theorized, he would have enough atomic material inside the palladium for fusion. However, the experiments did not work, and Tanberg went on to other things.

Sensational News from Utah The modern flirtation with cold fusion began on March 23, 1989, when two chemists, Martin Fleischmann and Stanley Pons, announced in Utah they had created fusion at room temperature on a desktop, with nothing more sophisticated than palladium rods and beakers of heavy water. The timing of the announcement was fortuitous, as that night the *Exxon Valdez* ran aground off Alaska, creating a huge oil spill. Cleaner energy was a hot topic, and a media frenzy ensued.

Their news was broken in an unusual manner for a scientific discovery, by publication in the London *Financial Times* and the *Wall Street Journal* and in a hastily organized press conference. However, they did not publish any of their results in scientific journals, or give conference papers on their breakthrough either before or immediately after their announcement.

One possible reason for the hasty release of Fleischmann's and Pons's results is that other nearby scientists were also researching cold fusion, and about to publish. They may have felt they were in a race. However, the second group, at Brigham Young University, was looking at results barely above background radiation, and made no extravagant claims for their work.

The Fleischmann and Pons press conference sparked a wave of positive press coverage, and high-level funding negotiations. If the claim was true then cold fusion was worth billions. Big money was forthcoming for the big claims. However, what was missing was big evidence. Details of Fleischmann's and Pons's experiment were sketchy at best. Their press releases seemed more concerned with the economic potential of the discovery than the scientific specifics. The lack of information frustrated those in the wider scientific community, many of whom were rushing to reproduce the experiment, to confirm or refute the claims.

There had been a similar rush just two years previously when another unexpected scientific breakthrough, high-temperature superconductivity, had been announced. However, whereas the discoverers of that phenomenon, Georg Bednorz and Karl Mueller, had published their work in peer-reviewed journals, and shared information on all details of their experiment, Fleischmann and Pons did not. Within weeks of the announcement of high-temperature superconductivity, successful reproductions were reported everywhere the experiment was reproduced.

Other Scientists Unable to Confirm The lack of details from Fleischmann and Pons led to spy tactics being employed by physicists and chemists attempting to reproduce their work. Press photos and video coverage were closely examined for information. Groups of scientists began to recreate the work of Fleischmann and Pons as best they could.

Given the lack of details, the rush of initial confirmations that appeared was surprising. Scientists at Texas A&M and Georgia Tech announced positive results. From India and Russia came more confirmations. However, most groups that were to confirm the work would only confirm either the excess heat or the neutron emissions, but not both.

Then came the retractions. Experimental errors and inaccurate equipment had given rise to false positives in many cases. Texas A&M scientists had announced that not only had their reproduction with heavy water worked, but so had their control experiments using normal water and carbon rods. They eventually isolated the excess heat generated in their work to the experimental setup itself. After correcting their error they never saw excess heat from any further experiments. Georgia Tech scientists realized that they had not been measuring neutrons at all. Rather their detector was overly sensitive to the heat of their experiment. Other labs admitted they had had similar errors.

Soon after came a wave of negative announcements from many labs across the world. Even the Harwell lab in England, the only lab to be given the full details, from Fleischmann himself, and equipment from Utah including supposedly "working" cells, announced that it had failed to reproduce the experiments. It usually takes longer to refute a claim than to confirm it, and the slow process of careful scientific study does not always fit with popular attention spans and media coverage. Most of the negative results were not widely reported.

How the Utah Research Went Wrong Aside from the problems of reproduction, there were also growing concerns regarding the original experiments of Fleischmann and Pons. In essence their work was very similar to that of Tanberg more than fifty years earlier. Using heavy water and electrolysis, together with palladium rods, they hoped to force heavy hydrogen atoms close enough together to fuse. They ran several experiments using different-sized palladium rods, and noticed excess heat from at least one. They also appeared to get positive readings of neutron emissions from their experiment, which they interpreted as a sign that fusion was taking place.

The fusion of two deuterium atoms is one of the most studied of nuclear processes. Normally the nuclei of the atoms will repel each other, as they are both positively charged. However, if they can be forced close enough together they can fuse to form a new, heavier nucleus of helium. In the process excess energy is released,

and about half of the time a high-speed neutron. If fusion was really occurring in the Utah experiments then both helium and neutron emissions should have been obvious. There should also have been a large amount of gamma radiation; so much, in fact, that it would have been fatal. However, no precautions had been taken, and no health problems had arisen in the Utah lab. Pictures showing the two chemists holding one of their working cells in bare hands suggested to many scientists that something was wrong.

Fleischmann and Pons had not performed control experiments with normal water, which should have been standard practice. Indeed, this was to become a major sticking point for the cold fusion claims. Many later experimenters got the same readings for normal water, which would imply that what was occurring was not fusion, as there would not be enough nuclear material to form helium atoms, just as in the original Paneth and Peters experiments.

The little information that had been published contained many errors and omissions. There was a revelation that Fleischmann and Pons had doctored some of their results to make it seem that they were seeing gamma radiation, misrepresenting some of their data to make it seem significant. Further setbacks for the cold fusion claims mounted. Fleischmann and Pons admitted that their biggest claims for excess heat were from projected work that had never been completed.

Perhaps the most damning criticism was that of the missing neutrons. Fusion of the power they claimed should have produced a billion times more neutrons than detected. In effect even those they claimed to have detected were subject to doubt, as they were close to background levels, and could have been from experimental errors or background fluctuations. There was also a lack of other by-products from the supposed cold fusion. Fleischmann and Pons had claimed that inside their cells a new and unknown type of fusion was taking place that made much more helium than normal. However, their experiments showed much less helium than would have been expected even in conventional fusion, and that may have come from sources other than fusion.

An American government panel that investigated the claims of cold fusion pointed to a lack of control experiments, failure to address errors, and incorrect use of significant figures, among other problems. It was also noted by many that the original basis for the experiment was flawed. Heavy hydrogen molecules inside palladium are actually further apart from each other than they are in gaseous form, and so are less likely to fuse, not more so.

Cold Fusion Remains a Dream Despite all the refutations and flaws, Fleischmann and Pons continued to claim that they had produced fusion. The lack of reproducibility was attrib-

Stanley Pons (l) and Martin Fleischmann (r) testify at a congressional hearing about their cold fusion research (1989).
(Photograph by Cliff Owen. © Bettmann/Corbis. Reproduced by permission.)

uted to poor methods by other experimenters, caused by their having insufficient details or their poor understanding of the intricacies of the experiment (which were the fault of Fleischmann and Pons themselves). However, even those few who were privileged with all the technical details could not reproduce the results. Supporters of cold fusion have made various claims ever since, ranging from effects that are barely detectable, to claims of limitless energy production, and beyond. Some claim the process works with normal water, others that it can be used to transform any element into another. A number of cold fusion companies have been formed and have spent millions of dollars of investors' money in research and development. While amazing claims for the potential of cold fusion have been made, not one working device has ever been demonstrated or patented.

Supporters point to hundreds of reports of cold fusion, but forget the many retractions, the more widespread negative results, and the errors and misrepresentation in the original work by Fleischmann and Pons. Advocates are also quick to point out that "well-respected" laboratories have done cold fusion research. However, most quickly abandoned such work after obtaining only negative results. Those that persisted, including NASA and the Naval Research laboratories as well as at least two large Japanese corporations, have done so for reasons that have more to do with gambling than science. Cold fusion research is very cheap compared with many other "alternative fuels" experiments, but the potential payoffs would be astronomical, so even a minute chance is seen as worth pursuing by some. However, the vast majority of scientists place the probability at exactly zero, and call such research a waste of time and money.

Many casual followers of the cold fusion saga are struck by the claims of excess heat. Even if fusion is not occurring, surely the excess heat is still worth investigating? However, the experiments of Fleischmann and Pons were not closed systems, and the heat could easily have come from the atmosphere or electrical current in the apparatus. Their reports of the greatest amounts of excess heat were only in projected, not actual, experiments. And finally, if a nuclear process was not occurring then the claims of excess heat would violate one of the most ironclad principles of physics, the first law of thermodynamics and the conservation of energy. Fusion allows the transformation of mass into energy, a process most famously noted in Einstein's $E=mc^2$ equation. However, without such a transformation it is impossible to generate more heat from a system than is put into it (which is what makes perpetual motion impossible). In short, if fusion is not occurring in the so-called "cold fusion" experiments, then they cannot be producing excess heat, and there is no compelling evidence for fusion in any of the work done to date.

Cold fusion is impossible. It contradicts all that we know in nuclear physics, and no concrete evidence, theoretical or experimental, has ever been forthcoming. No single working device has ever been demonstrated, and there have been no explanations for the shortcomings in the work of Fleischmann and Pons or later followers. The only journals that publish papers on cold fusion today are pro–cold fusion journals, which seek to promote the field, rather than critically examine it. Cold fusion is a dream, a golden promise of a world with unlimited energy and no pollution, but a dream nonetheless. —DAVID TULLOCH

Further Reading

Beaudette, Charles G. *Excess Heat: Why Cold Fusion Research Prevailed*. South Bristol, ME: Oak Grove Press, 2000.

Close, Frank. *Too Hot to Handle: The Race for Cold Fusion*. London: W.H. Allen Publishing, 1990.

Huizenga, John R. *Cold Fusion: The Scientific Fiasco of the Century*. Oxford: Oxford University Press, 1993.

Mallove, Eugene F. *Fire from Ice: Searching for the Truth Behind the Cold Fusion Furor*. New York: John Wiley & Sons, 1991.

Park, Robert. *Voodoo Science: The Road from Foolishness to Fraud*. Oxford: Oxford University Press, 2000.

Peat, F. David. *Cold Fusion: The Making of a Scientific Controversy*. Chicago: Contemporary Books, 1989.

Scaramuzzi, F., ed. *Proceedings on the 8th International Conference on Cold Fusion, Lerici, Italy, May 21–26, 2000*. Bologna: Societa Italiana de Fisica, 2001.

Taubes, G. *Bad Science: The Short Life and Weird Times of Cold Fusion*. New York: Random House, 1993.

Historic Dispute:
Is Earth the center of the universe?

Viewpoint: Yes, early scientists believed that what appeared to be movement around Earth by the Sun and other entities was, in fact, just that.

Viewpoint: No, later scientists such as Nicolaus Copernicus and Galileo correctly realized that Earth moves around the Sun, not vice versa, and thus cannot be the center of the universe.

It is easy in our human nature to believe that we are the center of the universe. A newborn infant must learn through experimentation and sensation that he is a part of the world, not the entirety. Any parent can attest that as children mature, they must be helped to understand that the universe does not revolve around them and their needs and desires. Even as adults, we often struggle to see the world from a perspective other than our own.

Likewise, it was natural for ancient peoples to assume that the universe they observed was centered upon Earth. After all, they saw the Sun rise in the east every morning, and set in the west at night. They saw the stars and planets appearing to move across the sky. These patterns were repeated in cycles, and cycles implied revolution. There was no reason to question that what appeared to be movement around Earth was, in fact, just that. By the fourth century B.C., the Greeks had developed a picture of the stars as fixed on a celestial sphere that rotated around Earth, with the Sun, Moon, and planets moving independently beneath it.

In accordance with Greek philosophy, the orbits of the heavenly bodies were assumed to be a circle, regarded as a "perfect shape." As astronomers carefully recorded the apparent movements of the stars and planets, this conceptual model needed to be adjusted to account for the observations. The planets, named from the Greek word for "wanderers," were a particular problem, because sometimes they appeared to move backward. In the second century A.D., Ptolemy developed a complex system of circles within circles, called epicycles, which accurately reproduced the observed celestial patterns.

In the sixteenth century, the Polish scholar Nicolaus Copernicus proposed that the Sun was the stationary center of the universe, with Earth, planets, and stars moving around it. However, Ptolemy's model was more accurate in predicting the celestial movements, since that was what it had been designed to do, and this provided a powerful argument against the Copernican system. Another competing model was that of the Danish astronomer Tycho Brahe. The Tychonic system held that the Sun and Moon revolved around Earth, while everything else revolved around the Sun.

When the seventeenth-century Italian scientist Galileo Galilei built his telescope, he observed that the planet Venus showed phases, and deduced that it orbited the Sun. He also found four moons orbiting the planet Jupiter, conclusively disproving the idea that everything in the universe revolved around Earth. Although these observations were consistent with either the Copernican or Tychonic model, Galileo threw his lot in with Copernicus.

Nicolaus Copernicus
(Copernicus, Nicholas, engraving. The Library of Congress.)

Likewise, Tycho Brahe's assistant, Johannes Kepler, became an adherent of the Copernican model. After his mentor's death, and using Tycho's extensive and precise observations, Kepler developed a heliocentric system in which the orbits were elliptical rather than circular. This finally produced more accurate predictions than Ptolemy's epicycles and greatly advanced the Sun-centered (heliocentric) view.

The concept of ourselves as the center of the universe was no easier for human societies to give up than it is for individuals. Copernicus escaped censure because his theories weren't published until he was on his deathbed. Galileo, however, attracted the attention of the Inquisition, and was found guilty of heresy, forbidden to publish, and sentenced to house arrest for life. This was relatively lenient treatment for the time, perhaps meted out because of Galileo's age and poor health. In 1600, the philosopher Giordano Bruno had been burned at the stake for espousing the same ideas.

Moving the designated center of the universe from Earth to the Sun may have been the hardest step, but it was only the beginning. As astronomy progressed in the nineteenth and early twentieth centuries, we gradually became aware that our planet orbits an ordinary star tucked away in an arm of an ordinary spiral galaxy, of no particular prominence within an enormous universe. But most people don't find that discouraging. Though it turns out that we are not at the center of anything, the universe we know today has far more wonders than Ptolemy could have imagined. —SHERRI CHASIN CALVO

Viewpoint:
Yes, early scientists believed that what appeared to be movement around Earth by the Sun and other entities was, in fact, just that.

Throughout most of recorded history, man believed that Earth was the center of the universe. This belief, which we now call the geocentric theory of the universe, was so strong that the few who dared to challenge it were often persecuted or even killed for their heretical beliefs. The persecution suffered by Italian mathematician and astronomer Galileo Galilei during the early seventeenth century for expressing his views opposing the prevailing geocentric model is well known. On the other hand, few are familiar with the story of Italian philosopher Giordano Bruno. Bruno was burned as a heretic in 1600 for supporting the same position as Galileo, namely that the Sun was actually the center of the universe and Earth revolved around it while rotating on its own axis. For centuries it had been an integral part of man's belief system that Earth was the center of the universe. This belief was not easily overturned.

There were many reasons for man's conviction that a geocentric system described his universe. Mythology and religion played important roles, as did prevailing scientific theories. However, probably the oldest and most persuasive reason for believing that Earth was the center of the universe was common sense based on everyday observations.

The Geocentric Theory For an untold number of years, man had watched the Sun "rise" in the east every morning, move across the sky through the day, and "set" in the west. This simple motion repeated itself the next day, and the next, and the next, ad infinitum. Man had no reason to suspect that this daily motion was anything other than what it seemed, or that it had ever been different, or would ever change. Some explanations for this phenomenon were based on myths. For instance, one such myth envisioned the Sun dying every day only to be reborn the next day. However, the obvious logical explanation for the Sun's movement was that Earth is a stationary object, and the Sun revolved about it every day. It is comparable to looking out a window at a scene as it passes by one's field of vision. You may be moving past the stationary scenery, or you might be stationary while the scenery moves past your window. If you experienced no sensation of movement, the obvious conclusion would be the latter. Man experienced no sensation of movement on Earth; therefore, the conclusion was that the Sun moves while Earth remains stationary. Because similar observations were made of the motion of the Moon and the planets (although their motion was a bit more complicated), it was thought that Earth must be at the center of the universe. Then the heavenly bodies revolved about Earth. There was very little reason to suspect otherwise.

The ancient Babylonians observed and studied the motions of the heavens, even developing mathematical techniques to predict the motions of the heavenly bodies. However, it was the Greeks who first developed scientific theories concerning these motions. With only a few exceptions, the ancient Greek philosophers believed Earth was the center of the universe. One Greek philosopher, Eudoxus, proposed a rather complicated system of fixed spheres to which the Sun, Moon, the five known planets (Mercury, Venus, Mars, Jupiter, Saturn), and the stars were attached. With Earth fixed at the center, these spheres revolved and carried the heavenly bodies in a circular motion around Earth. By employing some rather sophisticated mathematics, Eudoxus was able to explain reasonably well the motion of the Sun and Moon, as well as the motions of the planets. However, his system was only partially successful in predicting the motion and location of the various heavenly bodies as they revolved about Earth. One reason for the popularity of Eudoxus' model was that it was adopted by Aristotle. Aristotle was a Greek philosopher whose teachings were extremely influential until the dawn of modern science.

Greek astronomers realized that observational discrepancies existed in the geocentric theory of the universe. The most obvious difficulty was the unexplained irregularities in the motion of the planets. Astronomers noted that the planets sometimes appeared to move in a direction opposite to that of their usual movement. This motion, called retrograde motion, presented a mathematical and physical puzzle that was tackled by many Greek astronomers. They constructed ingenious models that met with varying degrees of success to explain retrograde motion. Eudoxus' model of the universe, with its collection of concentric spheres, was useful in explaining retrograde motion for some, but not all, of the planets.

The puzzle of retrograde motion, as well as certain other incongruencies in Eudoxus' system, was eventually "solved" by the use of

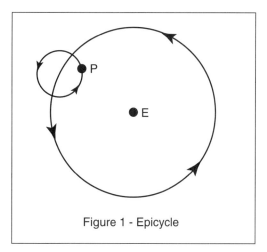

Figure 1
(Electronic Illustrators Group.)

Figure 1 - Epicycle

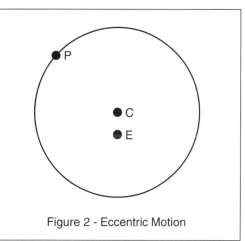

Figure 2
(Electronic Illustrators Group.)

Figure 2 - Eccentric Motion

epicycles. Essentially, an epicycle is a circle on a circle. The planet moves on a circle (called the epicycle) while this circle revolves around Earth on a larger circle (see figure 1). In this way, the planet appears to move backwards on occasion as it moves around its epicycle in the direction opposite to its motion on the larger circle. The use of epicycles helped preserve two of the primary tenets of ancient astronomy: the centrality of Earth and the belief in the perfection of uniform circular motion in the heavens.

The Ptolemaic Model The work of the second-century Greek astronomer Ptolemy represents the apex of the geocentric theory. In his work, entitled the *Almagest*, Ptolemy described a complicated system that was exceptionally accurate in its description of the motion of heavenly bodies. To do so, Ptolemy had to expand on Aristotle's rather simplistic description of circular motion. Ptolemy used two devices in his model in an effort to predict more accurately planetary motion: the previously mentioned epicycle and another device called eccentric motion. Eccentric motion was one in which the planet traveled around a circle whose center was not Earth (see figure 2). Although both epicyclical and eccentric motion had been proposed by Apollonius as

KEY TERMS

CRYSTALLINE SPHERES: The theory that the planets and stars orbit Earth attached to solid, transparent spheres.

EPICYCLE: A circle moving on the circumference of another circle. Used by Greek astronomers to accurately model planetary motions.

GEOCENTRIC: Earth–centered. The model of the universe generally accepted from ancient Greek times until the seventeenth century.

HELIOCENTRIC: Sun–centered. The model of the universe popularized by the writings of Copernicus during the Scientific Revolution.

RETROGRADE MOTION: The motion of the planets, as observed from Earth, in which the planet appears to change direction against the backdrop of the fixed stars.

STELLAR PARALLAX: The apparent change in position of a distant star, as observed from Earth, due to Earth's change in position as it orbits the Sun.

---

early as the third century B.C., it was Ptolemy who eventually used these two devices to construct a geocentric model that was successful in matching observational data. This system had the added benefit of providing an explanation for the varying length of the seasons, a feat earlier models had failed to accomplish.

The Ptolemaic model proved successful in predicting the motions of heavenly bodies and was the prevailing theory used by astronomers for centuries. However, the Ptolemaic model was not universally accepted. The eccentric motion violated the basic premise of uniform circular motion as prescribed by Aristotle. There were those, like the eleventh–century Muslim scientist Ibn al–Haytham, who tried to create models retaining the predictive powers of the Ptolemaic system without sacrificing the doctrine of uniform circular motion. Ultimately Ptolemy's model won the day, primarily due to its impressive accuracy.

Aristotelian Physics In addition to everyday observations, another argument for the centrality of Earth evolved from the physical theories of Greek philosophers, especially Aristotle. Aristotelian physics, which was the dominant paradigm until the Scientific Revolution, assumed the existence of five elements. Four of these elements, earth, water, air, and fire, formed the world and its surrounding atmosphere. The fifth element, the ether, was perfect and unchanging and formed the celestial bodies. In Aristotle's conception of the physical world, earth, as the heaviest element, naturally tended toward the center of the universe. Of course, this center of the universe was the center of Earth itself. Water, lighter than earth, also tended toward the center, gathering on top of the heavier earth. The lighter elements, fire and air, rose and collected above earth and water. Because the tenets of Aristotelian physics became so ingrained into society's picture of the universe, the concept of the centrality of Earth went essentially unchallenged. Astronomy began with this belief as a central assumption, and it was seldom questioned.

Later, in Europe, Aristotelian physics blended with Medieval Christianity to form a conception of the physical world that would dominate scientific thought until the work of Galileo, Sir Isaac Newton, and the other founders of modern science. Ideas such as the perfection of the heavens, the immobility of Earth, and the centrality of human creation all contributed to the pervading thought that Earth must be the center of the universe. The third century B.C. Greek mathematician/astronomer Aristarchus was labeled impious for placing the Sun at the center of the universe. Centuries later, Christians called upon the Bible to support their geocentric claim. They argued that Joshua commanded the Sun to stand still during a great battle so that his army might have more daylight in which to fight (Joshua 10: 12–13). The key to this passage was that Joshua did not command Earth to stand still, but rather the Sun. For the Sun to stand still implied that it must first be moving.

Allusions to ancient philosophers and to the Bible demonstrate that part of the reason for the acceptance of the geocentric model for so many centuries was man's preoccupation with authority. Whereas the Church was the ultimate authority in religious matters, Aristotle, Ptolemy, and other Greek thinkers were often considered the ultimate authority on scientific subjects. With a few adjustments made to their teachings to allow them to coexist with Christian doctrine, the science and philosophy of the Greeks was accepted almost without question.

The Heliocentric Theory Although the geocentric model of the universe dominated thought from ancient time through the seventeenth century, there were those who proposed the possibility of a Sun–centered, or heliocentric model. This model, with its requirement that Earth not only revolve about the Sun but also rotate on its own axis, was fraught with error, according to common opinion. First, argued the defenders of the geocentric model, if Earth moved man would have some sort of perception of that movement. If Earth were moving at the speed required to explain movements observed in the heavens, a strong wind would continually

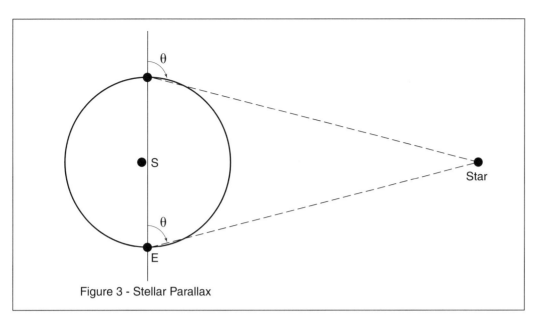

Figure 3
(Electronic illustrators Group.)

Figure 3 - Stellar Parallax

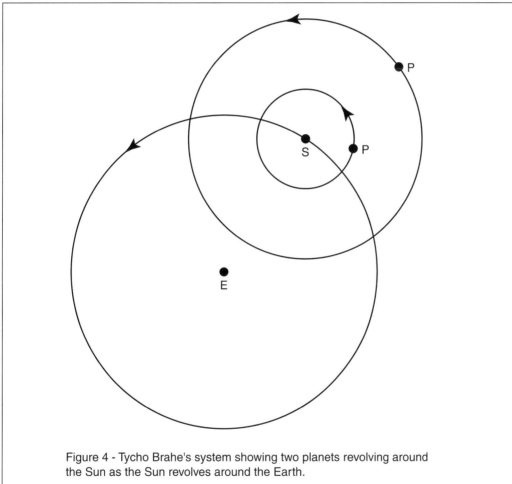

Figure 4
(Electronic Illustrators Group.)

Figure 4 - Tycho Brahe's system showing two planets revolving around the Sun as the Sun revolves around the Earth.

blow in the direction opposite to the motion of Earth. In addition, if one were to throw a stone straight up into the air, a moving Earth would cause the stone to fall some distance behind its original position. Everyday observations confirmed that none of these things happened. This was evidence in support of the geocentric model. Furthermore, it was ridiculous to assume, the argument went, that the heaviest element (earth) was propelled through the universe while the lightest (ether) remained motionless. The precepts of Aristotelian physics made such motion impossible. It was infinitely more logical that the heavy Earth was stationary while the light ether possessed the movement necessary to explain observable phenomena.

An even more sophisticated argument held that if Earth were revolving about the Sun, the motion of Earth should cause an apparent change in the position of the stars. This motion is called stellar parallax (see figure 3). Stellar parallax is not observable to the naked eye, or even through the first telescopes; therefore, proponents of the geocentric model argued that Earth was not moving around the Sun. Furthermore, if stellar parallax could not be observed due to the great distances involved, the universe would have to be much larger than anyone had imagined—too large, the geocentric theorists believed, to be a viable alternative.

Even some time after sixteenth–century astronomer Copernicus proposed his heliocentric model of the universe, most Europeans clung to the geocentric model. In answer to some of the questions raised by Copernicus' model, the Danish astronomer Tycho Brahe developed a new structure for the universe that was a compromise between the heliocentric model and the geocentric model. Brahe placed Earth at the center of the universe, with the Sun and the Moon revolving about it. However, instead of also requiring the other planets to revolve around Earth, in Brahe's model the planets revolved about the Sun as it revolved about Earth (see figure 4). This system seemed to encompass the physical and theological advantages of the geocentric model, as well as the observational and mathematical advantages of the heliocentric model. Brahe's complicated and rather illogical system serves to show just how far man would go in order to preserve the idea of geocentricity.

Eventually, all of the arguments used to defend the geocentric model of the universe were abandoned. The time it took to repudiate these arguments is a testament to the physical and astronomical systems devised to explain the world by the Greeks. It would take the complete overthrow of Aristotelian physics and Ptolemaic astronomy to finally nullify the geocentric theory. Yet, even today, we speak of the Sun "rising" and "setting" as if it moved rather than Earth. —TODD TIMMONS

Viewpoint:

No, later scientists such as Nicolaus Copernicus and Galileo correctly realized that Earth moves around the Sun, not vice versa, and thus cannot be the center of the universe.

The geocentric (Earth–centered) model of the universe was almost universally accepted until the work of astronomers Nicolaus Copernicus, Galileo Galilei, and Johannes Kepler in the sixteenth and seventeenth centuries. There were, however, a few radicals who proposed alternatives to the geocentric model in ancient times. For instance, followers of the Greek philosopher Pythagoras (to whose school the Pythagorean theorem is attributed) proposed that Earth revolved around a "central fire." Although this central fire was not the Sun, Pythagoras' theory was one of the earliest expressions of the novel idea that Earth might not be the center of the universe. Later, a few other Greek philosophers followed suit. In the fourth century B.C., Heracleides sought to resolve difficulties involved in the observations of Venus and Mercury by proposing that these two planets revolved around the Sun, while the Sun in turn revolved around Earth. Heracleides also suggested that Earth rotates. A little later, Aristarchus of Samos (third century B.C.) maintained that the Sun was the center of the entire universe and that Earth revolved around it. None of these theories, however, exhibited any marked influence on mainstream scientific thought. In spite of these heliocentric (Sun–centered) theories, the geocentric model reigned supreme thanks primarily to the philosophy and physics of Aristotle and the astronomical work of Ptolemy. It was not until the work of Copernicus many centuries later that a heliocentric model was seriously considered.

The Revolutionary Ideas of Copernicus
Nicolaus Copernicus (1473–1543) developed a heliocentric model of the universe and in the process initiated the Scientific Revolution. In his model, Copernicus maintained that Earth was not the center of the universe. Instead, Copernicus believed that Earth and the other planets revolved around the Sun. Although the notion that Earth was not the center of the universe presented many problems to sixteenth–century scientists and theologians, some of the advantages of the Copernican system over the Ptolemaic were readily apparent. Copernicus' system offered a simple explanation for many of the observed phenomena that could not be easily explained within the old system. Retrograde motion was one such phenomenon. Retrograde motion is the apparent change in direction that is observed in a planet's motion as it travels across the sky. The Ptolemaic system attempted to account for retrograde motion with epicycles. An epicycle is essentially a circle on a circle. According to the Ptolemaic system the planet moves on a circle (called the epicycle) while this circle revolves around Earth on a larger circle. With the Sun at the center of the universe, however, retrograde motion is easily explained. The apparent change in direction of the planet is a direct result of its orbit around the Sun (see figure A). Notice

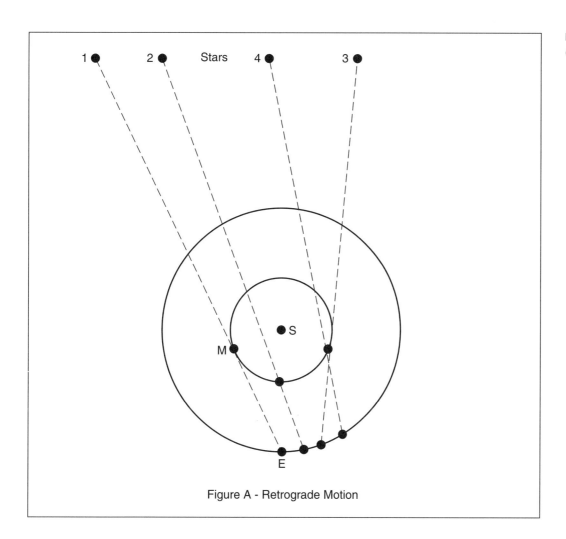

Figure A
(Electronic Illustrators Group.)

Figure A - Retrograde Motion

the position of the planet (Mercury, in this case) in relation to the fixed stars. Mercury appears to move in one direction from position 1 to 2 to 3 and then change direction as it moves to position 4. This movement was much more difficult to explain in the geocentric system of Ptolemy.

Another advantage of the Copernican system was its ability to simply and effectively pinpoint the relative position of the orbits of Mercury and Venus. In the old geocentric models it was never clear in which order the orbits of Mercury and Venus occurred. When the orbits of these inner planets were analyzed in the context of Copernicus' heliocentric model, their positions were, for the first time, unambiguous. The observations confirmed that the orbit of Venus was closer to Earth than that of Mercury.

The Advances of Brahe and Kepler
Although revolutionary in his ideas concerning the motions of the heavenly bodies, Copernicus remained a product of the medieval Aristotelian natural philosophy. In some ways, Copernicus' system, as explained in his famous work of 1543, *De Revolutionibus orbium coelestium* (On the revolutions of the heavenly spheres), was similar to the centuries–old model developed by Ptolemy. For instance, although Copernicus placed the Sun at the center of the universe, he retained the notion that the heavenly bodies were carried around on their revolutions by solid crystalline spheres. It was not until the work of two astronomers of the next generation, Tycho Brahe (1546–1601) and Johannes Kepler (1571–1630), that this theory was challenged. Brahe observed two occurrences in the heavens that cast serious doubt on the theory of crystalline spheres. First, he observed the birth, and later disappearance, of a new star (a nova). When Brahe was able to show that this new object in the sky came into existence beyond the orbit of the Moon, he challenged the belief that the heavens were perfect and unchanging. Secondly, Brahe calculated the path of a comet (a hazy gaseous cloud with a bright nucleus) and showed that it was moving across the heavens beyond the orbit of the Moon. In other words, its orbit would take the comet "crashing" through the crystalline spheres, an obvious impossibility. Brahe concluded that there were no physical spheres containing the orbits of the planets.

Kepler's contribution to the mounting evidence pointing toward the truth of Copernicus' theory came in the form of his three laws of

Galileo
(Galilei, Galileo, drawing. Archive Photos, Inc. Reproduced by permission.)

planetary motion. Kepler's first law states that the planets orbit the Sun following an elliptical path with the Sun at one focus of the ellipse. This revolutionary break with the tradition of circular motion allowed a simple geometrical model to explain the motions of the planets. No longer requiring awkward epicycles and eccentrics, elliptical orbits presented an elegant mathematical solution to a sticky problem. Kepler's other two laws are mathematical theories relating to the heliocentric model. His second law states that the orbits of planets sweep out equal areas in equal times, and his third law concludes that the square of the period of each orbit is proportional to the cube of the semimajor axis of the elliptical orbit—that is, one half the distance across the ellipsis at its widest point. The regularity that these three laws implied made the heliocentric model compelling to scientists who looked for order in the universe.

Interestingly, we remember Kepler's three laws of planetary motion but seldom hear of his other theories that did not stand the test of time. In Kepler's early work, *Mysterium Cosmographicum* (Cosmographic mystery), the astronomer defended the truth of Copernicus' heliocentric model by constructing his own model in which the orbits of the planets were separated by the five regular solids. A regular solid is one whose faces are all identical. For instance, a cube is a regular solid because its sides are all equal squares. Since only five such solids exist (cube, tetrahedron, dodecahedron, icosahedron, and octahedron), Kepler believed it was God's intention that they appear between the orbits of the six planets. Kepler argued that this was further proof of the heliocentric theory because, in the geocentric theory, the Moon was the seventh planet. If there are only five regular solids, they could not fit in between the orbits of seven planets. To Kepler, this was important evidence in favor of accepting the heliocentric model as God's divine plan.

The Discoveries of Galileo In one of the most important series of events in the Scientific Revolution, the Italian scientist Galileo Galilei (1564–1642) turned his newly acquired telescope toward the sky and discovered many wonders that would cause man to rethink his previous conceptions of the cosmos. One of Galileo's first discoveries was that the Moon had surface features such as craters. This discovery was in direct conflict with the Aristotelian view of heavenly bodies composed of a perfect and unchanging substance. The features of the Moon indicated that it might be composed of the same sort of common material as Earth.

Galileo also discovered, with the help of his telescope, that Venus went through observable phases just as the Moon. In the Ptolemaic system, the phases of Venus (undetectable without a telescope) would be impossible. If Venus orbited Earth, inside of the Sun's orbit, it would never be seen as full. The fact that Venus appeared in phases was a strong argument that it was revolving around the Sun.

Two discoveries made by Galileo with the help of his telescope changed the way man perceived the stars themselves. Galileo noticed that the planets appeared as solid discs with a well-defined outline when viewed through the telescope. The stars, on the other hand, continued to twinkle and resisted definition even when gazed upon through the telescope. Galileo concluded that the distance to the stars must be many times greater than the distance to the planets. This meant that the size of the universe was much greater than allowed by the Ptolemaic model. In addition, Galileo was able to observe that the Milky Way was actually composed of many individual stars, suggesting that the number of stars in the sky was much greater than had been previously believed.

Finally, a crucial discovery made by Galileo was the existence of the moons of Jupiter. The old paradigm maintained that Earth was the center of all revolving bodies. This centrality formed the very essence of Aristotelian physics and Ptolemaic astronomy. The existence of bodies revolving around a center other than Earth brought into question all of the previous assumptions upon which science was based. If

Earth was not the center of revolution for all heavenly bodies, then other tenets of ancient science might also be false.

In addition to the evidence for the heliocentric model discovered by Galileo with the aid of his telescope, the great Italian scientist also made an equally important contribution to the eventual acceptance of Copernicus' theory. If the heliocentric model were true, then the whole of physics, essentially unchanged since Aristotle, was in error. Galileo provided an alternate explanation for motion that did not require the philosophical conclusions concerning the primacy of Earth and its place in the center of the universe. In Aristotle's conception of motion, Earth must be at the center because it did not move. Motion, or lack of motion, was an inherent characteristic of a body, and Earth's lack of movement made it different from the continuously moving heavenly bodies. Galileo, on the other hand, argued that motion was only an external process and not an inherent characteristic of the body itself. Movement was not anymore an innate characteristic of the planets than lack of motion was innately inherent in Earth. Before the theory that Earth moved could be accepted, these important consequences regarding the nature of motion had to be explained.

Galileo also argued that a body in motion would continue in motion without the continuous influence of an outside force. In fact, it required an outside force to stop the body's motion. This is the concept that modern scientists call inertia. This conception of motion helped to answer one of the claims against Earth's diurnal, or daily, motion. Opponents of the heliocentric model claimed if Earth spun on its axis, a ball dropped from a tower would land some distance away from the base of the tower because Earth had moved beneath it. Galileo's answer was that the ball already possessed a motion in the direction of the spinning Earth and would continue with that motion as it fell, thus landing at the base of the tower.

One disturbing question that arose from Copernicus' theory was that of stellar parallax. If Earth did revolve about the Sun, the relative position of the stars should change as Earth moved. Unfortunately for Copernicans, this stellar parallax could not be observed. Today we know that stellar parallax can only be observed through telescopes much more powerful than those available to Galileo and his contemporaries. In fact, proof of stellar parallax was not supplied until the work of Friedrich Wilhelm Bessel in the nineteenth century. Bessel's work provided a final proof for the yearly motion of Earth revolving around the Sun.

It was also in the nineteenth century that a final proof of the rotation of Earth on its axis was supplied by the French physicist Jean–Bernard–Léon Foucault. Foucault suspended a large iron ball from a wire swinging freely from a height of over 200 feet. As the ball, known as "Foucault's pendulum," swung in the same vertical plane, Earth rotated beneath it. Foucault's pendulum is now a common exhibit at many modern science museums. A series of blocks standing in a circle around the pendulum are knocked over one by one as Earth rotates once in a twenty–four hour period.

Bessel's discovery of stellar parallax and Foucault's pendulum represented the final direct proofs of the two primary motions of Earth. These nineteenth–century events marked the end of a long process of discovery begun by Copernicus some four centuries earlier. —TODD TIMMONS

Aristotle
(Aristotle, lithograph. The Bettmann Archive. Reproduced by permission.)

Further Reading

A Brief History of Cosmology. <http://www-groups.dcs.st-and.ac.uk/~history/HistTopics/Cosmology.html>.

Copernicus, Nicolaus. <http://www-groups.dcs.st-and.ac.uk/~history/Mathematicians/Copernicus.html>.

Galilei, Galileo. <http://www-groups.dcs.st-and.ac.uk/~history/Mathematicians/Galileo.html>.

Greek Astronomy. <http://www–groups.dcs.st–and.ac.uk/~history/HistTopics/Greek_astronomy.html>.

Kepler, Johannes. <http://www–groups.dcs.st–and.ac.uk/~history/Mathematicians/Kepler.html>.

Kuhn, T. S. *The Copernican Revolution.* Cambridge, Mass: Harvard University Press, 1957.

Lindberg, David C. *The Beginnings of Western Science.* Chicago: University of Chicago Press, 1992.

Lloyd, G. E. R. *Early Greek Science: Thales to Aristotle.* New York: W. W. Norton, 1970.

———. *Greek Science after Aristotle.* New York: W.W. Norton, 1973.

Moss, Jean Dietz. *Novelties in the Heavens.* Chicago: University of Chicago Press, 1993.

Neugebauer, O. *The Exact Sciences in Antiquity.* New York: Dover, 1969.

Pannekoek, A. *A History of Astronomy.* New York: Dover, 1961.

Rossi, Paolo. *The Birth of Modern Science.* Oxford: Blackwell Publishers, 2001.

Westfall, Richard S. *The Construction of Modern Science.* Cambridge: Cambridge University Press, 1977.

INDEX

A

Abbey, David M., 138
Academy of General Dentistry, 175
Accelerators, Particle, **219–226**
    See also Cyclotrons
Adam and Eve, 207
Adams, Henry, 131
Adams, John Couch, *29*
    discovery of Neptune, 24–31
Adleman, Leonard, 165
AFC power plants. *See* Fuel cells
Aggressiveness and XYY karyotype, **111–118**
AIDS
    syphilis, 208, 211
    xenotransplants, 185, 192
Airy, George, 25–31
Airy model of isostasy, 60
Alkaline fuel cells. *See* Fuel cells
Allen, Carl, 42–43
Allotransplants. *See* Organ transplants
Alpha (space station). *See* International Space Station
American Dental Association, 175
American Museum of Natural History, 3, 4
American National Academy of Sciences, 180
American National Cancer Institute, 180
American Physical Society, 245
America's Cup (yachting), 98
Amino acids, 33–35
Aneuploidy, 111–112, 116
    See also Genetic disorders
Angular gyrus, 159–160
Animal cloning. *See* Cloning
Animal transplants. *See* Xenotransplants
Antineutrinos. *See* Neutrinos
Antinori, Severino, 194, 205
Apatosaurus, 129
Apollo spacecraft, 69, 70
Apollonius, 255–256
Arago, Dominique François Jean, 30
Archaea, 40
    See also Hyperthermophiles; Thermophiles
Ariane rockets, 19
Aristarchus of Samos, 256, 258
Armelago, George J., 213
Artificial intelligence, 86
Artistotle, *261*
    Aristotelian physics, 255, 256–257
Asilomar agreement, 89
Aspect, Alain, 242
Asteroid Belt, 6
Astrid (pig), 191
Astronomy
    dark matter and neutrinos, 229, 234
    forward modeling, 29
    geocentricism and heliocentricism, **253–261**
    Greek theories, 253, 255–257
    inverse perturbation theory, 29–30
    Neptune's discovery, **24–31**
    Pluto's status as planet, **1–7**
    solar neutrino problem, 228, 229, 231–232, 234
    supernovas and neutrinos, 228, 232
Asymmetric encryption, 165–166

Athletes and technology. *See* Sports and technology
Atomic structure, 219, 221, 231
Australia and space debris, 18
Automotive engines, **67–75**
Aviation Week & Space Technology (magazine), 10

B

B cells, 187
Baboon xenotransplants, 185, 191
Baby Fae, 185, 191
Bacon, Edmund, 136
Bacteria, 40
Bagger, Jonathan, 225
Bahcall, John, 231–232
Bailey, Leonard, 191
Baker, Brenda J., 213
Bakker, Robert T., 119–120, 121
Ballard (company), 70
Barger, Herbert, 137
Barksdale, James, 169
Barron, William, 148
Baseball and technology, 96
Battery-powered vehicles, 69
Beall, Donald R., 10
Beckett, Margaret, 105
Beckwith, Jon, 113
Bednorz, Georg, 250
Beipiaosaurus, 125
Bejel and syphilis, 209, 213
Bell, John, 237, 240, 242
Benton Hills (MO) earthquake faults, 54
Berlin Observatory (Germany), 28
Bessel, Friedrich Wilhelm, 261
Beta radiation and neutrinos, 227, 228, 231
Bill of Rights. *See* United States Constitution
Binning, Gerard, 91
Biomass power generation, 82–83
Birds and dinosaurs, 119–130
Blondlot, René, 249
Blood transfusions, 191
Bohm, David, 240, 242
Bohr, Niels, *239*
    quantum theory, 236
Boisselier, Brigitte, 194
Bones
    dinosaurs, 121, 123, *124*, 129
    syphilis, 210, 213–214, 217
Book, Stephen A., 17
Borg, Bjorn, 95
Bouvard, Alex, 30
Bovine spongiform encephalopathy, 107
Bowie, William, 60
Brahe, Tycho, 253–254, 258, 259
Brain and mathematics, 159–160, 162
Braun, Werner Von, 10
Brigham Young University, 245, 250
Brodie, Fawn M., 132
Brookhaven National Laboratory, 232
Brown, Louise, 194
Brown, Nick, 103
Bruno, Giordano, 254

263

BSE (bovine spongiform encephalopathy), 107
Building codes for seismic reinforcement, 49
Bumpers, Dale, 224
Burk, Dean, 178
Buses, Hybrid-powered (vehicles), 72
Bush, George W.
    cloning and stem cells, 195
    renewable energy, 79
    science budget forecasts, 226
    terrorism, 164
Butterworth, Brian, 162

C

Calculus instruction, **149–156**
Callender, James Thomson, 131, 135, 137
Calorimeters and cold fusion, 246–247
Cambridge Observatory (England), 27, 30–31
Canada and space debris, 19
Cancer
    fluoridation, 177–178
    power lines, 180
Capitan Bermudez (Argentina) and *Salyut 7* debris, 18
Capstone (company), 72
Car engines, **67–75**
Carbon dioxide on Mars, 44, 45–46
Carbon nanotubes, 86, 91
    See also Nanotechnology
Carbon tetrachloride in neutrino detection, 232
Carcinogens. *See* Cancer
Carr, John, 132, 138
Carr, Peter, 132, 134, 135, 137, 138
Carr, Samuel, 132, 134, 135, 137, 138
Carrel, Alexis, 191
Cathode-ray tubes, 220, 221
    See also Particle accelerators
Catholic church. *See* Roman Catholic church
Caudipteryx, 125
CDC (Centers for Disease Control and Prevention), 175, 182–183
Census Act (1954), 143
Census Act (1964), 143
Census Act (1976), 141, 143, 146–147
Census Bureau and statistical sampling, **141–148**
Census taking
    interviews, *145*
    privacy, 147
    statistical sampling, **141–148**
Center for Orbital and Reentry Debris Studies, 19
Centers for Disease Control and Prevention, 175, 182–183
Cerenkov radiation, 228–229, 232, 234
Ceres (asteroid), 6
CERISE (artificial satellite), 19
CERN (physics laboratory)
    Large Hadron Collider, 224
    neutrinos and mass, 229, 231, 232
    World Wide Web, 220, 223
Chaikin, Andrew, 43
Challenger (space shuttle), 18
Challis, James, 25, 27–28, 30–31
Chamberlin, T.C., 60
Chang, Michael, 95
Charles VIII (France), 210, 212
Charon (moon), 6
Cherenkov radiation, 228–229, 232, 234
Chestnut, Mary Boykin, 132
Childhood leukemia, 180
Chimpanzee xenotransplants, 191
China and greenhouse emissions, 79
Chinese language and mathematics, 161
CHOOZ (neutrino source), 229
CHORUS (neutrino detector), 229
Chromosomal abnormalities. *See* Genetic disorders
Circular particle accelerators. *See* Cyclotrons
Classroom instruction
    calculus, **149–156**
    minorities and mathematics, 149–150, 153, 154–155
    technology, 152–153, 154
Clinton, William Jefferson
    data encryption, 170
    sexual indiscretions, 133
Clipper Chip encryption, 168–169
Clonaid (company), 203

Cloning
    imprinting, 201, 204
    legal aspects, 194–195, 196, 201, 206
    moral and ethical aspects, 197–200, 205–207
    oocyte nucleus transfer, 195
    preimplantation genetic diagnosis, 200
    reproductive cloning, 196–202, 205–206
    risks and benefits, **194–207**
    sheep, *198*
    somatic cell nuclear transfer, 195–196, 198–201, 203–204
    therapeutic cloning, 198, 206
    See also Genetic engineering
Coates, Eyler Robert, 138
Cocke, John Hartwell, 135
Cold-blooded dinosaurs, **119–130**
Cold fusion, **244–252**
    calorimeters, 246–247
    gamma radiation, 246, 251
    peer review, 245–248, 250
    spectroscopy, 247
Cold War physics research, 225
Collision boundaries, 62
    See also Plate tectonics
Columbus, Christopher, *215*
    syphilis, **208–217**
Combating Terrorism Act (2001), 167
Comets
    forecast by Tycho Brahe, 259
    origins of life, 34, 37, 39
    Pluto, 3–4
    See also Meteorites
Communism and fluoridation, 174, 181
Computer Security Act (1987), 169
Computer viruses, 87
Computers in calculus instruction, 152–153, 154
Conrad, Charles "Pete," 14
Conservation of energy, 252
Conservation of momentum, 239
Constitutional rights. *See* United States Constitution
Continental crust
    composition, 58
    plate tectonics, 62
    See also Lithosphere
Continental drift, **57–64**
    See also Plate tectonics
Coolidge, Ellen Randolph, 132
Cooper, David K.C., 189
Copernicus, Nicolaus, *254*
    heliocentricism, 253–254, 258–259
Cosmos 954 (artificial satellite), 19
Cowan, Clyde, Jr., 228, 231
Criminality and XYY karyotype, **111–118**
Crocodiles, 123
Crust, Earth's, 58, 61–62
Cryptography. *See* Encryption
Curry, Kristina, 129
Cuvier, Georges, 119
Cyberwarfare, 166
Cyborgs, 89
Cyclotrons
    cancer radiotherapy, 222
    costs and benefits, **219–226**
    See also Particle accelerators

D

Dana, James Dwight, *60*
    continental permanence, 58–59
Dark matter and neutrinos, 229, 234
DARPA (Defense Advanced Research Project Agency), 92
d'Arrest, Heinrich, 25, 28, 29, 31
Darwin, Charles, 35
    land bridges, 59
    "little warm pond" theory, **33–40**
    origins of life, 33, 35
Data encryption. *See* Encryption
Davis, Gary, 138–139
Davis, Ray, 231–232
Dean, Henry Trendley, 174
DeBakey, Michael E., 10
Decennial Census Improvement Act (1991), 142–143, 146–147, 148
Defense Advanced Research Project Agency, 92

Definition of continuity (mathematics), 154
Definition of limits (mathematics), 154
Dehaene, Stanislaus, 162
Deinonychus antirrhopus, 120, 121, 127
Delay, Tom, 195
Delta rockets, 19, *21*
Democratic Party, 144
Dendrimers, 92
Denis, Jean-Baptiste, 184
Denmark and windmills, 82
Dental caries. *See* Tooth decay
Dental fluorosis, 175, 177, 178
Deoxyribonucleic acid. *See* DNA
Department of Commerce
 census, 144, 146–147
 data encryption, 170
 See also Census Bureau
Department of Commerce et al. v the United States House of Representatives et al. (1998), 144–145
Department of Defense, 226
Department of Energy
 alternative automotive power sources, 72
 budget forecasts, 226
 cost overruns, 224
Deuterium in cold fusion, 244–252
Diaz de Isla, Rodrigo Ruiz, 209, 211–212
Digital Encryption Standard, 170
Dimetrodon, 122
Dinosaurs
 behavior, 127
 bones, 121, 123, *124,* 129
 endothermy and ectothermy, **119–130**
 evolutionary pressures, 120, 122, 125–127
 gigantothermy, 126
 hybrid metabolism, 119–120, 129
 morphology, 120, 125
 organs and tissues, 123–125, 128–129
 polar regions, 120, 122–123
 predator-to-prey ratios, 120, 122
 teeth, 123
 Velociraptor, 129
Dioxins, 107
Diplo-Y syndrome. *See* XYY karyotype
Divergent boundaries, 62, 63–64
 See also Plate tectonics
DNA
 origins of life, 36
 testing in Jefferson-Hemings controversy, **131–139**
Dog xenotransplants, 191
Dolly (sheep), 194, 195, 196, 200
Donaldson, Alex, 104
Double planets, 6
Dr. Strangelove (motion picture), 181
Dragila, Stacy, 95
Drexler, Eric, 90
Drinking water
 chlorination, 181
 fluoridation, **173–183**
Dromaeosaurs, 125
Drugs, Performance-enhancing, 97
Du Toit, Alexander, 61, 62
Dual system estimates, 145–146, 147
Duve, Christian R. de, 37

E

Earth
 conditions for origins of life, 33–40
 continental drift, **57–64**
 crust, 58, 61–62
 early atmosphere, 35–36, 38
 formation, 38
 geocentric universe, **253–261**
 internal heat, 58, 61, 63
 internal structure, 60, 61–62
 rotation, 261
Earth-centered universe. *See* Geocentric universe
Earthquake-proof construction, 48–49
Earthquakes
 charting Earth's internal structure, 60, 63
 forecasting charts, *53*
 New Madrid Seismic Zone, **48–55**
 North America, *51*
Eccentric motion, 255–256

Ectothermy in dinosaurs, **119–130**
Edgett, Ken, 42, 43
Einstein, Albert, *241*
 brain, 159–160
 quantum theory, 236–243
 See also Einstein-Podolsky-Rosen paradox
Einstein-Podolsky-Rosen paradox, 238–240
 conservation of momentum, 239
 hidden variables, 236–243
El Paso (TX) Renewable Energy Tariff Program, 79
Electric generation
 power lines and cancer, 180
 sources, 81–84
 wind power, **76–84**
Electric-powered vehicles
 history, 67, 72, 73
 range, 73
 viability, 68–72, 73–74
Electron neutrinos. *See* Neutrinos
Embryos in cloning of humans, 201, 202, 206
Empiricism, 157, 158
Encryption
 Clipper Chip, 168–169
 constitutional rights, 168
 e-commerce, 166
 government keys, 167
 international aspects, 169–170
 keys, 165–166
 private strong encryption and risks, **163–171**
 RSA cryptography, 165–166
 symmetric encryption, 164–165
 trusted third parties, 167
Endeavor (space shuttle), 17, *22*
Endogenous retroviruses and xenotransplants, 186–187
Endothermy in dinosaurs, **119–130**
Engineering
 internal combustion engine alternatives, **67–75**
 nanotechnology, **86–92**
 sports and technology, **93–99**
 wind power, **76–84**
Enumeration for census, **141–148**
Environmental Protection Agency, 175
EPA (Environmental Protection Agency), 175
Epicycles, 253, 255–256
Epidemiology
 AIDS, 211
 biblical plagues, 216–217
 syphilis, **208–217**
 syphilis and AIDS, 208, 211
EPR paradox. *See* Einstein-Podolsky-Rosen paradox
ERVs (endogenous retroviruses), 186–187
Escalante, Jaime, 155
Eudoxus, 255
Eukarya, 40
Europa (moon), 40
European Laboratory for Nuclear Physics. *See* CERN (physics laboratory)
European Union and data encryption, 170
Eve (first woman), 207
Evolution
 dinosaurs, 120, 122, 125–127
 origins of life, **33–40**
EVOLVE space debris tracking model, 22
Extracorporeal liver perfusion, 189–190
Exxon Valdez (ship), 250

F

Faint Young Sun paradox, 45
Farlow, James, 130
Farms, *187*
FBI (Federal Bureau of Investigation), 166–167, 169–170
FDA. *See* Food and Drug Administration
Feathers on dinosaurs, 125
Federal Bureau of Investigation, 166–167, 169–170
Federalist Party, 131, 135
Fermi, Enrico, 227, 231
Fermilab
 MINOS neutrino experiment, 231
 particle accelerators, 224–225
Fertility treatments and cloning, 199
 See also In vitro fertilization
Feynmann, Richard
 nanotechnology, 90, 91

quantum theory, 236
technology and risk, 51
Finland and data encryption, 169
Fleischmann, Martin, *246, 251*
    cold fusion, **244–252**
Fluoridation
    carcinogens, 177–178
    communism, 174, 181
    discovery, 173–174, 179
    fluorosis, 175, 177, 178, 179–180
    protests, *177*
    risks and benefits, **173–183**
Fluoride, 173
    See also Fluoridation
Fluorosis, 175, 177, 178, 179–180
FMD (foot-and-mouth disease). *See* Foot-and-mouth disease
Food and Drug Administration
    fluoridation, 175
    xenotransplants, 185, 190
Foot-and-mouth disease
    containment options, **101–109**
    culling, 103, *104,* 105–106
    economic impact, 106–107
    United States, 106, *108*
    vaccination, 103–104, 107–109
45, X karyotype. *See* Turner's syndrome
47, XYY karyotype. *See* XYY karyotype
Fossils
    dinosaur bones and heart, *124*
    dinosaurs, 119–130
    origins of life, 35, 38
Foster, Eugene A., 132, 133–134, 137–139
Foucault, Jean-Bernard-Léon, 261
Fox, Sidney W., 35
France and data encryption, 169
Frascastoro, Girolamo, 209
Frechet, Jean M., 92
Freeh, Louis, *169*
    data encryption, 169
Freitas, Robert A., Jr., 90–91
Frosch, Paul, 101
Frosch, Robert A., 18
Fuel cells
    alkaline fuel cell power plants, 70
    motor vehicle engines, 68, 69–72, 73–74
    PEM fuel cells, 70
    spacecraft, 69, 70
Full-body running suits, 94, *98*
Fusion. *See* Cold fusion; Nuclear fusion

G

Galilei, Galileo. *See* Galileo
Galileo, *260*
    heliocentricism, 253–254, 258, 260–261
Galle, Johann, 25, 28, 29, 31
Galton, Francis, 111, 116
Gamma radiation from cold fusion, 246, 251
Garden of Eden, 207
Geist, Nicholas, 129
Gemini spacecraft, 69, 70
Genealogy and Jefferson-Hemings controversy, **131–139**
Genetic disorders
    aneuploidy, 111–112, 116
    cloning, 195–196, 200–201
    Klinefelter's syndrome, 112
    XYY karyotype and aggression, **111–118**
    See also Turner's syndrome; XYY karyotype
Genetic engineering
    cloning risks and benefits, **194–207**
    unintended consequences, 88, 89
    xenotransplants, 185
    See also Cloning
Geocentric universe
    development and rejection, **253–261**
    Ptolemaic model, 255–256
    spheres, 255
    Tychonic system, 253, *257,* 258
Geology
    earthquake risk, **48–55**
    Mars, 41–46
    Moon, 38
    mountains, 58
    plate tectonics, **57–64**
    uniformitarianism, 57, 60
Georgia Tech, 250
Geothermal energy, 83
    Guatemala, 83
Germany and data encryption, 169
Gibbons, Thomas, 135
Gigantothermy in dinosaurs, 120, 126
Gill, Ben, 103
Ginsburg, Ruth Bader, 144
Gisin, Nicolas, 242
Global Positioning System (GPS), 53–54
Goldin, Daniel, 12, 13–14
Golf and technology, 94–95, 95–96
Gondwanaland, 58–59
Gonorrhea, 214
Goodlatte, Bob, 169
Gore, Al, 166
Gotz, M.J., 117
GPS (Global Positioning System), 53–54
Gran Sasso National Laboratory, 231
Grandpa's Knob (VT), 76
Gravitational laws, 151
Gray goo, 86, 87, 90
    See also Nanotechnology
Great Britain. *See* United Kingdom
Greece and foot-and-mouth disease, 104
Green, Daniel W.E., 5–7
Greenhouse effect
    Earth and Mars, 44–45
    reduction, 79
Greenwood, Jim, 195
Greger, Paul, 105
Groundwater contamination and foot-and-mouth disease, 107
Grove, William, 69
Guirac tree, 212
Gullies on Mars, *42, 44*
Günzburg, Walter H., 190

H

Hale, Everett, 9
Hardy, James, 191
Harwell laboratory, 250
Hawaiian wind farm, *80*
Hayden Planetarium, 4
Head, Jim, 42
Heartland Institute, 74
Hearts of dinosaurs, 123–125, 128
Heavy water, 244–252
Heisenberg, Werner, 236
Heliocentric universe
    Copernican system, 258–259
    development and acceptance, **253–261**
    laws of planetary motion, 259–260
    relative motion, 256–257
Hemings, Betty, 137
Hemings, Beverly, 131–139
Hemings, Eston, 131–139
Hemings, Harriet, 131–139
Hemings, Madison, 131–139
Hemings, Sally
    descendants, *136*
    relationship with Thomas Jefferson, **131–139**
Heracleides, 258
Herschel, John, 31
Herschel, William, 26
Hersh, Reuben, 160
Hess, Harry, 64
Hidden variables, **236–243**
Higgs boson, 223
High-energy physics. *See* Particle physics; Physics
High-temperature superconductors, 250
Hijacking and data encryption, 164
Hind, John Russell, 27
Hitchcock, Edward, 120
HIV (human immunodeficiency virus), 192
Hocker, Lon, 226
Hoffman, David, 54
Hoffman, Paul Erich, 209
Holmes, Arthur, 61, 64
Holtz, Susan
    solar system size, 20
    space debris, 22–23

Homeothermy in dinosaurs, **119–130**
HTLV viruses, 187–188
Hubble Space Telescope, 21
Human cloning. *See* Cloning
Human Cloning Prohibition Act (2001), 194–195, 196, 206
Human Fertilisation and Embryology Act (United Kingdom, 1990), 195
Human Genome Project, 225
Human immunodeficiency virus, 192
Humane Society of the United States, 185
Humans and math, **157–162**
Hussey, Thomas, 25, 26
Hybrid electric vehicles, 68, 72, 74
Hydrogen fuel cells. *See* Fuel cells
Hydrogen in fuel cells, 69–72
Hydropower, 82
Hydrothermal vents, *39*
    origins of life, 34, 37, 39
Hypacrosaurus, 128–129
Hyperthermophiles, 36–37, 39–40
    See also Archaea; Thermophiles

I

Ibn al-Haytham, 256
IGF2R (insulin-like growth factor II receptor), 204
Iijima, Sumio, 91
IMB neutrino detector, 228, 232
iMEDD (company), 92
Immune system and xenotransplants, **184–192**
Imutran (company), 191
In vitro fertilization, 194, 203
    See also Infertility and cloning
Inertia, 261
Inertial homeopathy in dinosaurs, 120, 126
Infertility and cloning, 199
    See also In vitro fertilization
Information technology and encryption, **163–171**
Insulin-like growth factor II receptor, 204
Integrated coverage management, 142, 147
Inter-Agency Space Debris Coordination Committee, 22
Intermediate value theorem, 154
Internal combustion engines
    history, 67–68
    pollution, 68, 69, 72–73
    viable alternatives, **67–75**
International Astronomical Union, 1–7
International Fuel Cells (company), 70–71
International Space Station, 13
    cost, 9
    international cooperation, 11–12
    role in space exploration, **8–15**
    service module, *10*
    space debris, 17–18, 21
Internet
    CERN (physics laboratory), 220, 223
    risks and benefits, 203
Inverse perturbation theory, 29–30
Irvine-Michigan-Brookhaven neutrino detector, 228, 232
Isostasy, 58–59
    Airy model, 60
    Pratt model, 60

J

Jacobs, Patricia, 116
Japan and foot-and-mouth disease, 104
Japanese National Accelerator Facility, 230
Jefferson, Beverly, 135
Jefferson, Eston Hemings, 131–139
Jefferson, Field, 132, 134, 138
Jefferson, Isaac, 137
Jefferson, Israel, 132, 135
Jefferson, Martha (Martha Jefferson Randolph), 132, 133, 137
Jefferson, Martha Wayles, 131
Jefferson, Mary, 133
Jefferson, Randolph, 138–139
Jefferson, Thomas, *132*
    descendants, *136*
    relationship with Sally Hemings, **131–139**
Jefferson's Daughter (poem), 131
Jeffreys, Harold, 60, 61
Jewitt, David, 4–5
Jirtle, Randy, 204
Johnson, Michael, 94
Johnson Space Center, 21
Johnston, Arch, 52–53
Johnstone, E.C., 117
Joly, John, 58
Jones, Marion, *98*
Jones, Steven, 245
Jordan, Daniel P., 133
Joshua (biblical figure), 256
Jovian moons, 253, 260–261
Joy, Bill, 87, 90
Jupiter's moons, 253, 260–261

K

Kamiokande neutrino detector, 228, 232
    See also Super-Kamiokande neutrino detector
Kanamori, Hiroo, 55
Karyotype disorders. *See* Genetic disorders
KBOs (Kuiper Belt objects), 1–7
KEK particle accelerator, 230
Kepler, Johannes, 254, 259–260
Killian, J. Keith, 204
King, Jonathan, 113
Klinefelter's syndrome, 112
Koch, Robert, 213
Korea, South and foot-and-mouth disease, 104
Kosmos 1275 (artificial satellite), 21
Kuhn, Thomas, 29
Kuiper, Gerard, 4
Kuiper Belt, 1, 4–7
Kuiper Belt objects, 1–7

L

Laak, James Van, 18
Lackoff, George, 160
Land bridges, 59
Language and mathematics, 152, 157, 161–162
Lanza, Robert P., 189
Lanzerotti, Louis, 12
Large Hadron Collider, 224
Lawrence, E.O., *222*
    cyclotron, 220, 222
Le Verrier, Urbain, 24–31
Leatherback turtles, 128
Lebofsky, Larry, 4
Leprosy, 209, 212
Leshin, Laurie, 43–44
Leukemia, Childhood, 180
Leverrier, Urbain, 24–31
Levison, Hal, 5–6
Levy, Marlon, 190
l'Hôpital's rule, 154
Life, Origins of, **33–40**
Liquid air-powered cars, 67
Liquid Scintillating Neutrino Detector, 229
Lithopedion, 213
Lithosphere, 58, 61–62
    See also Continental crust; Oceanic crust
"Little warm pond" theory, **33–40**
Local hidden variables, 237, 240, 242
    See also Hidden variables
Locke, John, 157, 158
Loeffler, Friedrich, 101
Low Earth orbit
    space debris, 17–23
    space exploration, 14
Lowell, Percival, 2
Lowell Observatory, 2
LSND (Liquid Scintillating Neutrino Detector), 229
Lunar geology, 38
Luu, Jane X., 4
Lymphoma, 188

M

Macedonia and foot-and-mouth disease, 108–109
Mad cow disease, 107
Magnetism and continental drift, 63–64
Maiasaura, 127
Main, Ian, 54
Main Injector Neutrino Oscillation, 231

Major planets. *See individual planets by name*
Malin, Michael, 43
Mammals and dinosaurs, 119–130
Manned space stations, 8–15
    See also Space exploration
Maran, Stephen P., 22
Mariner 9 (Mars probe), 42
Mars Global Surveyor, 42–43
Mars Pathfinder (Mars probe), 43
Mars (planet)
    atmosphere, 43, 44–46
    canali, 42
    exploration strategies, 14–15
    gullies, *42*
    origins of life, 37, 40
    outflow channels, 42–43
    rocks, 43
    surface water, **41–46**
    valley networks, 42–43
Mars Rover (Mars probe), 43
Marsden, Brian, 4, 7
Mathematics
    abstraction, 157, 160
    brain, 159–160, 162
    calculus instruction and reform, **149–156**
    counting, 160–161
    definition, 158
    forward modeling, 29
    innate human capacity, **157–162**
    inverse perturbation theory, 29–30
    language, 152, 157, 161–162
    place value systems, 161
    RSA cryptography, 165–166
    subitization, 157, 158–159, 160–161
    symbolic representation, 157, 161
Matilda (sheep), *198*
Matthews, Drummond, 64
McKay, Frederick Sumter, 173–174, 179
Measles, mumps, rubella vaccine, 190
Mednick, S.A., 113, 117
Meiosis, 111–112
Memphis (TN), 49
Mercury (metal), 214
Mercury (planet), 3, 259
Meteorites
    carbonaceous chondrites, 38–39
    Martian, 40
    origins of life, 34, 37, 40
    See also Comets; Space debris
Mice, 196
Microgravity
    human health, 13
    research on *International Space Station,* 8
Microturbines, *71, 81*
    biomass power generation, 82–83
    hybrid electric vehicles, 72
Midmore, Peter, 107
Midwestern United States and earthquake risk, **48–55**
Milky Way (galaxy), 260
Miller, Dan, 148
Miller, Kevin, 161
Miller, Stanley, 33, 35, 38
Miller, Willoughby D., 173
Miller-Urey experiments, 33, 35, 38
Minor planets, **1–7**
MINOS (Main Injector Neutrino Oscillation), 231
Mir (space station)
    microgravity research, 12–13
    space debris, 21
Mirkin, Chad, 92
Mississippi River Valley and earthquake risk, **48–55**
Mitsubishi, 247
MMR (measles, mumps, rubella) vaccine, 190
Monitor lizards, 128
Monofilaments, 87
    See also Nanotechnology
Monticello, *136*
    Jefferson-Hemings controversy, 131–139
Moon
    craters and Galileo, 260
    geological record, 38
Moore, Gordon, 91
Moore's law, 91–92
Morley, L.W., 64

Motor vehicle engines, 67–75
Mueller, Karl, 250
Muon neutrinos. *See* Neutrinos

N

N-rays, 249
Nanodevices. *See* Nanotechnology
Nanosphere, Inc., 92
Nanotechnology
    benefits and dangers, **86–92**
    carbon nanotubes, 86, 91
    gray goo, 86, 87, 90
    monofilaments, 87
    nanoweapons, 87
    space exploration, 14
Nanotubes, Carbon, 86, 91
Nanotyrannus, 128–129
Nanoweapons, 87
Naples (Italy) and syphilis, 210
NASA (National Aeronautics and Space Administration)
    cold fusion, 252
    fuel cells, 69–70
    International Space Station, 12, 13–14
    privatization, 14–15
    space debris, 17–22
National Academy of Sciences, 142–143, 146–147
National Aeronautics and Space Administration. *See* NASA
National Cold Fusion Institute, 247
National Institute of Dental Research, 178
National Institute of Standards and Technology, 169
National Institutes of Health, 226
National Nanotechnology Initiative, 86
National Science Foundation
    calculus reform, 151
    seismic research, 55
National Security Agency, 168–170
Nature (journal)
    Jefferson-Hemings controversy, 132, 138–139
    n-rays, 249
Neptune (planet), *25*
    controversy over discovery, **24–31**
Neumann, John von, 240
Neuroscience and mathematics, 159–160
Neutrino Oscillation Magnetic Detector, 229, *233*
Neutrinos
    dark matter, 229, 234
    detectors, 227–234
    long baseline experiments, 230–231
    mass, **227–234**
    oscillation, 227–234
    solar neutrino problem, 227, 228, 231–232, 234
    Standard Model (physics), 228
Neutron radiation from cold fusion, 246, 251
New Madrid Seismic Zone
    plate tectonics, 62
    risk, **48–55**
    seismic activity, *51*
Newton, Isaac, 151
Newton's laws of gravity, 151
Noachis Terra (Mars), *42*
NOMAD (Neutrino Oscillation Magnetic Detector), 229, *233*
Non-local hidden variables, 237, 240, 242–243
    See also Hidden variables
North American seismic activity, *51*
North Carolina apportionment, 148
Northridge earthquake (1994), 54–55
NSA (National Security Agency), 168–170
Nuclear fusion, 244
    cold fusion, **244–252**
Nuclear weapons, 89
Nunez, Rafael, 160

O

Oberth, Werner, 10
Oceanic crust
    composition, 58
    plate tectonics, 62
    See also Lithosphere
O'Connor, Sandra Day, 144
O'Neill, Gerard, 14
Oort cloud, 4
Ophiolte suites, 63

Orbital debris. *See* Space debris
Organ transplants
    history, 184–185, 191
    organ donors, 188
    taboos, 189
    waiting lists, 188
    xenotransplants, **184–192**
Oriani, R., 246
Origins of life, **33–40**
Ornithomimus, 128–129
Ostrom, John H., 120, 121
Outflow channels on Mars, 42–43, 44
Oviedo y Valdés, Fernández de, 211–212
Owen, Richard, 119, 120
Oxygen
    isotopes in dinosaur bones, 123
    origins of life, 38
Ozone layer, 38

P

Paleomagnetism, 63–64
Paleontology
    continental drift, 59–60
    metabolism of dinosaurs, **119–130**
Palladium in cold fusion, 244–252
Pallas (asteroid), 6
Pan-American Foot-and-Mouth Disease Center, 101–102
Paneth, Fritz, 245, 249
Paracelsus, 209
Parietal lobes, 159–160, 162
Paris Observatory (France), 27
Park, Robert, 12–13
Parnevik, Jesper, 96
Partial functions (mathematics), 154
Particle accelerators, **219–226**
    See also Cyclotrons
Particle beams, 220, 221
Particle physics
    costs and benefits of research, **219–226**
    Einstein-Podolsky-Rosen paradox, 238–240
    hidden variables, **236–243**
    Higgs boson, 224
    neutrinos and mass, **227–234**
    quarks, 219
    See also Physics; Quantum theory
Pauli, Wolfgang, 227, 228, 231
Peer review of cold fusion, 245–248, 250
Pennington, Robert, 189–190
People for the Ethical Treatment of Animals, 185
Performance-enhancing drugs, 97
PERVs (porcine endogenous retroviruses), 185, 186–187, 190, 192
PETA (People for the Ethical Treatment of Animals), 185
Peters, Kurt, 245, 249
Phillips, 225
Photomultipliers, 228–229, 232, 234
Photons, 242
Photovoltaic power, 83–84
Phylogeny
    dinosaurs, 127–128
    tree of life, *36,* 40
Physics
    Aristotelian physics, 255, 256–257
    cold fusion, **244–252**
    costs and benefits of high-energy research, **219–226**
    neutrinos and mass, **227–234**
    Standard Model, 219, 228, 234
    Theory of Everything, 219
    See also Particle physics; Quantum theory
Pickering, William H., 2
Pigs, *187*
    transgenic, 191
    xenotransplants, **184–192**
Pinta (disease), 213
Planetesimals, 1–7
Planets. *See individual planets by name*
Planets, Minor, 1–7
Plate tectonics
    acceptance of theory, **57–64**
    collision boundaries, 62
    divergent boundaries, 62, 63–64
    earthquake risk, 50, 52
    mechanisms, 58, 61–64

    transform boundaries, 62
    See also Geology
Ploskoye (Russia), 19
Pluto (planet), *5*
    axial tilt, 3
    composition, 3
    discovery, 2
    luminosity, 4–5
    orbital eccentricity, 3, 6
    size, 3, 4, 6
    status as planet, **1–7**
Podolsky, Boris. *See* Einstein-Podolsky-Rosen paradox
Poikilothermy in dinosaurs, **119–130**
Polarization of photons, 242
Pole vaulting and technology, 94, 95
Poll, Nicholas, 212
Polling and accuracy, 141
Pollution
    foot-and-mouth disease, 107
    internal combustion engines, 68, 69, 72–73
    nanotechnology, 88
    space debris, 16, 18–19
    wind power, 76–77, 78–79
Polysomy Y. *See* XYY karyotype
Pons, Stanley, *251*
    cold fusion, **244–252**
Pontecorvo, Bruno, 228
Porcine endogenous retroviruses, 185, 186–187, 190, 192
Power lines and cancer, 180
PPL Therapeutics, 187
Pratt model of isostasy, 60
Price, D.A., 117
Prince Charming (Mars rock), 43
Private strong encryption. *See* Encryption
Proctor and Gamble, 178
Protarchaeopteryx, 125
Proteinoids, 35
Proteins, 35
    See also Amino acids
Ptolemy, 253, 255–256, 258, 259
Public health
    fluoridation, **173–183**
    risk-benefit evaluations, 174, 181
Pyrite, 37
Pythagoras, 258

Q

Quakeproof construction, 48–49
Quantum computers, 91–92
Quantum entanglements, 237, 239–242
    See also Quantum theory
Quantum mechanics. *See* Quantum theory
Quantum theory
    applications, 222
    Einstein-Podolsky-Rosen paradox, 238–240
    entanglements, 237, 239–242
    hidden variables, **236–243**
    neutrinos, 228
    particle physics, 222
    polarized photons, 242
    waves, 236–243
    See also Particle physics; Physics
Quantum waves, 236–243
    See also Quantum theory
Quarks, 219

R

Radioactivity
    cold fusion, 246, 251
    Earth's internal heat, 58, 63
    manned space flight, 13
    space debris, 18–19
Raeburn, Paul, 42
Raelians and cloning, 203
Randall, Henry S., 132
Randall, Willard, 139
Randolph, Martha Jefferson, 132, 133, 137
Randolph, Thomas Jefferson, 132, 134
Ratcliffe, S.G., 117
Reagan, Ronald, 12
Reemtsma, Keith, 191
Reines, Frederick, 228, 231

Relativity theory and quantum theory, 236–243
Renaissance Age and syphilis, 208–209
Renewable energy from wind power, **76–84**
Reptiles and dinosaurs, 119–130
Republican Party, 144
Retrograde motion, 255, 258–259
Retroviruses and xenotransplants, 186–187
Ribonucleic acid and origins of life, 36
Ricqlès, Armand de, 120
Rift valleys, 62, 63–64
    See also Plate tectonics
Rivest, Ronald, 165
RNA and origins of life, 36
Rocket fuel pollution, 19
Rockets and payload efficiency, 14
Roentgen, Wilhelm Conrad, 249
Rohrer, Heinrich, 91
Roland, Alex, 12, 14
Roman Catholic church
    cloning, 203
    geocentricism, 254, 256
Room-temperature fusion. *See* Cold fusion
Rose Center for Earth and Science, 3, 4
Rosebury, Theodore, 216–217
Rosen, Nathan. *See* Einstein-Podolsky-Rosen paradox
Ross, Donald, 191
Rothschild, Bruce, 214
Rothschild, Christine, 214
Royal Observatory (Greenwich, England), 27
RSA cryptography, 165–166
    See also Encryption
Ruben, John, 128–129
Running suits, Full-body, 94, *98*
Russell Sage Foundation, 147
Rutherford, Ernest, 220, 221

S

S-waves. *See* Seismic waves
Sachs, David, *189*
Safe Water Foundation, 178
Salmons, Brian, 190
Saloman, Daniel, 187
Salyut 7 (artificial satellite), 18
Sandford, Avery, 116
Saudi Arabia and space debris, *18*
Savannah River nuclear reactor, 231
Schaudinn, Fritz Richard, 209
Schiapparelli, Giovanni, 42
Schrödinger's Cat paradox, 236, 240
Schuchert, Charles, *61*
    land bridges, 59–60
Schulsinger, Fini, 117
Schwinger, Julian, *248*
    cold fusion, 248
Scipionyx samniticus, 129
SCNT (somatic cell nuclear transfer), 195–196, 198–201, 203–204
Security and Freedom through Encryption Act, 169–170
Segall, Paul, 53
Seidenberg, Abraham, 161
Seismic reinforcement, 48–49
Seismic waves
    charting Earth's internal structure, 63
    propagation in Midwest, 50
Sexually transmitted diseases
    AIDS, 208
    gonorrhea, 214
    syphilis, **208–217**
Shamir, Adi, 165
Sheep cloning, 196, *198*
Simian immunodeficiency virus, 192
Singh, Simon, 167–168
SIV (simian immunodeficiency virus), 192
Skeletal fluorosis, 175, 177, 178, 179–180
Skeletons. *See* Bones
Skylab (space station)
    debris, 18
    role in space exploration, 12
Slavery and Jefferson-Hemings controversy, **131–139**
Smalley, Richard, 91
Snider-Pellegrini, Antonio, 62
SNO (Sudbury Neutrino Observatory), 234
Soffen, Gerald A., 42

Solar neutrino problem, 227–228, 231–232, 234
Solar power, 83–84
Solar system
    geocentricism and heliocentricism, **253–261**
    Neptune's discovery, **24–31**
    Pluto's status, **1–7**
    size, 20
Somatic cell nuclear transfer, 195–196, 198–201, 203–204
Souter, David, 144
South Korea and foot-and-mouth disease, 104
Space debris
    Australia, 18
    dangers, **16–23**
    debris clouds, 19–20
    international policies, 22
    International Space Station, 13
    reentry dangers, 18–19
    Saudi Arabia, *18*
    size and speed, 17, 20
    statistical models, 17, 21–22
    Texas, *21*
    tracking, 17–18
Space exploration
    dangers from space debris, **16–23**
    International Space Station, **8–15**
    privatization, 14–15
    See also Manned space stations
Space shuttles
    o-rings and safety, 18
    space debris, 17, 21
Space Studies Board, 12
Space Surveillance Network, 21
Space tourism, 8
Speck, Richard, 113
Spectroscopy and cold fusion, 247
Spirochete bacteria. *See* Bejel; Pinta (disease); Syphilis; Yaws (disease)
"Spooky" forces. *See* Einstein-Podolsky-Rosen paradox; Quantum theory
Sports and technology, **93–99**
    baseball bats, 96
    golf, 94–95, 95–96
    integrity of the game, 96–99
    measuring systems, 96
    performance-enhancing drugs, 97
    pole vaulting, 94, 95, 96
    safety, 96–99
    tennis, 94, 95
    yachting, 98
Spotila, James, 128
SPS (Super Proton Synchrotron), 229, 231
SSC (Superconducting Supercollider), 220, 224–226
St. Louis Fault, 54
Standard Model (physics), 219, 228, 234
Stansberry, John A., 3
Staphylococcus aureus, 216
Stardust (space probe), 39
Starzl, Thomas, 191
Statistical sampling
    census and statistical sampling, **141–148**
    dual system estimates, 145–146, 147
    integrated coverage management, 142, 147
Statistics. *See* Statistical sampling
Steam cars, 67
Stellar parallax, *257,* 258, 261
Stem cells
    cloning, 195–196, 206
    organ transplants, 188
Sterile neutrinos. *See* Neutrinos
Stern, Alan, 3
Stevens, John Paul, 144–145
Stewart, Ian, 160
Subitization, 157, 158–159
Subterranean organisms, 39–40
Sudbury Neutrino Observatory, 234
Suess, Eduard, 58–59
Sun-centered universe. *See* Heliocentric universe
Super-Kamiokande neutrino detector, 228–232, 234
    See also Kamiokande neutrino detector
Super Proton Synchrotron, 229, 231
Superconducting Supercollider, 220, 224–226
Superconductors, High-temperature, 250
Supermale syndrome. *See* XYY karyotype
Supernova 1987A, 228, 232

Superposition. *See* Quantum theory
Surface water on Mars, **41–46**
    See also Water
Surrogate mothers, 205
Sutton, Philip R.N., 181
Swartz, Morris, 225
Swisher, Randall, 77
Synchrotron radiation, 220, 222
Syphilis
    AIDS, 208, 211
    bejel, 209, 213
    bones, 210, 213–214, 217
    genome, 217
    gonorrhea, 214
    leprosy, 209, 212
    mercury treatment, 214
    origins, **208–217**
    pinta (disease), 209, 213
    symptoms and progression, 209, 210–211, 214, 216
    unitary theory, 209–210, 214
    Vikings, 217
    yaws (disease), 209–210

T

T cells, 187
Taino, 212, 214
Tanberg, John, 249, 250
Tau neutrinos. *See* Neutrinos
Teaching. *See* Classroom instruction
Tectonic plates. *See* Plate tectonics
Tennis and technology, 94–95, 96
Terrorism and data encryption, 164
Texas
    space debris, *21*
    wind power, 79
Texas A&M, 250
Thecodont, 125
Theilgaard, Alice, 114–116
Thematic Apperception Tests, 114–115
Thermophiles, 34, 36–37, 39–40
    See also Archaea; Hyperthermophiles
Thescelosaurus, 123, *124,* 128
Thioesters, 37
Thomson, J.J., 220, 221
Tito, Dennis, 8
TNOs (trans-Neptunian objects), 1–7
TO66 (trans-Neptunian object), 5
Tombaugh, Clyde, *2*
    discovery of Pluto, 2
Tooth decay, *182*
    fluoridation, **173–183**
Toxic shock syndrome, 216
Trans-Neptunian Belt, 1, 4–7
Trans-Neptunian objects, 1–7
Transform boundaries, 62
    See also Plate tectonics
Tree of Knowledge, 207
Treponema pallidum. See Syphilis
Treponema pertennue. See Yaws
Tritium, 247
Tsiolkovsky, Konstantin, 10
Turbines. *See* Microturbines
Turner, Thomas, 135
Turner's syndrome
    cloning, 201
    genetic disorder, 112
Turtles, Leatherback, 128
TWA Flight 800 crash, 169
Tyrannosaurus rex
    bone structure, 123
    predator-to-prey ratio, 122

U

Uncertainty principle. *See* Quantum theory
Underground organisms, 39–40
Undersea volcanic vents. *See* Hydrothermal vents
Uniformitarianism, 57, 60
United Kingdom
    cloning, 195
    data encryption, 169
    foot-and-mouth disease, 101–109
    greenhouse emissions, 79
United Nations Committee on the Peaceful Uses of Outer Space
    space debris, 19
United States
    census and statistical sampling, **141–148**
    foot-and-mouth disease, 106, *108*
    greenhouse emissions, 79
    high-energy physics research policy, **219–226**
United States Constitution
    census, 142
    data encryption, 163, 168
United States Department of Commerce. *See* Department of Commerce
United States Department of Defense, 226
United States Department of Energy. *See* Department of Energy
United States Geological Survey, 49, 51, 54
United States House of Representatives et al., Department of Commerce et al. v. (1998), 144–145
United States Postal Service, 143
United States Public Health Service, 174, 177, 178
United States Space Command, 17
United States Supreme Court, 142, 143, 144–145, 147
University of Michigan Medical School, 92
University of Utah, 245, 250
Uranus (planet)
    discovery, 26
    gravitational influence from Neptune, 24–30
Urey, Harold, 33, 35, 38
Urey-Miller experiments, 33, 35, 38
Utah apportionment, 148

V

Vaccinations for foot-and-mouth disease, 103–104, 107–109
Valley networks on Mars, 42–43, 44, *45*
Variables, Hidden, **236–243**
Velociraptor, 129
Venus (planet)
    orbit, 259
    phases, 253, 260
Veterinary medicine and foot-and-mouth disease, **101–109**
Viking (Mars probe), 42
Vikings and syphilis, 217
Vine, Fred, 64
Viruses, Computer, 87
Viruses and xenotransplants, **184–192**
Volcanoes, 34
    See also Hydrothermal vents

W

Wachterhauser, Gunter, 37
Waldbott, George L., 180
Warlpiri tribe, 161
Warm-blooded dinosaurs, **119–130**
"Warm little pond" theory, **33–40**
Wassermann, August von, 209
Water
    contamination and foot-and-mouth disease, 107
    Mars, **41–46**
    origins of life, 33–40
Waves, Quantum, 236–243
    See also Quantum theory
Wayles, John, 137
Weakly interacting massive particles, 229
Wegener, Alfred, 59–64
Weiss, Robin, 187, 188, 190, 192
Wetmore, Samuel F., 132
Whatmore, A.J., 117
WHO (World Health Organization), 191–192
Wilcox, R.R., 217
Wild-2 (comet), 39
Williams, J. Leon, 173
Williams, Stan, 91–92
Willis, Bailey, 60, 61
Wilson, E.O., 111
WIMPs (weakly interacting massive particles), 229
Wind farms, 77–78, *80*
    forecasts, 77
    usage, 76
Wind power
    environmental considerations, 76–81
    history, 76–77, 78
    viability, **76–84**
    See also Wind turbines; Windmills

Wind turbines, 77–78
    history, 76–77
    See also Wind power; Windmills
Windmills
    Denmark, 82
    history, 76–77
    See also Wind power; Wind turbines
Witkin, Herman A., 117
Wood, R.W., 249
Woods, Tiger, 96
Woodson, Thomas C., 131–139
World Health Organization, 191–192
World Wide Web. *See* Internet
Wyss, Max, 54

X

X chromosome, 111, 112
X-rays, 249
Xenotransplants
    extracorporeal liver perfusion, 189–190
    history, 184–185, 191
    retroviruses, 186–187, 190, 192
    risks and benefits, **184–192**
    zoonoses, 186

XXY karyotype. *See* Klinefelter's syndrome
XYY karyotype
    aggression, **111–118**
    causes, 111–112, 116
    symptoms, 112, 116

Y

Y chromosome
    Jefferson-Hemings controversy, **131–139**
    XYY karyotype, 111, 112
Yachting and technology, 98
Yaws (disease)
    genome, 217
    syphilis, 209–210, 213
Yiamouyiannis, John, 178
YY syndrome. *See* XYY karyotype

Z

Zavos, Panayiotis Michael, 194
Zoback, Mark, 53
Zoonoses and xenotransplants, 186
Zubrin, Robert, 14
Zyvex (company), 90–91

ISBN 0-7876-5765-4